STRUCTURE AND PROPERTIES OF NEARBY GALAXIES

INTERNATIONAL ASTRONOMICAL UNION
UNION ASTRONOMIQUE INTERNATIONALE

SYMPOSIUM No. 77

HELD IN BAD MÜNSTEREIFEL, F.R.G., AUGUST 22–26, 1977

STRUCTURE AND PROPERTIES OF NEARBY GALAXIES

EDITED BY

ELLY M. BERKHUIJSEN AND RICHARD WIELEBINSKI

Max-Planck-Institut für Radioastronomie, Bonn, F.R.G.

D. REIDEL PUBLISHING COMPANY

DORDRECHT : HOLLAND / BOSTON : U.S.A.

1978

Library of Congress Cataloging in Publication Data

Main entry under title:

Structure and properties of nearby galaxies.
   (Symposium–International Astronomical Union: No. 77.)
   Bibliography: p.
   Includes index.
   1. Galaxies–Congresses. i. Berkhuijsen, Elly M., 1937–
ii. Wielebinski, R. iii. International Astronomical Union. iv. Series: International
Astronomical Union. Symposium No. 77.
QB857.S863   523.1'12   78-7292
ISBN 90-277-0874-6
ISBN 90-277-0875-4 pbk.

Published on behalf of
the International Astronomical Union
by
D. Reidel Publishing Company, P.O. Box 17, Dordrecht, Holland

All Rights Reserved
Copyright © 1978 by the International Astronomical Union

Sold and distributed in the U.S.A., Canada, and Mexico
by D. Reidel Publishing Company, Inc.
Lincoln Building, 160 Old Derby Street, Hingham,
Mass. 02043, U.S.A.

No part of the material protected by this copyright notice may be reproduced or utilized
in any form or by any means, electronic or mechanical, including photocopying, recording
or by any informational storage and retrieval system, without written permission from
the publisher

Printed in The Netherlands

TABLE OF CONTENTS

FOREWORD		xi
LIST OF PARTICIPANTS		xiii
INTRODUCTION / J.H. Oort		xvii

I		THE SMOOTH BACKGROUND	1
I.1		THE DISTRIBUTION OF LIGHT IN GALAXIES / K.C. Freeman	3

DISCUSSION I.1 — 10
    Problems in galactic dynamics studied by means of three-
       dimensional simulations / R.H. Miller — 11
    The kinematics of the lenticular galaxies NGC 1291 and
       NGC 1326 / U. Mebold — 14
    The dwarf-spheroidal galaxies near M31 / T.D. Kinman — 14

I.2   THE PAST HISTORY OF STAR FORMATION IN GALAXIES /
       B.M. Tinsley — 15

DISCUSSION I.2 — 19
    Multi-colour observations of NGC 5128 = Centaurus A /
       S. van den Bergh — 20

I.3   ROTATION CURVES IN THE OUTER PARTS OF GALAXIES FROM HI
       OBSERVATIONS / E.E. Salpeter — 23

DISCUSSION I.3 — 27
    HI sizes and rotation curves of some edge-on galaxies /
       R. Sancisi — 27
    The kinematics of a sample of about twenty spiral
       galaxies / A. Bosma — 28
    New optical rotation curve of M101 / G. Comte — 30

I.4   THE LARGE-SCALE RADIO CONTINUUM STRUCTURE OF SPIRAL
       GALAXIES / P.C. van der Kruit — 33

DISCUSSION I.4 — 45
    The thermal content of IC 342 / R. Wielebinski — 47
    Deconvolved single-dish radio observations of NGC 6946 /
       J. Pfleiderer — 47

I.5	THE LARGE-SCALE DISTRIBUTION OF RADIO CONTINUUM IN E AND S0 GALAXIES / R.D. Ekers	49
DISCUSSION I.5		51
	The gas content of lenticular galaxies / H. van Woerden	52
	HI observations of a large sample of elliptical galaxies / F.J. Kerr	53
	HI in NGC 1052 and NGC 4636 / J.S. Gallagher	54
	The discontinuity between elliptical and disk galaxies / M. Capaccioli	55
I.6	GLOBAL DYNAMICS OF THE INTERSTELLAR GAS, MAGNETIC FIELD, AND COSMIC RAYS / E.H. Levy	57
DISCUSSION I.6		62
	Comments on the radio continuum emission from normal disk galaxies / R.J. Allen	63
	On the nature of a galactic boundary layer flow / A.M. Waxman	64
	Optical polarization in galaxies / A. Elvius	65
II	SPIRAL STRUCTURE AND STAR FORMATION	67
II.1	THE EVOLUTION OF DISK GALAXIES / S.E. Strom and K.M. Strom	69
DISCUSSION II.1		92
	Drift and broadening of ageing spiral arms / R. Wielen	93
	Comments on NGC 3312, NGC 1291 and NGC 1079 / J.S. Gallagher	94
II.2	SOME COMMENTS ON RADIO OBSERVATIONS OF SPIRAL ARMS / W.W. Shane and J. Bystedt	97
DISCUSSION II.2		103
II.3	THE DYNAMICS OF THE SPIRAL GALAXY M81 / H.C.D. Visser	105
DISCUSSION II.3		112
II.4	A CONFRONTATION OF DENSITY WAVE THEORIES WITH OBSERVATIONS / A.J. Kalnajs	113
DISCUSSION II.4		125
	Discrete spiral modes in disk galaxies / J.W.-K. Mark	127
	Growth of spiral waves in disk galaxies / G. Bertin	128
	Galactic shocks in open-armed normal spirals and barred spirals / W.W. Roberts	129

TABLE OF CONTENTS

II.5	RADIO OBSERVATIONS OF MOLECULES IN NEARBY GALAXIES / J.B. Whiteoak	131
DISCUSSION II.5		137
	A first step toward the radial distribution of CO in M31 / L. Weliachew	137
II.6	HOT GAS IN THE GALAXY: HOW EXTENSIVE IS IT? / F.H. Shu	139
DISCUSSION II.6		145
III	NEARBY GALAXIES OF LARGE ANGULAR SIZE	147
III.1	RADIO CONTINUUM OBSERVATIONS OF M31 AND M33 / E.M. Berkhuijsen	149
DISCUSSION III.1		155
	Variation of spectral index across M31 / R. Beck	156
	M31 at 49 cm wavelength / J. Bystedt	157
III.2	VELOCITY DISPERSION IN THE BULGE OF M31; DYNAMICAL MODEL / G. Monnet, A. Pellet and F. Simien	159
III.3	THE SPIRAL STRUCTURE OF M31 / E. Athanassoula	163
DISCUSSION III.3		166
III.4	THE KINEMATICS WITHIN M31 / M.S. Roberts, R.N. Whitehurst and T.R. Cram	169
DISCUSSION III.4		172
	HI survey of M31 / T. Landecker	173
III.5	THE THREE-DIMENSIONAL DISTRIBUTION OF NEUTRAL HYDROGEN IN M31 / R.N. Whitehurst, M.S. Roberts and T.R. Cram	175
DISCUSSION III.5		179
	Anomalous motions of spiral arms in M31 / W.W. Shane	180
III.6	A WARP IN THE HI DISTRIBUTION AT THE EXTREME NE AND SW OF M31 / D.T. Emerson and K. Newton	183
DISCUSSION III.6		187
	New outer HI arms in M31 / R.D. Davies	188
III.7	THE LARGE-SCALE DISTRIBUTION OF HI IN M33 AND IC 342 / J.E. Baldwin	191
DISCUSSION III.7		195

III.8 A SENSITIVE SINGLE-DISH HI SURVEY OF THE GALAXY M33 /
W.K. Huchtmeier   197

DISCUSSION III.8   200
    Supernova remnants in M33 / S. D'Odorico   201

IV   NEARBY ACTIVE GALAXIES AND THEIR NUCLEI   203

IV.1   RADIO PROPERTIES OF ACTIVE NEARBY SPIRAL GALAXIES /
        A.G. de Bruyn   205

DISCUSSION IV.1   216
    The nucleus of NGC 1275 at 2.8 cm wavelength / E. Preuss   217
    HI in NGC 4258 / G.D. van Albada   218

IV.2   RADIO PROPERTIES OF THE NUCLEI IN ELLIPTICAL, SO AND
        SPIRAL GALAXIES / R.D. Ekers   221

DISCUSSION IV.2   223
    Compact components in active spiral and Seyfert galaxy
        nuclei and galaxy interactions / J.B. Carlson
        (presented by F.J. Kerr)   224

IV.3   INFRARED, OPTICAL, AND X-RAY PROPERTIES OF THE NUCLEI
        OF NEARBY GALAXIES / G. Burbidge   227

DISCUSSION IV.3   236

IV.4   EMISSION FROM THE NUCLEI OF NEARBY GALAXIES: EVIDENCE
        FOR MASSIVE BLACK HOLES? / M.J. Rees   237

DISCUSSION IV.4   242

V   THE OUTSKIRTS OF GALAXIES   245

V.1   GALAXY HALOES AND THE MISSING MASS PROBLEM / S. van den
        Bergh   247

DISCUSSION V.1   263
    Sensitive observations of HI envelopes of late-type
        galaxies / W.K. Huchtmeier   264

V.2   ON PERIPHERAL DYNAMICS / A. Toomre   267

DISCUSSION V.2   267

V.3   INTERACTING GALAXIES: THE KINEMATICS OF NGC 4038/39 AND
        THE HI BRIDGE BETWEEN M81 AND NGC 3077 / J.M. van der Hulst   269

DISCUSSION V.3   274
    Distributed HI in small groups of galaxies / R.D. Davies   274

	Tidal interaction and accretion in the galaxy pair NGC 1512 and 1510 / H. van Woerden	274
	Tidal interactions within the NGC 4631 group of galaxies / F. Combes	275
	HI observations of the M51 system / R. Giovanelli	276
	Gas distribution and velocity field of the barred spiral galaxy NGC 5383 / R. Sancisi	276
V.4	GALAXIES WITH LONG TAILS / F. Schweizer	279
DISCUSSION V.4		285
V.5	MATERIAL IN THE VICINITY OF GALAXIES / B.F. Burke	287
DISCUSSION V.5		291
V.6	HIGH-VELOCITY CLOUDS: GALACTIC OR EXTRAGALACTIC? / R. Giovanelli	293
DISCUSSION V.6		297
	High-velocity clouds and warping of the Galactic plane / J. Einasto	297
	Fine structure in the Magellanic Stream / I.F. Mirabel	298
V.7	DISCUSSION ON IRREGULAR GALAXIES	299
	HI observations of dwarf galaxies / T.D. Kinman	299
	Gas distribution, motions and dynamics for some dwarf-irregular galaxies / R.B. Tully	299
	Mass of clumps in the irregular galaxy Markarian 296 / J. Heidmann	300
	HI observations of the irregular galaxy NGC 4214 / T.L. Landecker	300
	Radio recombination lines in M82 / E.R. Seaquist	301
	Preliminary results from an Hα-[NII] line spectroscopic study of M82 / G. Comte	302
	A dust-scattering model of M82 / S.M. Scarrott	302
INDEX OF GALAXIES AND CLUSTERS OF GALAXIES		305

With the publisher's permission the following figures have been reproduced from "Astronomy and Astrophysics":
Figures 1, 2, 3 and 4 in Paper I.4 by P.C. van der Kruit;
Figure 2 in Paper III.1 by E.M. Berkhuijsen;
Figures 1, 2, 3, 4 and 5 in Paper IV.1 by A.G. de Bruyn.

FOREWORD

The IAU Symposium No. 77 on "Structure and Properties of Nearby Galaxies" was held in Bad Münstereifel, F.R.G., from 22-26 August 1977. The Symposium was financially supported by the IAU, the Deutsche Forschungsgemeinschaft and the Max-Planck-Gesellschaft.

The aim of the Symposium was to compare recent high-resolution radio continuum and HI-line results on nearby galaxies with both new and established optical material and with recent theoretical work. Over 100 astronomers actively working in these fields discussed the "state of the art" from their various points of view. The Symposium was sponsored by IAU Commissions 28 and 40. The Scientific Organizing Committee consisted of R.J. Allen (Chairman), K.C. Freeman, I.V. Gossachinski, G. Monnet, J.H. Oort, M.S. Roberts, S.E. Strom, A. Toomre and R. Wielebinski. In order to limit the scope of the Symposium the S.O.C. ruled that nearby galaxies have distances of 0.5 to about 15 Mpc; consequently topics mainly concerned structural features occurring on length scales ranging roughly from 0.1 to 50 kpc.

The Local Organizing Committee was formed by R. Wielebinski (Chairman), Elly M. Berkhuijsen, Gabriele Breuer (Conference Secretary) and W. Sieber. Especially during the Symposium the L.O.C. was greatly helped by N. Bartel, R. Beck, D.T. Emerson, H. Hoech, U. Klein, H. Kühr and Anne Vogt. Their help has been essential to the smooth run of the conference. We also wish to acknowledge the pleasant cooperation with the very able staff of the "Colonia Schulungszentrum" and its excellent facilities.

The Proceedings were made from camera-ready manuscripts according to the new IAU rules (1971, Transactions of the IAU vol. XIV B, p. 287). The speakers have prepared their own manuscripts. Manuscripts of invited papers are published as they were received, essentially in the order of their presentation at the Symposium. In a few cases references were retyped, and some minor corrections made. Because short contributed papers are considered to be part of the discussions their abstracts were retyped, but their titles were kept and are given in the Table of Contents. They were rearranged with part of the discussions in order to increase the coherence and readability. The discussions were produced from the discussion sheets received from the participants after the sessions and/or from the tape recordings. Answers from speakers were taken from the tape recordings or from their written answers sent to us upon request. We are to blame for any mistakes, omissions or inconsistencies that the reader may find.

The S.O.C. was ably assisted by Joke Nunnink. Vicky Sherwood mastered the difficult task of typing out the tape recordings. Gabriele Breuer not only solved many organizational details before and during the Symposium, but also typed the manuscripts of the discussions; her amazing skill and pleasant cooperation has been of invaluable help to us.

January 1978

Elly M. Berkhuijsen
Richard Wielebinski

LIST OF PARTICIPANTS

ALCAINO, G.   European Southern Observatory, Santiago, Chile
ALLEN, R.J.   Kapteyn Astronomical Institute, Groningen, NL
APPENZELLER, I. Landessternwarte, Heidelberg, FRG
ATHANASSOULA, E. Observatoire, Besancon, France
BALDWIN, J.E.   Cavendish Laboratory, Cambridge, UK
BERKHUIJSEN, E.M. Max-Planck-Institut für Radioastronomie, Bonn, FRG
BERMAN, R.H.   University of Reading, Reading, UK
BERTIN, G.   Massachusetts Institute of Technology, Cambridge, USA
BERTOLA, F.   Osservatorio Astronomico, Padova, Italy
BIERMANN, P.   Astronomisches Institut der Universität, Bonn, FRG
BOKSENBERG, A.   University College London, London, UK
BOSMA, A.   Kapteyn Astronomical Institute, Groningen, NL
BURBIDGE, G.   University of California San Diego, La Jolla, USA
BURKE, B.F.   Massachusetts Institute of Technology, Cambridge, USA
BYSTEDT, J.   Sterrewacht, Leiden, NL
CAPACCIOLI, M.   Osservatorio Astronomico, Padova, Italy
COMBES, F.   Observatoire, Meudon, France
COMTE, G.   Observatoire, Marseille, France
COURTES, G.   Laboratoire d'Astronomie Spatiale, Marseille, France
CRANE, P.C.   National Radio Astronomy Observatory, Green Bank, USA
DAVIES, R.D.   N.R.A.L., Jodrell Bank, Macclesfield, UK
DE BRUYN, A.G.   Hale Observatories, Pasadena, USA
DEKKER, E.   Sterrewacht, Leiden, NL
D'ODORICO, S.   Osservatorio Astrofisico, Asiago, Italy
DUVAL, M.F.   Observatoire, Marseille, France
EINASTO, J.   Struve Astrophysical Observatory, Toravere, USSR
EKERS, R.D.   Kapteyn Astronomical Institute, Groningen, NL
ELVIUS, A.   Stockholms Observatorium, Saltsjöbaden, Sweden
EMERSON, D.T.   Max-Planck-Institut für Radioastronomie, Bonn, FRG
FITZGERALD, M.P. University of Waterloo, Waterloo, Canada
FREEMAN, K.C.   Mt. Stromlo Observatory, Woden P.O., Australia
GALLAGHER, J.S. III Observatory, University of Illinois, Urbana, USA
GIOVANELLI, R.   Osservatorio Astronomico Universitario, Bologna, Italy
GROSBOL, P.   Astronomical Observatory, Copenhagen, Denmark
HAYLI, A.F.   Observatoire, Besancon, France
HEESCHEN, D.S.   National Radio Astronomy Obs., Charlottesville, USA
HEIDMANN, J.   Observatoire, Meudon, France
HUMMEL, E.   Kapteyn Astronomical Institute, Groningen, NL
HUNTER, C.   Florida State University, Tallahassee, USA
HUCHTMEIER, W.K. Hamburger Sternwarte, Hamburg, FRG
ISRAEL, F.   Owens Valley Radio Observatory, Pasadena, USA

KALNAJS, A.J.   Mt. Stromlo Observatory, Woden P.O., Australia
KELLERMANN, K.I.   National Radio Astronomy Observatory, Green Bank, USA
KERR, F.J.   University of Maryland, College Park, USA
KINMAN, T.D.   Kitt Peak National Observatory, Tucson, USA
LANDECKER, T.L.   Dominion Radio Astrophysical Obs., Penticton, Canada
LEVY, E.H.   University of Arizona, Tucson, USA
LIN, C.C.   Massachusetts Institute of Technology, Cambridge, USA
LINDBLAD, P.O.   Stockholms Observatorium, Saltsjöbaden, Sweden
LUIKEN, M.   University of Waterloo, Waterloo, Canada
MADORE, B.F.   Institute of Astronomy, Cambridge, UK
MARK, J.W.-K.   Massachusetts Institute of Technology, Cambridge, USA
MEBOLD, U.   Max-Planck-Institut für Radioastronomie, Bonn, FRG
MEZGER, P.G.   Max-Planck-Institut für Radioastronomie, Bonn, FRG
MILLER, R.H.   University of Chicago, Chicago, USA
MIRABEL, I.F.   N.R.A.L., Jodrell Bank, Macclesfield, UK
MONNET, G.   Observatoire, Marseille, France
OORT, J.H.   Sterrewacht, Leiden, NL
PARMA, P.   Laboratorio de Radioastronomia, Bologna, Italy
PEL, J.W.   Max-Planck-Institut für Astronomie, Heidelberg, FRG
PFLEIDERER, J.   Astronomisches Institut, Innsbruck, Austria
PRICE, R.M.   National Science Foundation, Washington, USA
REES, M.J.   Institute of Astronomy, Cambridge, UK
ROBERTS, M.S.   National Radio Astronomy Obs., Charlottesville, USA
ROBERTS, W.W. Jr.   University of Virginia, Charlottesville, USA
ROSADO, M.   Universidad Nacional Autonoma de Mexico, Mexico
ROTS, A.H.   Radiosterrenwacht, Dwingeloo, NL
SALPETER, E.E.   Cornell University, Ithaca, USA
SANCISI, R.   Kapteyn Astronomical Institute, Groningen, NL
SCARROTT, S.M.   University of Durham, Durham, UK
SCHMIDT-KALER, T.   Astronomisches Institut der Universität, Bochum, FRG
SCHWEIZER, F.   Obs. Interamericano de Cerro Tololo, La Serena, Chile
SEAQUIST, E.R.   University of Toronto, Toronto, Canada
SELLWOOD, J.A.   The University, Manchester, UK
SHANE, W.W.   Sterrewacht, Leiden, NL
SHOSTAK, S.   Kapteyn Astronomical Institute, Groningen, NL
SHU, F.H.   University of California, Berkeley, USA
SIMIEN, F.   Observatoire, Marseille, France
STAUDE, H.J.   Max-Planck-Institut für Astronomie, Heidelberg, FRG
STROM, S.E.   Kitt Peak National Observatory, Tucson, USA
TOOMRE, A.   Massachusetts Institute of Technology, Cambridge, USA
TULLY, B.   Institute of Astronomy, Honolulu, USA
VAN ALBADA, G.D.   Sterrewacht, Leiden, NL
VAN DEN BERGH, S.   David Dunlap Observatory, Richmond Hill, Canada
VAN DER HULST, J.M.   Kapteyn Astronomical Institute, Groningen, NL
VAN DER KRUIT, P.C.   Kapteyn Astronomical Institute, Groningen, NL
VAN DER LAAN, H.   Sterrewacht, Leiden, NL
VAN WOERDEN, H.   Kapteyn Astronomical Institute, Groningen, NL
VELUSAMY, T.   Radio Astronomy Center, Ootacamund, India
VIALLEFOND, F.   Observatoire, Meudon, France
VISSER, H.C.D.   Kapteyn Astronomical Institute, Groningen, NL
WARNER, P.   Cavendish Laboratory, Cambridge, UK

LIST OF PARTICIPANTS

WAXMAN, A. University of Chicago, Chicago, USA
WAYMAN, P.A. Dunsink Observatory, Castleknock, Ireland
WELIACHEW, L. Observatoire, Meudon, France
WESTERHOUT, G. U.S. Naval Observatory, Washington, USA
WHITEHURST, R.N. National Radio Astronomy Obs., Charlottesville, USA
WHITEOAK, J.B. CSIRO Radiophysics Division, Epping, Australia
WIELEBINSKI, R. Max-Planck-Institut für Radioastronomie, Bonn, FRG
WIELEN, R. Astronomisches Rechen-Institut, Heidelberg, FRG
WOODWARD, P.R. Sterrewacht, Leiden, NL
WRIGHT, M.C.H. University of California, Berkeley, USA
YUAN, C. City College of New York, New York, USA

During an afternoon break in the scientific program of IAU Symposium 77, Professor J.H. Oort (77 yr) visits the 100-meter radio telescope at Effelsberg in August, 1977.

INTRODUCTION

J.H. Oort
Leiden Observatory, Leiden, The Netherlands

Research on galaxies is only little more than half a century old. In my student days, around 1920, astronomers as well as students generally assumed that spiral nebulae were distant stellar systems, but interest was focussed on other things. The small significance of extragalactic research in the earlier years of this century is well illustrated by Newcomb-Englemann's standard work "Populäre Astronomie". In the more than 800-page edition of 1914 the galaxies were discussed in a section "Sternhaufen und Nebelflecke" as just another class of nebulae, in between the gaseous and the reflection nebulae. The possibility that the spirals might be independent stellar systems was not even mentioned. But only a few years later the subject became immensely interesting in two ways: V.M. Slipher's discovery that some nebulae had exceedingly large radial velocities was followed by W. de Sitter's suggestion in 1917 that these might be due to an expansion of the universe, as deduced from Einstein's theory of gravitation. The second way, full of speculations relevant to this Symposium, may be illustrated by Jeans' inspiring book "Problems of Cosmogony and Stellar Dynamics", published in 1919. Its chapter on "The Evolution of Rotating Nebulae", featuring the first Mt Wilson photographs of spiral and elliptical nebulae, contains the following statement about spiral arms: "That these arms really represent an ejection of matter from the central nucleus is almost proved by the two instances of M51 and M101 already discussed in §4. All this is quite in accordance with theory." As a physicist specialized in the kinetic theory of gases Jeans approached the subject largely from a gas-dynamical point of view. The "central nucleus" was a rotating mass of gas (tentatively identified with the elliptical and lenticular nebulae) which upon concentration becomes unstable and sheds material from its outer edge. Jeans remarked that condensations would be formed in the ejected arms, and he conjectured that the spiral nebulae would be huge swarms of stars, or "island universes".

A fantastic extension of observational data on galaxies followed upon the successful construction of the reflectors of the first half of this century. Directly related to our present topics were Hubble's classification system, now so monumentally presented in the Hubble

Atlas, Hubble's discovery of the remarkable similarity and structural simplicity of elliptical systems, his discovery of Cepheid variables in the Andromeda nebula and Baade's discovery in the dark war nights of 1942-45, of the two stellar populations, through which stellar ages and the birth of stars made their entrance in galactic and extragalactic astronomy.

After the early attempt by Jeans the theory of spiral structure has at present reached a first stage of completion, due to the development of the theory of waves in stellar systems, started by Bertil Lindblad in the twenties, but only now worked out satisfactorily by Lin and his co-workers. The proceedings of the Symposium show how much has been accomplished, both in the theory and in its confrontation with various kinds of observation. The most important of the latter are: (1) The first successful photometric measurement of spiral gravitational fields; (2) 21-cm line observations of the response of the interstellar gas to this field; and (3) the observation of star formation in the spiral shock. An important turnpike has been reached. The further stretch of the road, which should lead to insight into the origin and driving mechanism of the waves, is still full of obstacles. There is a feeling that deeper knowledge of bar structures might be an important prelude to such an understanding.

An unexpected development has occurred in the field of elliptical nebulae. Jeans, as we have seen, considered them as rotating masses of gas, flattened by their rotation. Such a model had to be abandoned when it became clear that they consisted of stars. But it seemed natural to retain the idea that they had originally been such rotating gaseous bodies, and that the stars formed within them had roughly conserved the flattened distribution of the gas. Recently, however, as pointed out in the first introductory report, evidence has been found that the majority of ellipticals may not be oblate spheroids, and that their evolution will therefore have been quite different from that of a rotating mass of gas.

A class of phenomena which could hardly have been imagined in my student days is that concerning the activity of galactic nuclei. Many phenomena connected with nuclear activity can best be studied in the nearby galaxies forming the theme of the present symposium. The subject is of much importance, not only for the secret of the nuclei themselves, but likewise for the effects which nuclear activity can apparently exert on the structure of the surrounding galaxy. These effects were mainly revealed by radio astronomy.

The final section of the Symposium dealt with the outermost regions of galaxies, and with their surroundings. It illustrates some of the information obtained in recent years about the vast complexes of intergalactic gas drifting around and doubtlessly continually falling into the large spiral galaxies of our vicinity. The subject is gaining importance on account of its consequence for the general evolution of galaxies and for the dynamics of their outskirts.

I THE SMOOTH BACKGROUND

"When you subtract everything that doesn't look smooth, you are left with a smooth background"

G. Westerhout at buffet dinner

THE DISTRIBUTION OF LIGHT IN GALAXIES

K.C. FREEMAN
Mount Stromlo and Siding Spring Observatory,
Research School of Physical Sciences,
The Australian National University.

The distribution of light in galaxies is their most obvious and fundamental observable property. Hopefully it gives us some insight into their structure and dynamics. In this talk I will review some recent work on ellipticals and disk galaxies. In summary, the luminosity distributions for both these classes have several complexities, the dynamical significance of which is not yet clear. Both classes turn out to have roughly constant mean surface brightness. This may result from selection. However, if it is real, then it is important dynamically, and I will discuss this question at the end.

1. ELLIPTICALS

Ellipticals have a wide range of absolute magnitudes, and it seems clear that the structure changes with magnitude. The normal giant ellipticals ($M_B \leq -18$) have roughly similar radial surface brightness distributions $I(r)$. These $I(r)$ distributions can mostly be represented fairly well by King's (1966) models with $\log(r_t/r_c) \approx 2.2$. These models were originally constructed to represent globular clusters. They are a one-parameter family of truncated isothermal spheres, whose central concentration increases with increasing values of r_t/r_c (tidal radius to core radius): the limit is the isothermal sphere itself. Alternatively, two empirical laws are often used. One is de Vaucouleurs' well known $r^{\frac{1}{4}}$ law

$$\log I(r) \propto r^{\frac{1}{4}}$$

which has no free parameters. The other is the truncated Hubble law

$$I(r) = I_o(r+\beta)^{-2}\exp[-(r/\alpha)^2]$$

introduced by Oemler (1976), where α and β are length scales. This is again a one-parameter (α/β) family of increasing central concentration, like the King family.

There is evidence now that the $I(r)$ profiles for normal ellipticals may depend on their environment. For example, Strom's (1977) photometry of ellipticals in the Coma cluster shows that the brighter systems ($M_B<-21$) in the outer parts of the cluster have more extended envelopes than the ellipticals in the inner parts. This is probably due to tidal processes. Similarly the dwarf ellipticals ($M_B>-16$) are significantly less centrally concentrated than the giants, and this is again presumably tidal. Their $I(r)$ distributions are well fitted by King models with $\log(r_t/r_c)$ in the approximate range 0.5 to 1.5. On the other hand, the cD ellipticals, which dominate some galaxy clusters, have very extensive outer envelopes. These may result from accumulated cluster debris, or from the cD galaxies themselves forming through mergers of smaller galaxies. This question is not settled yet.

Now we come to the shape of ellipticals. The belief is that ellipticals are oblate spheroidal systems, flattened presumably because they are rotating. With the assumption of oblateness, we can infer the true distribution of axial ratio from the apparent distribution. It turns out that the true distribution is peaked at axial ratio 0.6, with very few truly spherical systems, and none flatter than axial ratio 0.3. However reality seems more complicated. (i) The isophotal eccentricities of ellipticals change with radius. King's results, quoted by Wilson (1975), show that $e(r)$ can be monotone increasing or decreasing, or take a maximum or minimum at intermediate radii. There is also evidence that the position angle of the major axis changes with radius. (ii) Observations of rotation curves for several flat ellipticals, by Bertola and Capaccioli (1975) and Illingworth (unpublished), give peak rotational velocities much smaller than expected. There are several possible interpretations of (i) and (ii), taken together. They certainly suggest that rotation and flattening are not uniquely related, and that ellipticals may be triaxial or even prolate. This now has some theoretical support. Binney (1976) has shown how the dissipative collapse of a galaxy can lead to an eccentricity independent of rotation, and Miller's recent N-body results show how triaxial systems can form during the collapse of a stellar system. Observations of the distribution of apparent axial ratio for ellipticals are not inconsistent with the concept that ellipticals are prolate systems. A true distribution peaked near axial ratio 1.8, with few systems more elongated than axial ratio 2.5, and again with very few truly spherical

I.1 THE DISTRIBUTION OF LIGHT IN GALAXIES

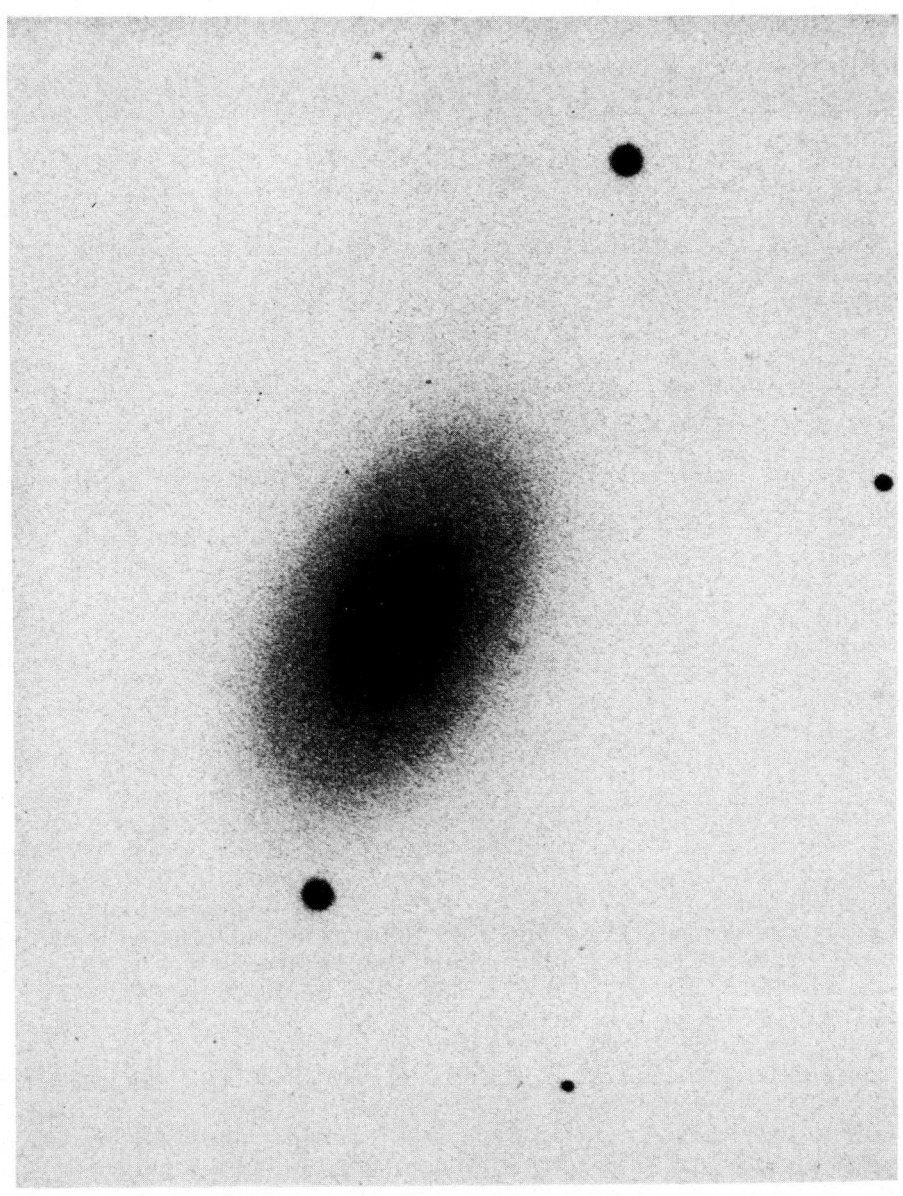

Figure 1: Yellow light photograph of the S0 galaxy NGC 1553. Note the prominent lens.

systems, represents the data just as well as the distribution of oblate systems discussed earlier.

2. DISK GALAXIES

The two-component (bulge+disk) structure of these systems is now well known. The bulge surface brightness follows the $r^{\frac{1}{4}}$ law, at least in the inner parts, and the disk surface brightness decreases exponentially with radius. The bulge to disk ratio varies greatly: for example, see Freeman (1974), Figure 1. Systems like M33 and NGC 300 are almost pure exponential disk, while galaxies like NGC 4594 are dominated by their bulge. This simple picture provides a first order description of disk galaxies, but again there are some complications.

The first complication is the lens. Figure 1 shows a yellow light photograph of the inner parts of NGC 1553. This is an S0 galaxy with a particularly prominent lens. The lens appears as the annular region of approximately uniform surface brightness. This structure is common in spirals and S0s: for examples, see NGC 210 and NGC 5101 in the Hubble Atlas. The lens is almost certainly a substructure of the disk, because a similar flat component is clearly visible in several edge-on S0s, like NGC 4762 (see the Hubble Atlas). Dynamically the lens component is not yet understood.

There is some interesting structure in the bulges of edge-on systems. These often show the effects of either differential rotation or the highly non-spherical gravitational potential of the disk or both. In particular, the bulges of systems like NGC 128 (see the Hubble Atlas) appear box-like or peanut shaped: this probably results from differential rotation, although no good models are yet available to support this.

A recent thesis by Vassiliki Tsikoudi at the University of Texas illustrates some of these features of disk galaxies particularly clearly. She made deep surface photometry of three edge-on S0s (NGC 4762, 4111 and 3115) to study the structure of the bulge and disk. We can summarise the main important results. (i) Along the _major_ axis of these galaxies (ie along the edge-on disk), the surface brightness does not decrease smoothly with radius, as we would expect for a purely exponential disk. Two or three plateaux are seen, the inner one corresponding to the obvious lens component in photographs. These plateaux appear symmetrically on both sides of the nucleus, so they are most likely to be annular structures. (ii) Along the _minor_ axis, the bulge follows the $r^{\frac{1}{4}}$ law in the inner parts, but further out there is an excess of light above this law. The difference, (observed - $r^{\frac{1}{4}}$) itself decreases exponentially with height z above the plane. (iii) For NGC 4762, which has a very small bulge, it is possible to study the vertical structure of the disk itself. For small z (a few hundred parsecs), the surface brightness is a gaussian function of z; for larger z it is well represented by an exponential.

Two comments. (i) The z-structure of both the bulge and the disk shows an exponential decrease of surface brightness with z for large z. This makes sense. We know that elliptical galaxies are well represented by the King models, which are weakly truncated isothermal spheres. We know also that the isothermal <u>sheet</u> has an exponential decrease of density with height for large z. So, if the bulge and the disk are also approximately isothermal, then the effect of the rather flat equipotentials of the disk itself will lead to an exponential structure at large z, as observed. (ii) Some new results on the radial distribution of globular clusters in M87, NGC 4594 and the Galaxy (Harris and Smith, 1976; Wakamatsu, 1977; de Vaucouleurs, 1977) show that the surface number density of globular clusters also follows the bulge's $r^{\frac{1}{4}}$ law. This very clearly identifies the halo population (as defined by the globular clusters) with the outer bulge population.

3. THE CONSTANT SURFACE BRIGHTNESS PROBLEM

The problem here is that both ellipticals and disk galaxies appear to have approximately constant mean surface brightness (different values for the two classes). Is this real, or is it just the result of selecting galaxies of approximately the same mean surface brightness for surface photometry ?

First I should summarise briefly the observational evidence. (i) For elliptical galaxies, Fish (1964) showed directly that the total luminosity is proportional to the effective area. Also, Faber and Jackson (1976) and others have shown that the total luminosity is proportional to v^4, where v is the stellar velocity dispersion. Again, simple dynamical arguments show that this implies constant mean surface brightness. (ii) For the <u>disk</u> component of disk galaxies, the surface brightness follows the exponential law $I(r) = I_o \exp(-\alpha r)$. Freeman (1970) showed that, for systems with detailed surface photometry at that time, I_o was approximately constant, at 21.6 ± 0.3 B mag arcsec^{-2}, after correction for inclination and galactic absorption. Most recent deep photometry of disk galaxies confirms this result, except for dwarf systems which have fainter I_o.

Kormendy (1976) argues that this apparent constancy of I_o has no physical significance, but results merely from the contribution of the bulge in the outer parts of the galaxy where the exponential disk seems best defined. This seems unlikely, however, because (a) the constant I_o is seen also for systems like M33 and NGC 300 which have very weak bulges, and (b) much of Kormendy's photometry was for galaxies with prominent lenses, and did not always go faint enough to define the exponential disk properly.

This constancy of mean surface brightness (which also implies constancy of mean surface density because the M/L ratios are now known not to vary greatly) is dynamically important if it is real. For ellipticals, Fish showed that it is associated with a particular potential energy - mass relation, which contains useful information about the energy radiated away during the collapse phase, before star formation occurred. For the disk galaxies, the constant I_o means that the disk angular momentum is proportional to the 7/4 power of its mass, and this constrains angular momentum acquisition theories. De Vaucouleurs (1974) warned, however, that this constancy of mean surface brightness may just be a selection effect, associated with choosing galaxies against the night sky background, and Disney (1976) has recently given a compelling theoretical argument to support this view.

Disney assumes that galaxies are chosen for photometric programs by their radius (or area) at some limiting isophote corresponding to the detection level on photographic plates. For conventional blue light photographs against a dark sky, this isophote level is approximately $\mu = 24$ B mag per square arcsec. For a given total galaxy luminosity, there is some value of the mean surface brightness which maximises the radius or area of the galaxy within the $\mu = 24$ isophote. The particular value depends on the law for the radial surface brightness distribution: for $\mu = 24$, this corresponds to an exponential disk with $I_o = 21.6$, as observed, <u>and</u> also to an $r^{1/4}$-law system with the mean surface brightness observed by Fish. The picture is then that disk galaxies, for example, have a wide range of true I_o-values at a given total luminosity: only those with $I_o \approx 21.6$ are chosen for photometry because they are the largest in radius. Although Disney's argument strongly suggests that selection effects act in this way, because it reproduces the observed mean surface brightnesses so well, I think this may be fortuitous, for the following reasons.

(i) The preferred value of I_o depends on the value of μ set by the emulsion - night sky brightness combination. As mentioned above, $\mu \approx 24$ for blue IIa-O exposures against

1.1 THE DISTRIBUTION OF LIGHT IN GALAXIES

a dark night sky, and the corresponding preferred value of I_o is about 21.6, as observed. Until recently, most galaxy photographs, including the Palomar Sky Survey, had approximately this value of μ. Now there is available the UK 48-inch Schmidt survey of the Southern sky, on IIIa-J emulsion, and this has $\mu \approx 26$. The largest galaxies at a given total luminosity then have $I_o \approx 23.7$. So, if Disney's hypothesis is correct, there should be many giant galaxies of relatively low surface brightness appearing on this survey, which does not appear to be so. The UK Schmidt survey does show many low surface brightness galaxies, but HI observations of these by members of the UK Schmidt unit and others in Australia suggest that they are mostly hydrogen-rich dwarfs.

(ii) The predicted value $I_o = 21.6$ for $\mu = 24$ depends also on the <u>exponential</u> form of the light distribution for disk galaxies. Although some systems have almost pure exponential disks, the surface brightness in most disk systems, at the $\mu = 24$ level, is dominated by the bulge or the lens. So if Disney's assumption is correct, that galaxies are chosen by their size at this level, then most are selected mainly by the size of their bulge or lens, for which the luminosity profile does <u>not</u> have the exponential form. The theory would then predict a value of I_o different from 21.6.

(iii) The observed I_o values for the exponential disks of disk galaxies are approximately constant <u>only</u> after correction for inclination and galactic absorption. The sample with <u>corrected</u> $I_o = 21.6 \pm 0.3$ included systems with <u>uncorrected</u> I_o values between 19.6 and 23.1. Obviously it is the uncorrected values that appear in selection procedures.

(iv) Although a particular mean surface brightness is preferred in Disney's picture, the selection effect is not strong enough to exclude from observation small systems of high surface brightness. Not many of these are known among the nearby galaxies; although the compact galaxies were possible candidates, recent work by Kormendy (1977) on red compacts and Rodgers (unpublished) on blue compacts shows that their surface brightnesses are fairly normal.

In summary, although there are certainly low surface brightness dwarf disk and elliptical galaxies, it appears unlikely to me that the apparent constant mean surface brightness (within each class) for giant galaxies results from selection effects. However this can now be tested directly, by choosing galaxies from the deep survey according to specific selection rules.

REFERENCES

Bertola, F. and Capaccioli, M.: 1975, Astrophys. J. 200, 439.
Binney, J.: 1976, Monthly Notices Roy. Astron. Soc. 177, 19.
de Vaucouleurs, G.: 1974, IAU Symposium No. 58, page 3.
de Vaucouleurs, G.: 1977, Astron. J. 82, 456.
Disney, M.J.: 1976, Nature 263, 573.
Faber, S. and Jackson, R.: 1976, Astrophys. J. 204, 668.
Fish, R.: 1964, Astrophys. J. 139, 284.
Freeman, K.C.: 1970, Astrophys. J. 160, 811.
Freeman, K.C.: 1974, IAU Symposium No. 58, page 129.
Harris, W. and Smith, M.: 1976, Astrophys. J. 207, 1036.
King, I.R.: 1966, Astron. J. 71, 64.
Kormendy, J.: 1976, Ph.D. thesis, Caltech.
Kormendy, J.: 1977, Astrophys. J. 214, 359.
Oemler, A.: 1976, Astrophys. J. 209, 693.
Strom, S. and Strom, K.: 1977, Preprint.
Wakamatsu, K-I.: 1977, Publ. Astron. Soc. Pac. 89, 267.
Wilson, C.: 1975, Astron. J. 80, 175.

DISCUSSION FOLLOWING REVIEW I.1 BY K.C. FREEMAN

TOOMRE: I would suggest that we deal with ellipticals first and with disks later.

STROM: I would like to point out that our analysis of $\sim$ 100 galaxies in the Coma cluster shows (1) the same range of $\varepsilon(r)$ characteristics (anything you can dream up!) as you report from Ivan King's sample, (2) in $\sim$ 10 - 15 per cent of our sample, a rotation of position angle of the major axis of the elliptical isophotes as one proceeds outward. These results as well raise the spectre of triaxial ellipsoids and/or bars in E-galaxies.

MILLER: In a paper in press we argue that ellipticals flatter than E2 are probably prolate objects. This is based on dynamical arguments, in that we have been unable to construct systems like ellipticals with axis ratios less than 2/3 that remain axisymmetric (oblate). Flatter systems are "unstable" to nonaxisymmetric disturbances, and ultimately form bar-like or triaxial forms that rotate about a short axis. Even so, ellipticals can show remarkably rapid rotation without much flattening. We have models with Ostriker-Peebles $t = T_{rot}/|W|$ of 0.17 that are nearly spherical, and systems with t as great as 0.27 that remain oblate. The system with t = 0.27 had about an E2 profile when viewed across the rotation axis. Systems that were spun even faster became bar-like.

A remarkable variety of forms has been obtained. For example, we have seen systems that resemble the Dedekind ellipsoids — figures stationary in space, but ellipsoidal in shape, supported by special particle motions.

HUNTER: My analysis referred to by Dr. Freeman is based on models for which the mass distribution function is assumed to depend only on the energy and the angular momentum about the axis of symmetry. Hence the possible effects of an extra integral of motion are ignored. The dynamical quantity that is significant in producing a distribution of oblateness (or prolateness) is the difference between the mean square velocity in the circular direction and that in the radial direction. Knowledge of this quantity alone does not allow any mean circular velocity of rotation to be deduced, because of the well-known insensitivity of the mass distribution to reversals in the directions in which orbits are described.

EKERS: The elliptical galaxies which have powerful radio lobes have a well defined axis which is presumably related to the ejection of the material which powers the radio source. If the apparent distribution of radio - optical major axes is analysed with the assumption that the elliptical galaxies are oblate and rotating about their minor axis then the radio axes are randomly distributed with respect to the rotation axes. A result which is somewhat difficult to reconcile with the constancy of the radio axis in time. However if the elliptical galaxies are prolate the analysis will be different and it is likely that the data is consistent with radio lobes aligned with the rotation axis.

MILLER: PROBLEMS IN GALACTIC DYNAMICS BY MEANS OF THREE-DIMENSIONAL SIMULATIONS

Computer simulations of galaxy models by means of n-body integrations are the only tools available for many problems in galactic dynamics. Long-range effects and collective effects can be followed into nonlinear ranges. A fully three-dimensional form is required to prevent unrealistic restrictions, especially if we hope to understand how galaxies can develop the remarkable symmetries shown by many observed objects. Large numbers of particles are required to separate dynamical and relaxation time scales. A fully three-dimensional simulation has been in operation on the ILLIAC IV computer at NASA-Ames Research Center for the past year. Simulations were based on 50 000 to 120 000 particles.

Motion pictures were shown for several classes of problems. These include: (1) Collapsing configurations started from a rigidly rotating sphere of uniform density. Within two free-fall times, runs started with different initial rotation speeds and different velocity dispersions all formed a prolate bar that rotated about a short axis. Some runs passed through intermediate stages in which a ring appeared briefly while others briefly showed sheets. These intermediate forms were too short lived to be of astronomical importance, but the bar is a long-lived form that may be important astronomically. (2) Particle motions in the bar. These bars are peculiarly stable, and may well represent barred spirals or prolate elliptical galaxies. There is pronounced streaming in the direction of rotation. Orbits have been studied, and rotation curves suitable for comparison with observation have been obtained. (3) Flattening of rotating systems. This investigation was undertaken to determine how flat an elliptical might be and still remain oblate. The flattest found so far is E2. These centrally condensed systems can rotate remarkably rapidly and still retain nearly spherical form. (4) Galaxy formation by collapse of gaseous spheres with star formation. Stars continue to form at about the same rate (relative to the amount of gas available from which to form them) and to deplete the gas to unacceptably low levels (1-2%). (5) Collision of pairs of galaxies. The transfer of energy from orbital motion to internal degrees of freedom is large, and leads to ejection of a surprisingly large number of stars. Orbital angular momentum is rapidly depleted in deeply penetrating collisions by means of ejected particles.

VAN DEN BERGH: Can one exclude the possibility that galaxies with "peanut-like" nuclear bulges are, in fact, objects in which we are looking <u>almost</u> along the axis of a bar?

FREEMAN: No, I don't think we can exclude that.

MARK: For axisymmetric fluid equilibria in differential rotation (specifically previously applied to rotating stars, Mark, J.W.-K., 1968, Ap.J. 154, 627), I have found that some of them are flatter at the polar axis than at some positive radii nearer the equator. These equilibria are stable as far as we know and they need only a two or three to one variation in rotation frequency from center to surface. Bulges of galaxies might well be exhibiting similar equilibria whose edge-on view would then be more box-shaped (versus spheroidal) or even bi-lobed (peanut-shaped). Presence of the flat disk might accentuate such behaviour and make them more observable.

BALDWIN: If one analyses the surface brightness in "peanut" structures as a function of R at constant z on the assumption of cylindrical symmetry, does it lead to a physically sensible distribution of luminosity? For example, is it everywhere positive?

FREEMAN: I have no idea.

EINASTO: Dr. Freeman indicated that in some galaxies (example VII Zw 303) the disk or lens should have a hole at the center. We have studied the mass distribution in our Galaxy (Tartu Astr. Obs. Teated No. 54, 3, 1976) and in M31 (Tartu Astr. Obs. Preprint A-4, 1977) with particular emphasis of the structure of the disk. For both galaxies two models have been calculated: (a) with a normal exponential disk, and (b) with a disk having a hole at the center. In the first case it is impossible to obtain a model circular velocity curve with a minimum at $\sim$ 2 kpc, in the second case the calculated curve represents well the observed minimum. It is well known that in Sa and Sb galaxies the hydrogen avoids central regions and forms a ring-like structure. If our interpretation of the velocity curve is correct, then we come to the conclusion that a gaseous disk from which disk stars have been formed has always had a ring-like structure. This means that all the proto-galactic matter with small angular momentum has been used to form the bulge and the halo of Sa and Sb galaxies and no gas with small angular momentum has been left over to form the central part of the gaseous disk.

BERMAN: I have recently examined a 3-dimensional computer model of a galaxy with a disk component, with a cutout center and a massive spherical bulge. This general structure lasts for many galactic years. The qualitative shape of the rotation curve agrees with the above shown by Dr. Einasto.

GALLAGHER: In regard to Dr. Freeman's comments about the universality of $B(0) \sim 21.5$ for exponential disks, one way to select low surface brightness disks is to look for galaxies which seem to have excessive HI content for their optical characteristics. One such object is NGC 6902, a southern Sb which was previously classified as S0 due to the low surface brightness of the disk. However, surface photometry from CTIO 4 m plates shows $B(0) \sim 21.8$. Thus where very different selection criteria were used to choose a galaxy for surface photometry, we get the usual result. It is also interesting to note that due to the large size of 6902, the galaxy is very luminous ($M_B \sim -22$ for $H_0 = 50$ km s^{-1} Mpc^{-1}), and it therefore does seem possible that we select against low surface brightness galaxies in searching for luminous systems.

WIELEN: In 1970, you were forced to introduce exponential disks of type II, which do not extend right into the center of the galaxy. You have not mentioned these objects today. What is your present opinion about these objects?

FREEMAN: The point was that if you extrapolated the exponential disk you had a hole at the center in the distribution of the surface brightness. I am sure now that that is due to the lens; all we do is taking a spheroid, adding on an exponential disk, and adding a lens structure on top of that.

WIELEN: Is the lens really a new, independent component of a galaxy, or can it be simply a perturbation of the disk? For example, is the thickness of the lens different from that of the disk? Perhaps the lenses are just positive deviations of surface density from the ideal exponential disks while the type II-disks represent the negative perturbations.

VAN WOERDEN: Do the brightness profiles of lenses indeed have their maximum away from the center? Hence, are lenses toroidal rather than spheroidal?

FREEMAN: Yes, probably, but it is in principle impossible to work it out. To do that the contribution of the bulge should be subtracted from the observed profile but we have no a priori knowledge about what the bulge is like.

MEBOLD: THE KINEMATICS OF THE LENTICULAR GALAXIES NGC 1291 AND NGC 1326

21-cm line observations of the galaxies NGC 1291 and NGC 1326 (van Woerden et al. 1976, P.A.S.A. 3, 68) with the Parkes 64-m telescope (HPBW = 15') have been used to determine the position angle PA(HI) of the maximum velocity gradient of the HI gas distribution. This position angle is of particular interest because the position angles PA(R) and PA(L) of the two main constituents of these galaxies, the outer ring and the inner lens-bar system, differ by about $90°$ for NGC 1291 and by about $60°$ for NGC 1326 (cf. the photographs and the more detailed discussion in Mebold et al. 1978).

We find that the HI gas distribution is sufficiently extended in both NGC 1326 (HPW of HI gas = 4 min) and NGC 1291 (HPW = 16 min) that a velocity gradient can be determined. For NGC 1291 we further find that a minimum of about 50% of the HI gas must be, and a maximum of 100% may be, located in the outer ring or even further out. For NGC 1291 we find that PA(HI) = $85° \pm 10°$ is inconsistent with PA(L) $\sim 170°$, but is consistent with PA(R) $\sim 80°$. However, PA(R) is ill defined because the outer ring is nearly face-on (inclination $i \leq 10°$). For NGC 1326 we find PA(HI) = $270° \pm 15°$ which again is inconsistent with PA(L) $\sim 30°$, but is consistent with PA(R) $\sim 90°$.

We conclude that the lens in both galaxies is either a triaxial spheroid rotating about one of its minor axes or a rotational ellipsoid seen partially edge-on and rotating about an axis which is not aligned with that of the outer ring and the HI gas distribution.

KINMAN: THE DWARF SPHEROIDAL GALAXIES NEAR M31

The distribution of stars in the three dwarf spheroidal galaxies discovered near M31 by van den Bergh (1972, Ap.J. 171, L31) have been derived from counts by L.L. Stryker and the author using IIIa-J plates taken at the prime focus of the KPNO 4-m reflector. Although the UBK7 Wynne triplet corrector has an unvignetted field of radius ~ 25 arcmin, seeing-dependent corrections to the star counts must be applied which can critically affect the distributions obtained for radial distances exceeding 10'. These corrections are particularly important where, as here, the brightest stars are close to the plate limit (B ~ 23), and the systems have a limiting radius (r_t) $\sim 15'$. The determination of the core radius (r_c) requires the faintest limiting magnitude that is consistent with a resolution which minimizes the confusion from crowding. This was achieved by the microphotometry of the three best plates of each galaxy. The array of densities from each plate was combined on a computer and adjusted to increase the contrast in a selected range of densities before being converted back to a photograph which was optimized for star counting. Provisional results for Andromeda II show $r_c \sim 1.2'$, and $r_t \sim 15.5'$, which correspond to 240 pc and 3.0 kpc at a distance of 700 kpc. The central surface brightness is ~ 24.7 V-mag per square arcsec giving an apparent V-magnitude of 13.0 and $M_V \sim -11.3$; this is similar to Leo I in our Galaxy (Hodge 1971, Ann. Rev. A.A. 9, 35).

THE PAST HISTORY OF STAR FORMATION IN GALAXIES

Beatrice M. Tinsley
Yale University Observatory

1. THE "STANDARD SCENARIO" - AN IDEALIZATION

It is generally believed that most galaxies have similar ages, and that the age differences of their most visible stellar populations are due to a diversity of evolutionary stages. The presence of old stars in even late-type galaxies rules out an age sequence increasing from irregulars through S0's; dynamical constraints prevent transitions of isolated disc systems into ellipticals; and the gas contents of typical galaxies provide enough fuel for a long life of star formation at current rates (Baade, 1963; Roberts, 1963, 1975; King, 1971; Sandage, 1975). More specifically, models for the evolution of photometric properties show that integrated colors of galaxies are consistent with the following hypothesis:

> Normal galaxies have the same age and stellar initial mass function (IMF), and have monotonically decreasing star formation rates, but the time scale for star formation to decline varies along the sequence of morphological types

(Tinsley, 1968; Searle et al., 1973; Larson and Tinsley, 1974). I shall refer to this hypothesis as the "standard scenario" for the evolution of galaxies. Its interest is twofold: in correlating star-formation time scales with structure in a way that calls for dynamical explanation, and in providing an idealized "smooth background" to the complicated real scene of star formation in galaxies. This paper will address the second point.

The colors of ordinary galaxies agree remarkably well with the predictions of this idealization, as illustrated simply by UBV colors in Figure 1. The data points are for a sample selected on the basis of morphological "normality", as detailed in Larson and Tinsley (1978; to be referred to as LT). The line is an eye-estimated mean, and it corresponds closely to the theoretical 2-color locus of models conforming to the standard scenario. Indicated on the figure are limiting

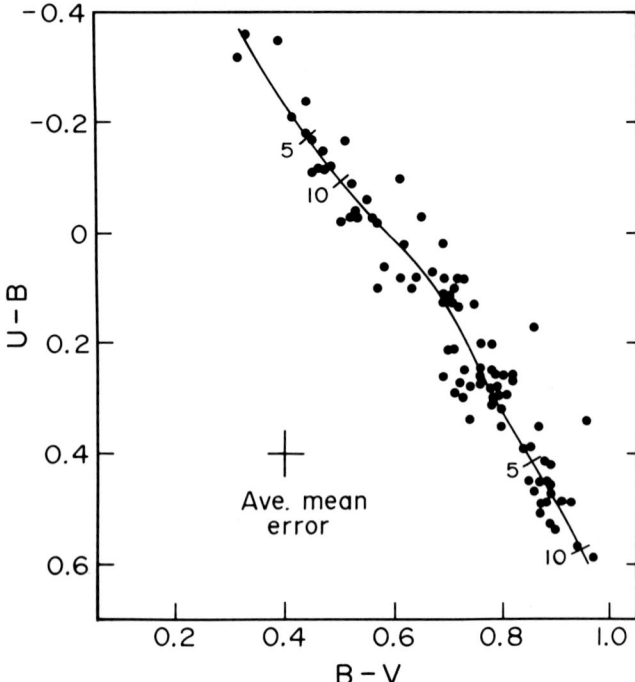

Figure 1. Colors of morphologically "normal" galaxies; cross marks indicate limiting colors of monotonic models with ages 5 and 10×10^9 yr.

colors for such models with ages 5 and 10×10^9 yr; the blue limit at each age is for constant star formation, the red limit is for a brief initial burst, and models with monotonically decreasing rates (of any time-dependence) have intermediate colors. It can be seen that few points deviate by more than the observational errors from the mean line, in accordance with our hypothesis. (The data points have not been adjusted for metallicity differences, which tend to scatter colors parallel to the mean line and probably account for the unusually blue top few points [LT].) Only stars above about 0.4 $M_\odot$ contribute significantly to the light, but if galaxies have the same IMF for less massive stars, the models predict that their mass-to-luminosity ratios should increase with color along the mean line by an overall factor $\sim$ 10. This indeed seems to be the case (Sargent and Tinsley, 1974; LT).

In spite of such successes, the standard scenario is obviously not a unique explanation of the broad photometric properties and gas contents of normal galaxies. The next section discusses some of the growing evidence that it is at best a very oversimplified view of galactic evolution.

2. SOME VARIANTS FOUND IN NEARBY GALAXIES

2.1. Variations in the initial mass function

The IMF is known to vary, at least on small scales within galaxies. Freeman (1977) has reported values of the power-law slope (x) in globular clusters and young LMC clusters ranging from 0.2 to > 3! Field stars in the solar neighborhood have $x \sim 1.3$ for most masses, and ~ 0.3 between 0.5 and 1 $M_\odot$, but there are wide variations among small-scale regions of star formation (associations and open clusters). Van den Bergh (1976) and Schweizer (1977) have noted an unusual paucity of O stars in the star-forming knots in NGC 4594, and Larson (1977) has discussed evidence that the most massive stars may form only in regions of large-scale dynamical disturbance or shock fronts. On the other hand, the data in Figure 1 are not consistent with entire normal galaxies having x as great as 3 or as small as 0.2 for stars above 1 $M_\odot$; and the giant-dominated spectra of elliptical galaxies and the central bulges of spirals show that $x \lesssim 1$ for their stars of 0.5 - 1 $M_\odot$ (Whitford, 1977; Tinsley and Gunn, 1976). Thus in spite of localized variations that are of great interest for understanding star formation, the IMF of nearby field stars may be a valid average for most galaxies.

2.2. Non-monotonic star formation rates

Ongoing bursts of star formation greatly affect the colors of some galaxies (Searle et al., 1973; Huchra, 1977; LT). Moreover, sudden enhancements in the star formation rate seem to be correlated with violent dynamical interactions: LT found that interacting galaxies have a much wider color distribution than the sample in Figure 1, and that the most anomalous colors can be interpreted as due to conversion of up to 5% of the mass of a system - i.e. the whole interstellar gas content of a typical spiral - into stars in times as short as a few times 10^7 yr. Bursts of this strength may not be very rare. Toomre (1977) and Vorontsov-Velyaminov (quoted in discussion after Toomre's paper) have estimated that 15% or more of all galaxies have undergone collisions or mergers in their lifetimes. Examples of systems with colors suggesting that a large burst of star formation occurred a few times 10^8 yr ago are NGC 2623 (Arp 243) and NGC 7252 (Arp 226), which Toomre cites as prospects for ongoing mergers because of their extensive tails. It is startling to realize that after only $\sim 10^9$ yr galaxies like these will resemble ordinary ellipticals in both morphology and colors! Some nearby peculiar galaxies such as M82 and NGC 5128 may be undergoing bursts of star formation induced by encounters with clouds of intergalactic gas (Larson, 1976a; Solinger et al., 1977), which again may be a fairly common interruption to the smooth conversion of a galaxy's gas supply into stars. In summary, the frequency of present-day interactions and the short time scale for obvious effects of these to die away indicate that many so-called normal galaxies could have had very disturbed pasts. Star formation may, in many cases, be "monotonic" only in averages over judiciously chosen time intervals.

2.3. Age differences

A number of candidates for young galaxies have been discussed in the literature (Sargent and Searle, 1971; Larson, 1976a), but in all cases 90% of the mass could be in old stars masked by the "terrific splash" (Baade, 1963) of a younger burst, and the chaotic appearance could be due to interactions. There are, however, significant large-scale age differences among stellar populations within galaxies. For example, Demarque and McClure (1977) find that almost all disc stars in the solar neighborhood are $\lesssim 1/3$ times as old as some globular clusters, and more generally, collapse models (Larson, 1976b) predict that star formation should reach its peak rate much later in the outer parts of discs than in the central regions. Only in integrated light that includes the central bulge are galaxies likely to appear coeval.

3. THE OUTLOOK

Altogether, the apparent success of the "standard scenario" in accounting for integrated UBV colors of most galaxies (Fig. 1) is due to the insensitivity of these colors to parameters other than the total age and the star formation rate in the last 10^9 yr (LT). Population syntheses based on far more detailed photometric data are, unfortunately, not decisively more sensitive to details of past evolution (although they can reveal some interesting components of a galaxy's population). Two general problems are: (1) the light is dominated by stars in late stages of evolution that are poorly understood and insensitive to the ages of their precursor stars; and (2) colors and spectral features are sensitive to chemical abundances, which vary in overall level and in relative proportions of heavy elements (Faber, 1977). Obviously, collisions and other vagaries play havoc with attempts to predict chemical abundances using evolutionary models! We shall therefore have to rely on detailed observations of nearby galaxies to tell us about complications that are smoothed over beyond detection in most cases.

In spite of its superficiality, the simple picture provided by the standard scenario for galactic evolution remains valid and suggestive in that correlations among properties of typical galaxies are consistent with the main variable along the Hubble sequence being the time scale for star formation. Details that are accessible for only a few galaxies point to the frequent occurrence of large deviations from such features of the idealized picture as a monotonic star formation rate, isolated evolution, or an invariant IMF, and some of the most interesting problems ahead are to document and understand such deviations.

I am indebted to Dr. Morton S. Roberts for presenting this paper at IAU Symposium No. 77. This work was supported in part by the National Science Foundation (Grant AST76-16329) and by the Alfred P. Sloan Foundation.

REFERENCES

Baade, W.: 1963, 'Evolution of Stars and Galaxies', C. Payne-Gaposhkin (ed.), Harvard University Press, Cambridge.
Demarque, P. and McClure, R.D.: 1977, in B.M. Tinsley and R.B. Larson (eds.), 'The Evolution of Galaxies and Stellar Populations', Yale University Observatory, New Haven, p. 199.
Faber, S.M.: 1977, in B.M. Tinsley and R.B. Larson (eds.), 'The Evolution of Galaxies and Stellar Populations', Yale University Observatory, New Haven, p. 157.
Freeman, K.C.: 1977, in B.M. Tinsley and R.B. Larson (eds.), 'The Evolution of Galaxies and Stellar Populations', Yale University Observatory, New Haven, p. 133.
Huchra, J.P.: 1977, Astrophys. J., in press.
King, I.R.: 1971, Publ. Astron. Soc. Pacific 83, 377.
Larson, R.B.: 1976a, Comments Astrophys. and Space Phys. 6, 139.
Larson, R.B.: 1976b, Monthly Notices Roy. Astron. Soc. 176, 31.
Larson, R.B.: 1977, in B.M. Tinsley and R.B. Larson (eds.), 'The Evolution of Galaxies and Stellar Populations', Yale University Observatory, New Haven, p. 97.
Larson, R.B. and Tinsley, B.M.: 1974, Astrophys. J. 192, 293.
Larson, R.B. and Tinsley, B.M.: 1978, Astrophys. J., in press (LT).
Roberts, M.S.: 1963, Ann. Rev. Astron. Astrophys. 1, 149.
Roberts, M.S.: 1975, in A. Sandage, M. Sandage, and J. Kristian (eds.), 'Galaxies and the Universe', University of Chicago Press, Chicago, p. 309.
Sandage, A.: 1975, in A. Sandage, M. Sandage, and J. Kristian (eds.), 'Galaxies and the Universe', University of Chicago Press, Chicago, p. 1.
Sargent, W.L.W. and Searle, L.: 1971, Comments Astrophys. and Space Phys. 3, 111.
Sargent, W.L.W. and Tinsley, B.M.: 1974, Monthly Notices Roy. Astron. Soc. 168, 19P.
Schweizer, F.: 1977, preprint.
Searle, L., Sargent, W.L.W., and Bagnuolo, W.G.: 1973, Astrophys. J. 179, 427.
Solinger, A., Morrison, P., and Markert, T.: 1977, Astrophys. J. 211, 707.
Tinsley, B.M.: 1968, Astrophys. J. 151, 547.
Tinsley, B.M. and Gunn, J.E.: 1976, Astrophys. J. 203, 52.
Toomre, A.: 1977, in B.M. Tinsley and R.B. Larson (eds.), 'The Evolution of Galaxies and Stellar Populations', Yale University Observatory, New Haven, p. 401.
van den Bergh, S.: 1976, Astron. J. 81, 797.
Whitford, A.E.: 1977, Astrophys. J. 211, 527.

DISCUSSION FOLLOWING PAPER I.2 WRITTEN BY B.M. TINSLEY

BIERMANN: I would like to mention that at Bonn we have made both theoretical calculations and radio continuum observations for Markarian

galaxies. The evidence supports the viewpoint that bursts of star
formation play an important role in the evolution of galaxies (Astron.
Astrophys.: 1977, 60, 353).

VAN DEN BERGH: The interstellar gas in NGC 5128 is loaded with dust
and the HII regions in it emit strong [NII]. This suggests that the
gas in this galaxy contains material that has been processed in stars
i.e. it is probably not pristeen intergalactic material. Note that
NGC 5128 is an isolated object so one would not expect to observe
material swept out of S0 galaxies such as is found in rich clusters.

GALLAGHER: I think that you have to be careful about the abundances in
the gas in NGC 5128. There is a band of gas across NGC 4278 (Knapp,
Kerr and Williams 1977, preprint), but little dust is seen. This
suggests that star formation in accreted gas could provide the heavy
elements, even though the gas was initially metal poor.

M.S. ROBERTS: We should be cautious in invoking intergalactic HI since
we have yet to find such gas except for apparent tidal remnants and the
gas in the Sculptor group. The latter is much more complex than
described in the discovery paper by Mathewson et al.

VAN DEN BERGH: MULTI-COLOUR OBSERVATIONS OF NGC 5128 (= CENTAURUS A)
   In cooperation with R.J. Dufour ultraviolet, blue and yellow plates
obtained with the CTIO 4-m telescope have been combined to produce both
real-colour (Dufour and Martins 1976, J. Appl. Photogr. Eng. 2, 93) and
false-colour photographs of NGC 5128. Each plate used for this work
was calibrated using both sensitometer spots and photoelectric UBV
photometry at over one hundred positions within NGC 5128 (van den Bergh
1976, Ap.J. 208, 673). Inspection of the real- and false-colour photographs and of composite images produced with the KPNO Image Picture
Processing System shows the following:
(1) The main body of NGC 5128 has a brightness profile that obeys de
Vaucouleurs' $\sigma \propto R^{-\frac{1}{4}}$ law for elliptical galaxies.
(2) The outer isophotes of NGC 5128 are quite highly flattened and have
an axial ratio $b/a \sim 0.7$. The major axis of these isophotes is almost
perpendicular to the equatorial dust band.
(3) Active star formation along the edges of this dust band manifests
itself by the presence of numerous blue stars, clusters and associations. Some of the reddened stars embedded within the dust band have a
distinctly orange colour on true-colour prints. The integrated UBV
colours of points within the dust band show that large numbers of hot
young stars must be embedded within dust clouds.
(4) The distribution of young stars (see Figure 1) and of HII regions,
which were observed with the Yale 1-m telescope, are most easily understood in terms of star formation within a doughnut-shaped volume. This
suggests that the dark band that appears to cross NGC 5128 is, in
reality, the front side of a ring or disk of absorbing material.
(5) No optical radiation is seen at the positions of the inner lobes of
the radio source Centaurus A.

Fig. 1. Ultraviolet plate (103aO + UG2) of NGC 5128 obtained by Dufour in seeing 1-2" with the CTIO 4-m telescope.

ELVIUS: Could the Hα regions (and regions of star formation) which are supposed to be seen through the bulge of NGC 5128 perhaps be connected with the dust clouds seen projected against the bulge?

VAN DEN BERGH: It is always difficult to interpret a three-dimensional structure that is seen on a two-dimensional picture. Continuity arguments and the absence of young stars associated with the "high latitude" dark clouds suggest that the high latitude HII regions and young stars lie behind the main body of NGC 5128, i.e. they form the backside of a ring- or doughnut-shaped region of star formation.

OORT: Have you an explanation of the north-south asymmetry in the number of dark patches projected on the bulge of NGC 5128?

VAN DEN BERGH: I have no idea.

"Curiously, one giant elliptical rotates as expected, and that is NGC 3557."

A. Toomre in Discussion I.1

ROTATION CURVES IN THE OUTER PARTS OF GALAXIES FROM HI OBSERVATIONS

Edwin E. Salpeter
Cornell University, Ithaca, New York

Abstract: 21cm observations at the Arecibo Observatory for 9 edge-on spiral galaxies are described. Flat rotation curves are found in most cases.

INTRODUCTION

I mainly report on 21cm observations carried out at Arecibo of the outermost parts of spiral galaxies in order to derive rotation curves. The data is most clearcut for late-type spirals since the neutral hydrogen tends to extend out further in these cases. The optical emission is then dominated by the disk component and the surface brightness (Freeman 1970) decreases approximately exponentially, $\sigma_L(r) \propto \exp(-\alpha r)$, with r the distance from the center in the galactic plane. The most dramatic question is whether the ratio of the surface density $\sigma_M(r)$ of total gravitational mass to surface brightness $\sigma_L(r)$ is typically independent of r (and of order 5 in solar units) as suggested by Baldwin (1975). Our data corroborates the suggestion by Roberts (1975, 1976) that the ratio σ_M/σ_L in fact increases markedly with r in the outer regions of many spiral galaxies.

If $\sigma_M(r)/\sigma_L(r)$ were constant, the circular rotation velocity $V_{rot}(r)$ of hydrogen in the disk (if velocity dispersion is neglected) would be a unique function of αr. This function (Monnet and Simien 1977) has a rather flat maximum but already decreases quite rapidly for αr between about 5 and 7. With optical radius R_{opt} defined as $0.5 D_o$, taken from de Vaucouleur et al (1976), Freeman (1970) gives $\alpha R_{opt} \sim 3$ to 3.5 and the rotation velocity should have dropped below the maximum by $r \sim 1.5 R_{opt}$ and even more by $r \sim 2 R_{opt}$. There is one clearcut case where the rotation curve is of just that form, the Sab (or Sb) galaxy M81 (Gottesman and Weliachew 1975, Rots 1975). However, there were (even before Arecibo) a few equally clearcut cases where the rotation velocity is flat in the vicinity of $r \sim 1.5 R_{opt}$, especially the Sb galaxy M31 (Roberts et al 1975, 1978) and the Scd galaxy NGC2403, which indicate σ_M/σ_L increasing with radial distance r. The outer regions of galactic disks can be warped by $\gtrsim 20°$ (Sancisi 1976) and the interpretation of rotation curves is clearcut only when a galaxy is viewed almost edge-on ($\cos 20°$ is close to unity, but $\cos 80°/\cos 60°$ is not). Furthermore, to obtain a rotation-velocity accurately one requires good signal to noise ratios to measure the outer edge of the spectral profile (and not merely its peak). For this reason we surveyed at Arecibo a number of large, nearby, late-type galaxies which are seen almost edge-on.

THE ARECIBO OBSERVATIONS

We selected 14 spiral galaxies of type Sb to Sm accessible to the 21cm beam of the upgraded 1000-foot dish at the Arecibo Observatory, with $R_{opt} > 2!7$ and disk inclination less than 40° away from edge-on (four of these 14 galaxies are two galaxy-pairs). IC1727 is too perturbed by its larger partner (NGC672) and four of the single galaxies showed too little hydrogen emission to be useful. Of the 9 remaining galaxies, NGC4517, 4527 and 7331 showed approximately flat rotation curves, but the emission extended only to $r \sim 6'$. The remaining six galaxies were detected out to $r > 10'$, but the radii reported by Krumm and Salpeter (1977) were too large because of sidelobes and beam-smearing (see also Sancisi 1978).

An example of the raw Arecibo data is illustrated in Fig. 1 (for the large, almost completely edge-on Sb galaxy NGC4565) which shows clearly a feature probably present in other galaxies as well: The hydrogen intensity fluctuates and decreases slowly at first and then drops suddenly. For NGC4565 the drop (on the N side) occurs near $11!6$ (the profile at $14!5$ is mainly the $\sim 10\%$ sidelobe of the emission at $8!7$) and the hydrogen extends at least to $10' = 1.8 R_{opt}$. For (one side of) NGC 925, 4559 and 4631 the minimum extent is about $7' = 1.5 R_{opt}$, $8' = 1.8 R_{opt}$ and $11' = 2.0 R_{opt}$. The outer edge of each spectral profile for NGC4565 in Fig. 1 is close to 1460 km s^{-1} and the flat rotation-curve (Fig. 2) gives a minimum total mass of about $6 \times 10^{11} M_\odot$ (assuming a distance of 20 Mpc). NGC925, 4559 and the W side of 4631 also show flat rotation-curves and $\sigma_M(r)/\sigma_L(r)$ must increase appreciably for $r \gtrsim 1.5 R_{opt}$ (the data is poorer for NGC672 and 4656 which are disturbed by their companions). On the E side of NGC4631 (closest to NGC4656) the emission for $r > 15'$ appears at much lower rotation velocities, but this may come from a disturbed extended halo.

A recent extensive survey, carried out at Arecibo on early-type galaxies for a different purpose (Krumm and Salpeter, 1978, in preparation), is also of some relevance. Of about 80 galaxies surveyed, 5 showed sufficiently extended hydrogen emission (tc $r \sim 3'$ to 6') to give rudimentary rotation curves: NGC3626 (S0$^+$), 4324 (S0$^+$) and 4698 (Sab) suggest flat rotation curves (but only the last with any reliability), whereas NGC3623 (Sa) shows a slightly decreasing rotation curve. Only one galaxy, NGC3593 (S0a) which is $\sim 1°$ away from NGC3623, displays a sudden decrease in rotation velocity on one side--rather similar to NGC4631.

I.3 ROTATION CURVES IN THE OUTER PARTS OF GALAXIES

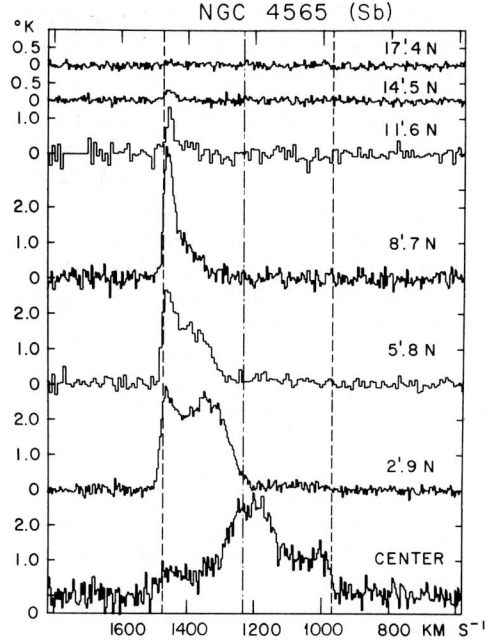
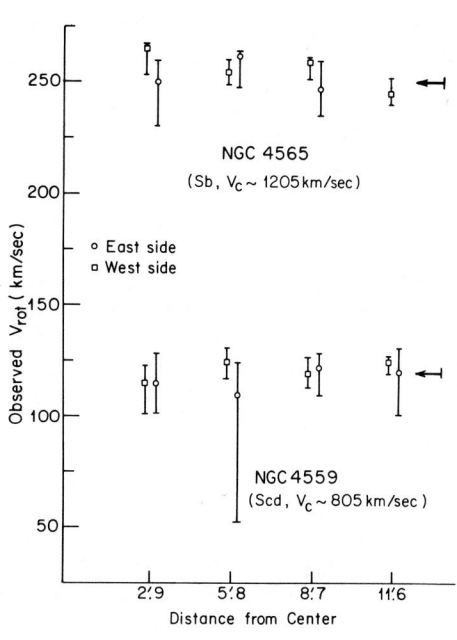

Fig. 1: Spectral profiles at various distances from the center along a semi-major axis.

Fig. 2: Rotation curves for two spiral galaxies.

DISCUSSION

M31 and the best few Arecibo cases show that at least some spiral galaxies have an increasing ratio $\sigma_M(r)/\sigma_L(r)$ of mass-density to surface brightness. Although M81 provides a counter-example, I want to argue that this is the rule rather than the exception: Huchtmeier (1975) gives a compilation of rotation curves for 17 galaxies and about half of these show flat or increasing rotation curves. The other half show a decreasing rotation velocity after reaching a maximum some distance r_{max} from the center, but the numerical value of r_{max} is crucial: If $\sigma_M(r)$ were indeed proportional to $\sigma_L(r)$, then r_{max} would have to be $\sim 0.6\ R_{opt}$. In reality, except for M81 and one or two other cases, r_{max} is much larger; for NGC3109, for instance, $r_{max} \sim 19'$, $R_{opt} \sim 6'$ and V_{rot} at $r \sim 40'$ is still larger than at $r \sim R_{opt}$.

It is not clear whether the constancy of σ_M/σ_L for M81 is connected with it being of early type or with having close companions. For the galaxies with increasing $\sigma_M(r)/\sigma_L(r)$, it is also not known how much further the "invisible halo" (Ostriker and Peebles 1973) extends than the hydrogen boundary at $\sim 2\ R_{opt}$. The physical form of the "invisible" mass need not be particularly exotic: White dwarfs and low mass stars ($M < 0.08\ M_\odot$) which cannot burn hydrogen cool to quite low light-to-mass ratios in 10^{10} years. An "early stellar population III" which formed mainly into stars with $M < 0.08\ M_\odot$ and/or $M > 1.5\ M_\odot$ (which end up as black holes, neutron stars or white dwarfs) would be effectively invisible. An initial mass function in the outer disk weighted more towards $M \sim 0.2\ M_\odot$ probably also gives sufficiently large σ_M/σ_L.

If V_{rot} were exactly constant, the (volume) mass density $\rho_M(r)$ would be proportional to r^{-2} and the mass $M(r)$ contained inside radius r would be proportional to r. This law must break down at very small and very large r, but for a number of galaxies V_{rot} is remarkably constant over quite a dynamic range of distances r. This is illustrated not only by the Arecibo data but by more accurate recent HI data on M31 (Roberts and Whitehurst 1978, this volume) and by some recent optical rotation curves (V. Rubin, unpublished). The precision of the constancy of V_{rot} is probably not an accident of the formation history of a galaxy, but some servomechanism was at work. One such possibility comes from an instability (Lovelace and Hohlfeld 1977), which is related to the derivative of $V_{rot}(r)$.

As mentioned, the outer radius R of the "invisible stellar population III" is not known, but for galaxies in clusters tidal breakup must limit R since $M \propto R$ and tidal effects depend strongly on M and R. "Population III" debris, stripped from the outermost layers of many galaxies, may accumulate towards the center of a cluster. Similar stripping may occur from individual galaxies in double or triple systems, such as M81.

I am indebted to N. Krumm, who carried out much of the work reported here, and to National Science Foundation Grant AST 75-21153.

REFERENCES

Baldwin, J. E.: 1975, in "Dynamics of Stellar Systems", IAU Symposium No. 69, A. Hayli (ed.), Reidel, Dordrecht.
de Vaucouleurs, G., de Vaucouleurs, A., and Corwin, H. G.: 1976, Second Ref. Cat. of Bright Galaxies, Univ. of Texas Press, Austin, Texas.
Freeman, K. C.: 1970, Astrophys. J. 160, 811.
Gottesman, S. T. and Weliachew, L.: 1975, Astrophys. J. 195, 23.
Hutchtmeier, W. K.: 1975, Astron. Astrophys. 45, 259.
Krumm, N. and Salpeter, E. E.: 1977, Astron. Astrophys. 56, 465.
Lovelace, R. V. and Hohlfeld, R. G.: 1977, Cornell Univ. CRSR Report 661.
Monnet, G. and Simien, F.: 1977, Astron. Astrophys. 56, 173.
Ostriker, J. P. and Peebles, P. J.: 1973, Astrophys. J. 186, 467.
Roberts, M. S.: 1975, in "Dynamics of Stellar Systems", IAU Symposium No. 69, A. Hayli (ed.), Reidel, Dordrecht.
Roberts, M. S. and Whitehurst, R. N.: 1975, Astrophys. J. 201, 327.
Rots, A. H.: 1975, Astron. Astrophys. 45, 43.
Sancisi, R.: 1976, Astron. Astrophys. 53, 159.
Sancisi, R.: 1978, this volume.

DISCUSSION FOLLOWING PAPER I.3 GIVEN BY E.E. SALPETER

BALDWIN: On the assumption of constant mass-to-light ratio in an exponential disk the circular velocity at one Holmberg radius is about 0.85 of that at the peak of the rotation curve. Do you agree then that for data at up to one Holmberg radius we are talking about an expected drop of only 30 km/s from the peak?

SALPETER: Assuming a pure exponential disk for NGC 4565, one has to guess its scale length from Freeman's measurements on other galaxies. I use correlations with de Vaucouleurs "corrected" radii (rather than Holmberg radii) and find a drop to less than 0.85 times the maximum velocity at a 10' distance. However, if your guess of the scale length is more correct than mine you would have to add a nuclear mass component to avoid too low a velocity at small distances (2', say), and this in turn would enhance the velocity drop at 10'. The velocity drop should be even more apparent for the W side of NGC 4631, for M31 and for Huchtmeier's best cases.

STROM: Is there any evidence for or against the presence of a "halo" of neutral hydrogen surrounding NGC 4565 or is the gas restricted to the disk?

SALPETER: The Arecibo beam cannot resolve the disk thickness and we have not looked far from the disk. The Westerbork data shows a warped disk but, as far as I know, no halo.

SANCISI: HI SIZES AND ROTATION CURVES OF SOME EDGE-ON GALAXIES

Two of the galaxies studied by Krumm and Salpeter (1977, A.A. 56, 465) at Arecibo, NGC 4565 and 4631, have also been investigated with the Synthesis Radio Telescope at Westerbork (WSRT). The WSRT data do not show the large extent of the HI disks and the flat rotation curves reported by these authors.

NGC 4565. The HI layer of this galaxy shows a large bending in the outer parts (Sancisi: 1976, A.A. 53, 159). On the western side the HI emission extends out to 11!5 radius, not to 14!5 as reported by Krumm and Salpeter. Beyond 11!5 the WSRT maps, even when convolved to the Arecibo resolution, do not show any signal: the 3σ upper limit is about 12 mJy, or $N_H = 7 \times 10^{19}$ cm^{-2} for a 27 km/s wide channel and the 0!8 × 1!9 WSRT beam, or correspondingly 16 mJy and 1.5×10^{19} cm^{-2} for the 3!2 × 3!2 Arecibo beam. At these positions the WSRT and Arecibo data have similar sensitivities.

The rotation curve is flat out to about 8' ($\approx$ 50 kpc if H = 55 km/s/Mpc). Between 8' and 11!5 it is not well defined because of the bending, but may drop off by 10 or 20 km/s. The maximum rotational velocity is 250 ± 10 km/s. For comparison the Holmberg radius is 10', and the de Vaucouleurs radius is 8'.

NGC 4631. This system is in close interaction with NGC 4656. The distribution of HI around NGC 4631 and between the two galaxies is quite complex (Weliachew et al.: 1977, A.A., in press). The HI disk of NGC 4631 on the western side extends to about 10' (= 15 kpc) from the center, not 20' as reported by Krumm and Salpeter. At 14!5 the Arecibo data show a peak intensity of about 100 mJy, whereas the WSRT 3σ upper limit at the same resolution is 20 mJy. At this position the WSRT observations have a factor 2 better sensitivity. The corresponding 3σ limits on the column density are 6×10^{19} cm^{-2} for a 27 km/s wide channel and a 48" × 89" beam, or 1.5×10^{19} cm^{-2} for the 3!2 × 3!2 Arecibo beam. At distances of 20' the WSRT data are not adequate for a comparison.

The rotation curve is approximately flat out to 9!5 (∼ 14 kpc) on the west side, i.e. out to the Holmberg radius. Beyond this distance no HI emission is detected. On the east side the rotation curve is flat out to about 6'. Beyond this limit the radial velocity drops off in rough agreement with Krumm and Salpeter's results. However this gas does not seem to be part of the NGC 4631 disk (Weliachew et al. 1977). The maximum rotational velocity is 150 ± 10 km/s.

Some spurious signal due to side lobes is clearly present in the Arecibo observations of these two galaxies, and may explain all or most of the discrepancy with the WSRT results. The possibility of sufficiently bright and extended emission (angular size > 15'), which could have escaped detection in the WSRT observations, seems to be ruled out by the agreement of the WSRT values of flux density with single dish measurements.

To sum up: (1) The extent of the HI disks at the detection limit of the present Arecibo and Westerbork observations ($1-5 \times 10^{19}$ cm^{-2}) is significantly less (25 to 50 percent) than reported by Krumm and Salpeter. The HI does not extend much farther than the Holmberg radius. It is possible, of course, that these objects have significantly larger HI sizes at lower densities. (2) The rotation curves are flat out to the edge of the bright optical disk well inside the Holmberg radius. At larger distances from the center they are poorly defined or not determined at all because no HI is detected.

These conclusions are also valid for NGC 5907 and 4244 (Sancisi, in preparation), and for NGC 891 (Sancisi et al.: 1975, in La Dynamique des Galaxies Spirales, p. 295). Other galaxies have been known to have flat rotation curves inside the Holmberg radius. The results from these edge-on galaxies, therefore, do not seem to bring any significantly new evidence on the dynamics and mass distribution in galaxies.

BOSMA: THE KINEMATICS OF A SAMPLE OF ABOUT TWENTY SPIRAL GALAXIES

We briefly describe the results of a study aimed to relate the morphology of spiral galaxies to dynamical parameters as inferred from e.g. rotation curves. 21-cm HI line data with high spatial resolution (ratio Holmberg radius/beamsize larger than ∼ 5) have been collected for about twenty galaxies of various morphological types, either from the literature or from still unpublished recent studies using the Westerbork telescope.

The radial velocity fields of these galaxies have the characteristic pattern expected for an inclined disk in differential rotation, but most of them show in addition specific patterns of noncircular motions, and small-scale irregularities. Often there is a strong resemblance in the type of noncircular motions among different galaxies, although their amplitude varies. These types of noncircular motions constitute not only dynamical problems by themselves, but also restrain the derivation of rotation curves.

We briefly discuss the main types of noncircular motions:

(1) Motions associated with spiral arms. Visser (this volume) has constructed a model to describe these motions in M81 and has corrected the rotation curve for their effects. For other galaxies showing hints of similar effects we have made no correction.

(2) Large-scale symmetric deviations. In many galaxies the kinematical major axis changes its position angle as function of radius, but the velocity field still has a central symmetry.

  a. In cases where this major axis change is in the inner parts usually there is a misalignment with the major axis of some of the structures seen on optical photographs. We then suspect an oval or bar-like distortion to be present in the potential field of the disk. The effect on the rotation curve is probably not so large; we did not correct for this effect.

  b. In cases where the major axis change is in the outer parts the major axis of the inner parts of the optically visible disk is usually well aligned. In these cases we suspect the plane of the galaxy to be warped. Models have been constructed using concentric tilted rings (cf. Rogstad et al.: 1974, Ap.J. 193, 309), with circular motion in each ring independent of its orientation.

Some ambiguity exists in the proposed split up (the warped barred spiral), and the assumptions concerning the resulting rotation curves are not backed up by an adequate theory.

(3) Large-scale asymmetries. These occur mainly in the outer parts (M81, M101), and can usually be attributed to tidal interaction with a neighbouring galaxy. We have used a rotation curve for the symmetric inner parts only.

Apart from these problems, instrumental limitations hamper the determination of the rotation curves in the inner parts of most galaxies. Sometimes rapid variations in radial velocities occur there on the scale of one beamsize, sometimes no HI gas is detected. These problems can be partly solved using optical data; however, for still $\sim$ 40% of the galaxies in our sample the rotation curve in the central parts is not well determined.

We have collected rotation curves for 23 galaxies, of which 11 are warped, 3-5 have oval distortions or bars, and 4 have large-scale asymmetries. Most rotation curves do not decline very fast, if at all, past the turnover radius, and none of them show the Keplerian drop-off in the outer parts. We have calculated mass models using both the thin-disk method described by Nordsieck (1973, Ap.J. 184, 719), and the spheroid(s)+disk model fitting described by Shu et al. (1971, Ap.J. 166, 465). These methods are perhaps not justified in view of massive haloes, but they are relatively simple.

The distribution of mass surface density, $\sigma_m(r)$, with radius r shows that larger galaxies have a slower decline of $\sigma_m(r)$ with r, and that probably earlier type spirals have a steeper increase of $\sigma_m(r)$ towards the center. No unique correlation with Hubble type has been found, partly because of the difficulties mentioned above. The curves of mass within a cylinder of radius r, $M(r)$, derived using Nordsieck's method, do not converge to a final value at the radius of the last measured point R_o. Comparison with the results from the mass model fits shows that usually 20% - 40% of the mass in the model fit lies beyond R_o.

The "total" mass-to-light ratio $(M(R_o)/L_B)$ turns out to be independent of Hubble type, colour and luminosity, with typical values between 5 and 15 (H = 75 km/s/Mpc). For a number of galaxies, most of them not being warped, we have calculated the radial distributions of the mass-to-light ratio and of the total mass-to-HI gas mass ratio. M/L has a tendency to increase with radius, M/HI has a tendency to drop off to a more or less constant, but for each galaxy different, level. Given the uncertainties arising from noncircular motions, beamsmoothing, thin disk approximations etc., these results should be considered as very tentative.

A detailed account of the work summarized here will be presented elsewhere (Bosma, 1978).

COMTE: NEW OPTICAL ROTATION CURVE OF M101

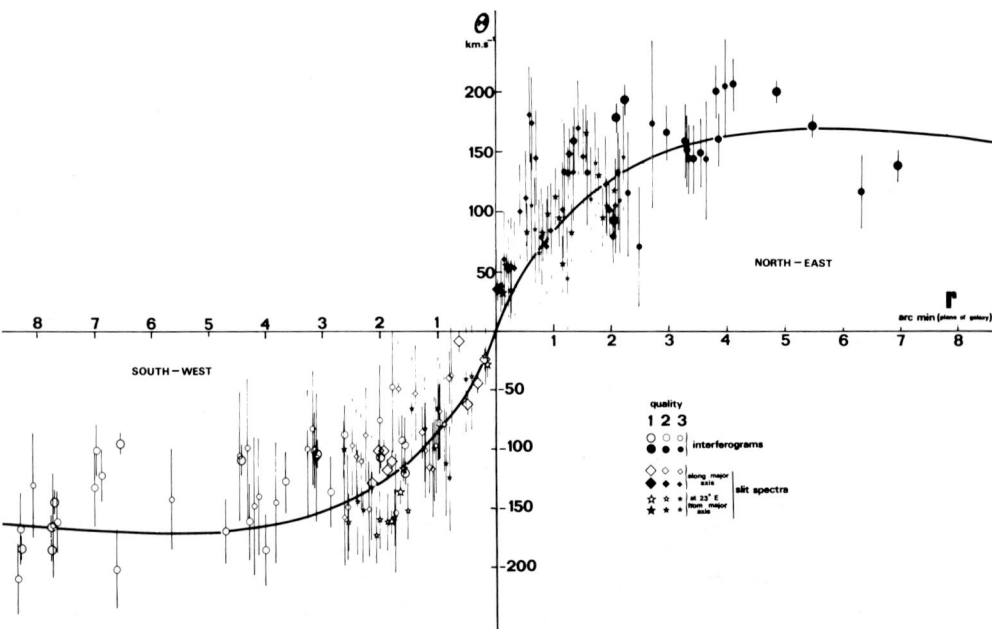

Rotation curve of the galaxy M101 derived from a velocity field obtained by interferometry and spectroscopy of the Hα line, across a 7' radius in the optical disk. The rotation velocities have been computed owing to a systemic velocity of 238 km/s, a major axis position angle of 35° and an inclination of 27°. A mass model, based on the photometric parameters of M101, built for a pure exponential disk in circular rotation with a blue M/L ratio of 4 and a total disk mass of $\sim 10^{11}$ M$_\odot$ at 6 Mpc, fits correctly the data (solid line) with a turnover velocity of 170 km/s at 5.'5 from the center. The blue M/L ratio of the bulge has been estimated to be ~ 1.5.

FITZGERALD: Work being done by Jackson, Moffat and myself indicates that the rotation curve for the Milky Way is probably also flat out to about $R \simeq 18$ kpc from the galactic center. This work is based on observations made of very small young open clusters and stars in HII regions. Distances are determined by cluster fitting where possible, and spectroscopic parallaxes for cases in which no main sequence of zero age is determined. Radial velocities are taken from the literature for Hα, and measured from various image tube plates for stars. A plot of $|\omega - \omega_0|$ versus $(R - R_0)$ for this new data suggests V(R) increases beyond the sun (R > 10 kpc). However, there is sufficient observational error that a flat curve with V = 250 km/s beyond R = 10 kpc is consistent with the data, whereas the Schmidt model is definitely excluded.

K.C. Freeman: "How many unambiguous S0s have double radio sources?"

R.D. Ekers: "None! But how many unambiguous S0s are there?"

Discussion I.5

THE LARGE-SCALE RADIO CONTINUUM STRUCTURE OF SPIRAL GALAXIES

P.C. van der Kruit
Kapteyn Astronomical Institute, University of Groningen,
The Netherlands

1. INTRODUCTION

This review concerns the large-scale structure of radio continuum emission in spiral galaxies ("the smooth background"), by which we mean the distribution of radio surface brightness at scales larger than, say, 1 kpc. Accordingly the nuclear emission and structure due to spiral arms and HII regions will not be a major topic of discussion here. Already the first mappings of the galactic background suggested that there is indeed a distribution of radio continuum emission extending throughout the Galaxy. This conclusion has been reinforced by the earliest observations of M31 by showing that the general emission from this object extended over at least the whole optical image. More recently, van der Kruit (1973a, b, c) separated the radio emission from a sample of spiral galaxies observed at 1415 MHz with the Westerbork Synthesis Radio Telescope (WSRT) into a nuclear, spiral arm and "base disk" component, showing that the latter component usually contains most of the flux density. This latter component is largely non-thermal and extends over the whole optical image (see also van der Kruit and Allen, 1976). Clearly it is astrophysically interesting to discuss the large-scale structure of the radio continuum emission.

The study of the spatial and frequency dependence of non-thermal radio continuum surface brightness aims at a number of questions, the most important ones being: (a) Where do the relativistic electrons, which are one component of the cosmic rays originate? (b) How strong are the magnetic fields and what is their distribution, orientation and origin? (c) Can the electrons be contained in the galaxy and if so, how?

From the observational point of view there are a number of difficulties which are partly the reason for the relatively slow progress in these areas. The first is one of observational selection in the sense that the best studied galaxies have a larger brightness temperature in the disk and are usually giant spirals. The effects of this selection are unclear but should always be kept in mind. A second problem is that even though our interest is in the smooth background we do need high-resolution observations. This is because we not only want to correct, if necessary, for smaller-scale structure (nucleus, background sources, bright HII regions) but we also want to study the background emission on scales of -say- the optical scale length. On the other hand, the severe requirements on the surface brightness sensitivity restrict effective studies to lower frequencies, where beamwidths are larger. This poses the problem that at some frequencies at least there are serious effects owing to the absence of short spacings in the synthesis observations required to attain the necessary resolution.

It also is important that the observations have a large dynamic range, since the outer disk features or the halos often have surface brightnesses of one percent or less of the brightest parts. A dynamic range of at least a factor 100 seems usually attainable at the synthesis instruments. Finally, a most serious problem is the necessary separation of thermal and non-thermal emission especially when variations of non-thermal spectral index with radius and hence origin and diffusion models for the electrons are studied. In fact, Baldwin (1976) has indicated that in the galactic background at 1.4 GHz as much as 30% is probably thermal.

2. RADIAL DISTRIBUTIONS OF NON-THERMAL RADIO EMISSION

2.1. Spiral galaxies with relatively bright disk emission

Of the larger Sc galaxies in the northern sky M51 and NGC 6946 have the highest brightness temperatures at radio wavelengths and have consequently been studied in most detail. Van der Kruit, Allen and Rots (1977) studied NGC 6946 with the WSRT at 610, 1415 and 4995 MHz (in wavelength respectively 49, 21 and 6 cm). Their maps showed that the galaxy exhibits large-scale structure, but that details are also present. Combining the measurements at the three frequencies they were able to show that most of the small-scale structure is due to thermal emission from large HII complexes and a relatively strong background source.

A main objective in that study was to obtain an accurate radial distribution of continuum emission and its spectral index and consequently much attention has been paid to corrections for absence of small baselines and for the different beam shapes at 21 and 49 cm. Although often done in practice it is in theory not sufficient to use a similar set of baselines at the two frequencies when there is a change of spectral index. This is because the morphology changes as a function of frequency and hence the visibility function at the two frequencies is different. One should therefore be careful when interpreting <u>large</u> variations in spectral index if they derive from this procedure.

Van der Kruit et al. referred the observations of NGC 6946 to an identical (Gaussian) beam after decomposing the maps into point sources, correcting in this way also for missing short spacings. The maps were averaged in rings which are circular in the galactic plane using orientation parameters from a kinematical HI study. The resulting distribution, which at 49 cm extends to 1.2 Holmberg radii, could be fitted well by exponential functions with a scale length quite similar to that of the optical light.

I.4 THE LARGE-SCALE RADIO CONTINUUM STRUCTURE OF SPIRAL GALAXIES

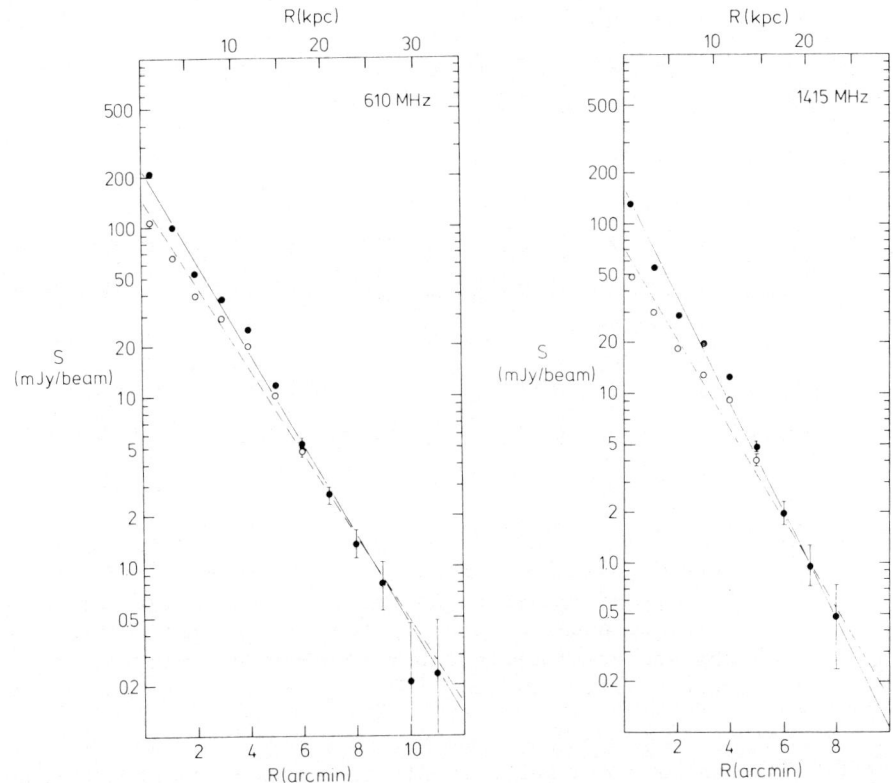

Fig. 1 The radial distribution of radio brightness temperature in NGC 6946 derived after averaging the maps in circular rings in the galactic plane. The open circles are points corrected for the predicted contribution of thermal emission. The HPBW in the maps from which these data are produced was about 1' at both frequencies. (Van der Kruit et al., 1977)

The spectral index varied from -0.5 in the central region to about -1.0 at the last points (see fig. 2). From various lines of arguments they showed that all of this change could be attributed to a radially varying contribution of thermal emission from the HII regions, so that there is no evidence at least over the brighter part of the disk that the spectral index of the non-thermal radio emission changes with radius.

A similar set of observations was obtained by Segalovitz (1976, 1977a) of M51. This system is known to have strong spiral ridges in the radio continuum interpreted as density-wave compression regions (Mathewson et al., 1972). Allen (1975) had earlier pointed out that the general radial fall-off of brightness temperature indicated here also an exponential disk with a scale length typical for late-type spirals.

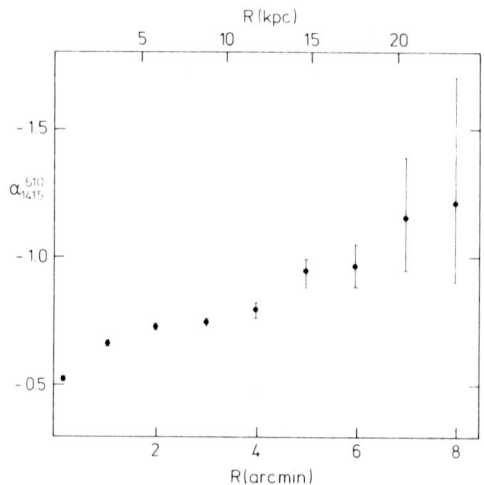

Fig. 2 The radial variation of spectral index in NGC 6946 derived from the points in fig. 1 without correction for thermal emission. As in fig. 1 the error bars relate to uncertainties in the zero-levels in the maps and are therefore not independent (van der Kruit et al., 1977).

Segalovitz (1977a) analysed his measurements in a manner similar to that described above and concluded that there was a real variation of non-thermal spectral index with radius from -0.65 in the central region to -0.8 at the outer edges. Van der Kruit (1977) disagreed with that conclusion and in particular on the basis of Hα flux densities measured by Tully (1974) and of various measurements of Balmer decrements in the HII regions concluded that this change should be attributed to thermal emission. The corrected radial distribution of non-thermal radio emission in M51 would then agree very well with that of the optical light, both being exponential with the same scale length over the main disk.

It should be noted that in NGC 6946 and M51 respectively about 30% and 15% of the total flux density at 21 cm is thermal, which is in agreement with the integrated radio spectra and in particular with recent measurements of the total flux densities at 5 GHz made at Effelsberg (von Kap-herr and Wielebinski, private communication).

An unpublished analysis of the WSRT measurements of M101 (Israel et al., 1975) shows that these observations are consistent with this interpretation but the serious effects of missing short baselines at 1415 MHz make this conclusion very uncertain. Harten (private communication) performed a similar study of IC 342 and found a very strong steepening of the spectrum between 21 and 49 cm with radius. However his maps are derived from similar baseline distributions at the two wavelengths and

1.4 THE LARGE-SCALE RADIO CONTINUUM STRUCTURE OF SPIRAL GALAXIES   37

consequently suffer the difficulties mentioned above. In particular because no attempt could be made to correct for missing short spacings his results may only be correct for small-scale structure.

There are only tentative results for a few more galaxies. For example in NGC 2841 the spectral index is roughly constant with radius, but in NGC 7331 it steepens somewhat with radius (Hummel and Bosma, private communication). The spectrum of NGC 4736 also steepens somewhat with radius (de Bruyn, 1977a), while changes in spectral index are also evident across the disk of NGC 4258 (de Bruyn, 1977b). In all these systems the observations are at least consistent with the view that the change in spectral index is due to thermal emission and that the non-thermal emission is distributed like the optical light. We stress this similarity and not the approximate exponential nature of the distributions in M51 and NGC 6946. Observations of edge-on systems are in general agreement with this conclusion (see section 3a). It has been noted before (Lequeux, 1971; van der Kruit, 1973a, b) that radio continuum distributions are qualitatively similar to the optical appearance.

2.2. Spiral galaxies with weak disks

For galaxies with faint disks information can only be obtained for those with a large angular size, because then the signal-to-noise ratio can be improved by smoothing the maps. However, especially for Sc galaxies the problem of contamination by thermal emission becomes much more serious. The emission of M33 at 21 cm is dominated by thermal emission (Israel and van der Kruit, 1974) so that present studies of this system may be irrelevant to the matters discussed here.

The best remaining galaxies with faint disks suitable for study then are the Sb galaxies M31 and M81. (Note that our Galaxy also has a faint radio disk with a surface brightness similar to these two systems (see also Baldwin, 1976).) Pooley (1969) first mapped M31 and found that the emission at 408 and 1417 MHz was strongly concentrated towards the broad ring of HII regions although the spectrum was clearly non-thermal. This behaviour was studied in more detail at 2.7 GHz (11 cm) with the Effelsberg telescope (Berkhuijsen and Wielebinski 1973, 1974). The radial profile is characterized by a minimum at about 4 kpc followed by a broad maximum at 9 kpc with an exponential fall-off in the outer regions (Berkhuijsen, 1977). The HII regions also peak radially at about 9 kpc.

The same behaviour is found in M81 (van der Kruit, 1973a; von Kapherr et al., 1975; Segalovitz, 1977b) with the minimum at 2 kpc and the maximum at about 5 kpc. The radial distribution of HII regions and possibly also of non-thermal emission in our Galaxy qualitatively behaves similarly.

The full-resolution Westerbork map of M81 (Segalovitz, 1977b) at 21 cm shows radio emission near the spiral arms which Segalovitz attributes to shocks in a galactic density wave. He notes that the spectral index is rather uniform starting in the inner minimum to the outer regions. The radio spectrum is clearly non-thermal. The same general characteristic follows from the detailed work by Berkhuijsen (1977) on M31. From this study it also follows that the thermal contribution to the radio emission in the arm regions is not negligible.

In the outer regions beyond the broad maximum the radio continuum and blue light fall-offs in M31 can both be described by exponential functions with similar scalelengths (Berkhuijsen, 1977). This same property approximately holds for Segalovitz's radial distribution in M81. From this it seems that the major difference between these galaxies and those described above lies in the deficiency of radio emission and HII regions in the central regions. It has been noted by Berkhuijsen and by Segalovitz that this may be related to the low density of neutral hydrogen in the same inner regions. In this respect our Galaxy also behaves similarly, even when the distribution of CO and that of H_2 inferred from this is added.

2.3. The origin of the cosmic rays

Lequeux (1971) suggested on the basis of the general similarity of the radio continuum extent in external spiral galaxies with that of the extreme population I (in particular the HII regions) that the supernovae of type II associated with this young population are the prime source of cosmic rays. Van der Kruit and Allen (1976) however pointed out that a relation with the general optical light rather than the HII regions seemed more likely.

The detailed analysis of NGC 6946 (van der Kruit et al., 1977) and M51 (van der Kruit, 1977) clearly indicate that the non-thermal emission is in the radial direction distributed similarly as the total stellar component and certainly different from that of the HII regions. Indeed, if the observed change in spectral index is due to mixture with thermal emission it follows that the thermal and non-thermal emission cannot be distributed in the same way. It also is a general property of spiral galaxies that the HI falls off much slower than the radio continuum brightness, so the non-thermal emission certainly does not correlate with the extreme population I. Van der Kruit et al. noted that the radial distributions of non-thermal emission and optical light are also similar to those derived statistically for supernovae (irregardless of type!) in galaxies. They concluded that it is then reasonable to take the view that indeed supernovae, their associated remnants and/or pulsars are important sources of cosmic rays but that then supernovae of both types have to be considered.

The above discussion then argues against a unique relation of the non-thermal radio emission and the young population I. It is evident from discussions of supernova statistics (e.g. Tammann, 1977) that only supernovae of type II (SNII) belong to the young population I. The progenitors of SNI are part of the older disk population (possibly in binaries). They may not be related to recent star formation unless in the interarm regions star formation deficient in the most massive stars occurs (Tinsley, 1977). The occurrence rates and energetics of the two

I.4 THE LARGE-SCALE RADIO CONTINUUM STRUCTURE OF SPIRAL GALAXIES

types of supernovae by themselves already indicate that if the origin of cosmic rays is associated with supernova activity, there is no reason to presume an exclusive relation to the young population I. On the other hand the evidence discussed here is also consistent with the view that cosmic rays originate in any constituent of the disk that is distributed like the total stellar disk population (e.g. flare stars). It is of course also possible that the distributions of magnetic field, lifetimes of the electrons and cosmic-ray sources conspire in such a way that a spurious relation with total star light results.

We now have to discuss how the radial profiles of the galaxies discussed in section 2.2. fit in the picture outlined above. From that discussion it follows that these systems deviate only in the central regions where a minimum in radio continuum, HI and HII is found. In these central minima the HI density is very low; much lower than found at comparable radii in e.g. NGC 6946 and M51. Segalovitz (1977b) in his study of M81 suggested that in these regions the extremely low gas density inhibits star formation but is not preventing shocks from forming (thin dust lanes are observed). This already might lead to reduced production of cosmic rays. Also Ekers (private communication) has suggested that the broader z-distribution of mass at these radii due to the bulge component might allow the cosmic rays to expand quickly in the z-direction. Finally in these regions of very low gas density the magnetic field could also be weaker than elsewhere in the disk.

The difference of these systems therefore is restricted to a relatively small central region and possibly related to extremely low gas densities. Note that in our Galaxy there appears a minimum in SNR's in the central region (e.g. Berkhuijsen, 1977), although there are obvious selection effects. It should be noted that those electrons that radiate in the minimum are probably accelerated there rather than diffusing from the nucleus or from the maximum further out. This follows from the observation of Segalovitz in M81 that the spectrum in the minimum is not steeper than that in the broad maximum (a significant thermal contribution would make the spectrum even steeper in the maximum).

An important inference from this model has to do with the diffusion of the cosmic rays in galactic disks. Clearly a systematic change of spectral index with radius is most easily interpreted as due to energy losses of the electrons while diffusing through the disk. The correction for thermal emission discussed above is therefore very critical. After concluding that thermal emission is negligible in M51 and that the observed variation of spectral index is entirely due to the non-thermal emission, Segalovitz (1977c) has constructed detailed models fitting the observations in M51. In his model the diffusion coefficient is $\sim 10^{29}$ cm^2 sec^{-1}, the leakage time of electrons out of the disk and its associated magnetic field is $\sim 3 \times 10^7$ years and the slope of the energy injection spectrum of the electrons $\gamma_0 \sim 2.2$, while the source function is strongly concentrated towards the inner regions.

Van der Kruit (1977) has constructed similar models for the case of constant spectral index assuming that the sources of the cosmic rays varied as an exponential disk with the optical scalelength and a constant magnetic field strength. Then the diffusion coefficient is $\sim 10^{29}$ cm^2 sec^{-1}, the leakage time has to be $\sim 10^7$ years or less and the injection spectrum has $\gamma_o \sim 2.6$. Note that these parameters do not differ much from those of Segalovitz in spite of the two greatly different geometries, and cannot alone be used to choose between the models. However, since the case for the observed change of spectral index being due to thermal emission is very good the last mentioned model is strongly preferred. The parameters derived are reasonable compared to what they are estimated to be in our Galaxy (Parker, 1976). Unfolding of galactic γ-ray observations also gives evidence that the distribution of relativistic electrons closely follows that of total stellar mass and of supernovae (Dodds et al. 1975; Stecker, 1977). Note that in both models the electrons escape from the disk in about 10^7 years.

Some comments will be made on the general brightness of disk emission. This can vary considerably from galaxy to galaxy (at 21 cm from T_b = 10 K to T_b < 0.1 K) and is not strongly correlated with Hubble type and colour and correlates only weakly with integrated optical magnitude (van der Kruit, 1973c; Ekers, 1975). Although severe selection effects exist there still is an apparent relation between the power of the nuclear radio source and the average brightness temperature of the disk emission (see for example van der Kruit and Allen, 1976). The evidence mentioned above that the spectral index of the disk emission does not change with radius is good evidence against a model in which the nuclei contribute significantly to the cosmic rays in the disk, since such a model would be qualitatively very similar to the one developed by Segalovitz for M51.

Also the above discussed association of the origin of cosmic rays with the total disk population suggests that the simple assumptions in Biermann's (1976) calculations are not justified, since he directly links the strength of the radio continuum emission to the very recent formation of massive stars. The absence of correlation of disk strength with Hubble type (and colour index) furthermore argues for at least another parameter to control the synchrotron volume emissivity. Since the latter depends most sensitively on the magnetic field strength this might in fact be a dominant factor. In this respect Pacini's (1975) suggestion that pulsars are a source of galactic magnetic fields is important to note, but difficult to quantify in the models.

3. Z-DISTRIBUTIONS AND RADIO HALOES

3.1. Observations of some edge-on galaxies

The most detailed investigation at present has been that of NGC 891 with the WSRT (Allen, Baldwin and Sancisi, 1977). They observed at the three wavelengths of 6, 21 and 49 cm and were also careful to produce maps with identical beam shapes and to correct for missing spacings. They separated the emission into two components. There is a thin disk in the equatorial plane of which at least two-thirds is non-thermal at 6 cm

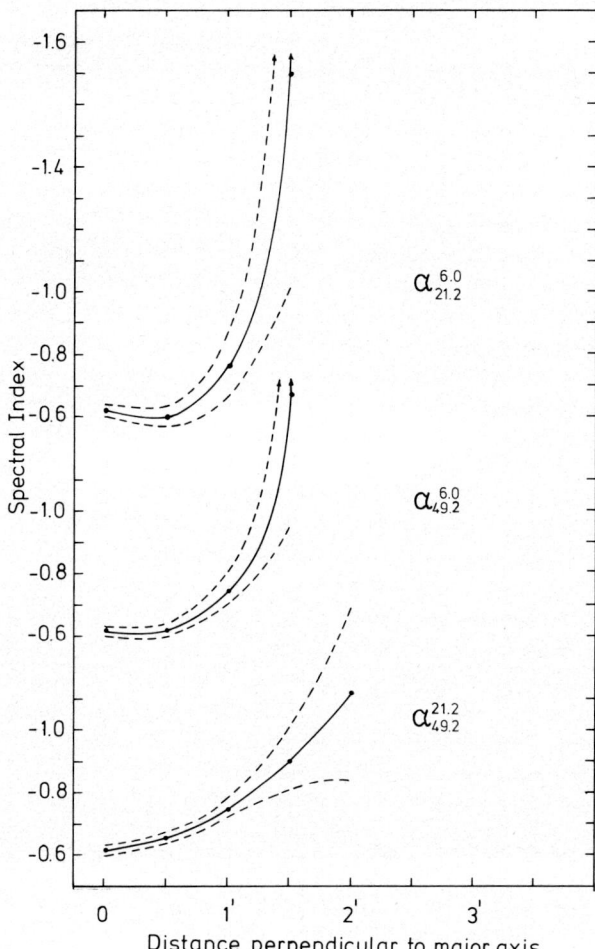

Fig. 3. The variation of spectral index with z in the edge-on galaxy NGC 891. The HPBW in this direction was 1' in all maps and the wavelengths of observation are indicated. At the assumed distance of 14 Mpc, 1' corresponds to 4 kpc. (Allen et al. 1977.)

and a "thick disk" with axis ratio 3.5 : 1 which is non-thermal. The emission has been detected up to about 6 kpc above the plane. In the plane the spectral index where measurable is constant with radius (up to 12 kpc from the centre, which is only half the optical extent), but the spectrum steepens at distances of more than 2.5 kpc out of the plane (see fig. 3).

The steepening of the radio spectrum with z is also found in a few

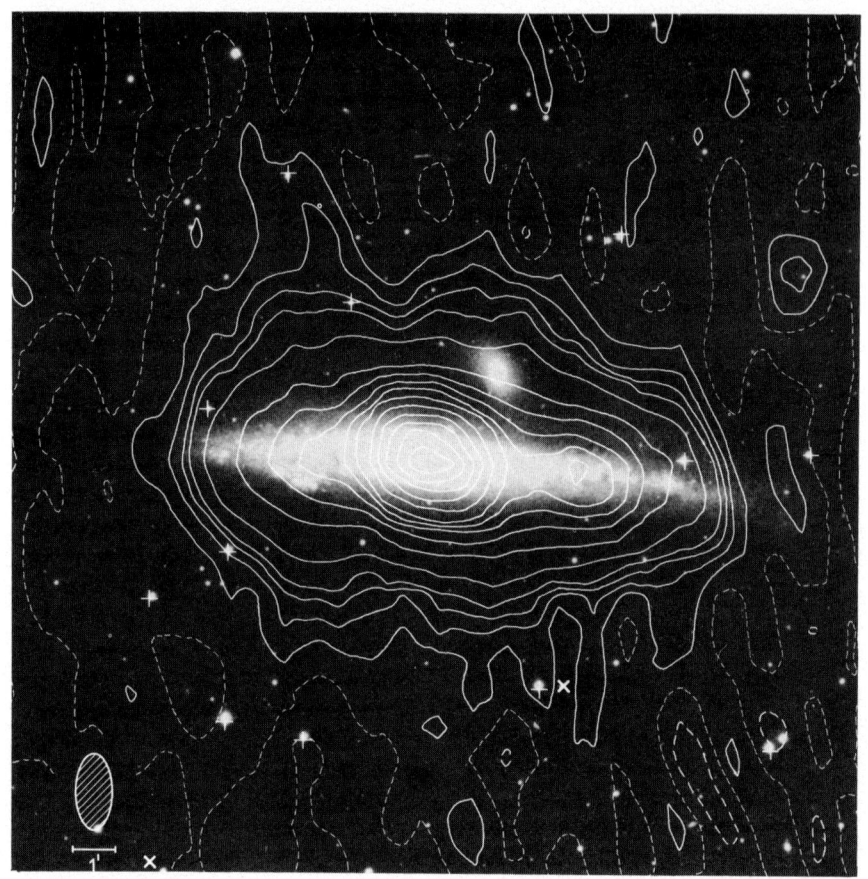

Fig. 4. The radio halo in NGC 4631 observed at 610 MHz (49 cm). The scale and HPBW are indicated in the lower-left corner. (Ekers and Sancisi, 1977.)

more edge-on galaxies observed in Westerbork such as NGC 3556 (de Bruyn and Hummel, private communication), where the emission apparently extends to at least 2 kpc above the plane (see also van der Kruit, 1973b). The spectrum in the disks of NGC 3556 and 5907 steepens with radius which can conceivably be explained by mixture with thermal emission.

Ekers and Sancisi (1977) presented evidence for a flattened radio halo in NGC 4631 with an axis ratio of about 2 : 3 (see Fig. 4). The volume emissivity is comparable to that for the proposed halo in our Galaxy. There is a definite spectral steepening with distance above the plane. Since steepening occurs in all cases still at $z \geq 1$ kpc, mixture with thermal emission cannot be invoked here as an explanation.

3.2. The radio haloes of our Galaxy and M31

Historical notes on these haloes can be found in van der Kruit and

Allen (1976). Webster's (1975, 1977) work has produced good evidence of
a halo in our Galaxy. In view of what has been mentioned above his
assumption that its spectrum is steeper than that in the disk seems well
justified. In Webster's models the volume emissivity is about 30 times
less than in the disk.

Wielebinski (1976) has recently reviewed the evidence for a radio
halo in M31. He compared in particular a recent 408-MHz Bonn map of M31
and surroundings with the distribution of 4C and 5C3 sources. This shows
that structure in the contours such as the spurs near the minor axis are
due to the distribution of these sources on the sky. The Bonn map however
does show a broad region of excess emission around M31, but Wielebinski
attributes it completely to the 5C3 sources. This seems invalid because
his integrated flux density of the region is derived with respect to a
chosen zero level in his map which then already contains the general
distribution of background sources. The smooth excess emission can in
principle also be due to the galactic background, but the data are also
consistent with a smooth halo around M31 with a flux density equal to
or less than that of the disk. The question of radio halo versus structure
in the galactic background can best be resolved by spectral index studies,
especially if a low frequency (maybe 100 MHz) is chosen.

3.3. Containment of cosmic rays

We will briefly discuss the question of containment of cosmic rays
in view of the observational material reviewed above. Allen et al.
(1977) interpreting the z-distribution of the radio emission in NGC 891
suggested that the observations can be understood in a model in which
the electrons diffuse out of the disk. The slope of the electron energy
spectrum steepens with z due to the energy losses and this effect plus
a fall-off of magnetic field strength with z would indicate a propagation
time of 4×10^7 years or more for the electrons to reach a height of
4 kpc above the plane. Note that the models described in section 2 for
the radial radio continuum distributions also require the electrons to
escape from the disk on timescales of the order of 10^7 years.

As is discussed above whenever a spiral galaxy has a strong radio
disk and its orientation is sufficiently edge-on there appear to be
z-extensions up to at least 2 kpc accompanied by a steepening of the
spectrum with z. The most straightforward conclusion from this is that
at least in these brighter systems the electrons are not contained in
the disk, but diffuse into the halo. There containment times would be
very long and the magnetic fields weak.

4. CONCLUSIONS

It is emphasized that conclusions following from the above discussion should be regarded tentative since they derive from a small and
somewhat biased sample. From the observations these are: (a) Often
the observed spectra in disks of spiral galaxies steepen with radius,
but most of this is probably due to mixture with thermal emission.
(b) The non-thermal brightness distribution in the disk correlates with

the total stellar disk population rather than the young population I.
(c) Extensions in the z-direction are observed to at least 2 kpc and
sometimes up to 6 kpc above the plane. (d) The radio spectra steepen
with z and this cannot be due to thermal emission. For the origin and
diffusion of cosmic rays the conclusions are: (e) Cosmic rays originate
in the disk but are not exclusively related to young population I.
Sources of cosmic rays can be supernovae of both types, their remnants,
pulsars and/or any constituent distributed like the total stellar disk.
(f) Relativistic electrons diffuse out of the disks after travelling not
very far (say 1 kpc in about 10^7 years) from their places of origin.
(g) Nuclei do not appear to be significant contributors to cosmic rays
in galactic disks.

ACKNOWLEDGEMENTS

I am grateful to various colleagues, in particular Drs. R.J. Allen,
A. Bosma, R.D. Ekers, R. Harten, K. Hummel, R. Sancisi, A. Segalovitz
and R. Wielebinski for discussions on these subjects over the last
years and for providing information prior to publication. Many of the
observations described here have been done with the Westerbork Synthesis
Radio Telescope which is financially supported by the Netherlands Foundation for the Advancement of Pure Research.

REFERENCES

Allen, R.J.: 1975, La Dynamique des Galaxies Spirales (C.N.R.S.), p.157
Allen, R.J., Baldwin, J.E., Sancisi, R.: 1977, Astron. Astrophys.
    (in press)
Baker, J.R., Haslam, C.G.T., Jones, B.B., Wielebinski, R.: 1977,
    preprint
Baldwin, J.E.: 1976, The Structure and Content of the Galaxy and Galactic
    Gamma Rays, ed. C.E. Fichtel and F.W. Stecker (Goddard), p. 206
Berkhuijsen, E.M.: 1977, Astron. Astrophys. 57, 9
Berkhuijsen, E.M., Wielebinski, R.: 1973, Astrophys. Lett. 13, 169
Berkhuijsen, E.M., Wielebinski, R.: 1974, Astron. Astrophys. 34, 173
Biermann, P.: 1976, Astron. Astrophys. 53, 295
Bruyn, A.G. de: 1977a, Astron. Astrophys. 54, 491
Bruyn, A.G. de: 1977b, Astron. Astrophys. 58, 221
Dodds, D., Strong, A.W., Wolfendale, A.W.: 1975, M.N.R.A.S. 171, 569
Ekers, R.D.: 1975, Structure and Evolution of Galaxies, ed. G. Setti
    (Reidel), p. 217
Ekers, R.D., Sancisi, R.: 1977, Astron. Astrophys. 54, 973
Israel, F.P., Goss, W.M., Allen, R.J.: 1975, Astron. Astrophys. 40, 421
Israel, F.P., Kruit, P.C. van der: 1974, Astron. Astrophys. 32, 363
Kap-herr, A. von, Jones, B.B., Wielebinski, R.: 1975, Astron. Astrophys.
    41, 115
Kruit, P.C. van der: 1973a, b, c, Astron. Astrophys. 29, 231, 249, 263
Kruit, P.C. van der: 1977, Astron. Astrophys. 59, 359
Kruit, P.C. van der, Allen, R.J.: 1976, Ann. Rev. Astron. Astrophys. 14,
    417
Kruit, P.C. van der, Allen, R.J., Rots, A.H.: 1977, Astron. Astrophys.
    55, 421

Lequeux, J.: 1971, Astron. Astrophys. 15, 42
Mathewson, D.S., Kruit, P.C. van der, Brouw, W.N.: 1972, Astron. Astrophys. 17, 468
Pacini, F.: 1975, Origin of Cosmic Rays (Reidel), p. 371
Parker, E.: 1976, The Structure and Content of the Galaxy and Galactic Gamma Rays, ed. C.E. Fichtel and F.W. Stecker (Goddard), p. 320
Segalovitz, A.: 1976, Astron. Astrophys. 52, 167
Segalovitz, A.: 1977a, Astron. Astrophys. 54, 703
Segalovitz, A.: 1977b, Astron. Astrophys. 55, 203
Segalovitz, A.: 1977c, preprint
Stecker, F.W.: 1976, The Structure and Content of the Galaxy and Galactic Gamma Rays, ed. C.E. Fichtel and F.W. Stecker (Goddard), p. 357
Tammann, G.A.: 1977, Supernovae, ed. D.N. Schramm (Reidel), p. 95
Tinsley, B.M.: 1977, Supernovae, ed. D.N. Schramm (Reidel), p. 117
Tully, R.B.: 1974, Astrophys. J. Suppl. 27, 415
Webster, A.: 1975, Monthly Notices Roy. Astron. Soc. 171, 243
Webster, A.: 1977, preprint
Wielebinski, R.: 1976, Astron. Astrophys. 48, 155

DISCUSSION FOLLOWING REVIEW I.4 GIVEN BY P.C. VAN DER KRUIT

TOOMRE: What spectral index would we *expect* from synchrotron theory, i.e., if thermal effects were absent?

VAN DER KRUIT: Such a prediction depends on the energy distribution of the cosmic ray electrons and on how fast they escape.

OORT: Do not possible changes in the magnetic field strength play an important part beside the density and propagation of the cosmic ray particles?

VAN DER KRUIT: Of course, the magnetic field must be playing a role, but we can only guess at its strength and at how it may vary across a galaxy. The main point I wanted to make is that the distribution of nonthermal emission seems to correlate with the older population to some extent rather than with the very young population. It would seem to be a mere coincidence if this apparent correlation were caused by a systematically varying magnetic field and a source function following the young population.

BURKE: After correction of the variation of spectral index for HII contamination in the disks of the galaxies you have considered, how uniform is the nonthermal spectral index? It does seem surprising that the injection and loss mechanisms are so uniform throughout the disk.

VAN DER KRUIT: If you correct as well as you can for HII contamination in M51 you find a nonthermal spectral index of -0.80 ± 0.05 at all radii; thus across this galaxy it is constant to within the errors.

VAN DER LAAN: I would like to make two remarks:
(1) The hope of using the continuum intensity distributions to sort out basic questions of cosmic ray physics, concerning their origin and propagation, is frustrated by the R-dependent nonthermal/thermal radiation mix. The prediction of free-free emission from optical line emission data is very uncertain, beset with problems of nonuniformity in absorption, of filling factor uncertainty and of very low brightness emission from large areas. The best way to determine the true nonthermal emission distribution is to measure the continuum emission at $\lambda \sim 2$ cm, where it is almost purely thermal, and subtract it from lower frequency array maps. Only the Effelsberg telescope can make the required thermal emission maps.
(2) The lack of λ-dependence of the nonthermal intensity in the R-direction indicates that cosmic ray production is widespread in the disks and diffusion is primarily in the z-direction. The λ-dependence of intensity in the z-direction is consistent with this. We should keep in mind in any quantitative treatment that the spectral steepening may be enhanced by a decrease of magnetic field strength with increasing z, so that it takes higher and higher energy electrons to emit at a given frequency.

BIERMANN: I would like to make two comments:
(1) You mentioned my models for the radio continuum emission from galaxies. These models were constructed with the assumption that only massive stars make supernovae and supernovae make nonthermal radio radiation, because that is the most simple assumption conceivable. It is obvious that more complicated models could be constructed and the observations may require more complicated assumptions.
(2) The spectral index map of M51 shows quite strong variation across the disk explainable as you say by variation of the thermal contribution. Then the steepest spectral index on the map corresponds to the nonthermal index, according to your arguments constant across the disk after subtraction of the thermal component. However, the outermost spectral index is steeper than about unity which would seem to be consistent with diffusion and losses of cosmic ray electrons being important contrary to your conclusions. In view of the large errors involved I do not see that you can really exclude a variation of the nonthermal spectral index across the disk.

VAN DER KRUIT: No, you cannot exclude that. Both a constant nonthermal spectrum as well as a steepening spectrum is consistent with the data.

BALDWIN: The nonthermal spectral index you deduce for the outer parts of spirals is close to 1. Are the total flux densities at low frequencies consistent with this value?

VAN DER KRUIT: Yes, they are, but the flux densities at low frequencies are rather uncertain. In the literature the total flux densities at high frequencies are also unreliable. However, the well determined values at 6 cm that have recently been obtained in Effelsberg are in full agreement with our predictions based on the observed spectral index and the important contribution of thermal emission.

WIELEBINSKI: THE THERMAL CONTENT OF IC342

The Scd galaxy IC342 is a very good case to demonstrate the presence of significant thermal emission in a normal galaxy. Recent radio continuum observations of Baker et al. (1977, A.A. 59, 261) and Harten (1977, in prep.) give well calibrated high resolution maps at widely separated frequencies as well as total flux values at a number of frequencies. The spectral index for frequencies below 1 GHz is $\alpha = 1.2$ ($S \propto \nu^{-\alpha}$), in fact greater than for any other normal galaxy. The spectral index reduces to $\alpha = 0.7$ for the frequency range 2.7 to 4.8 GHz. The reality of this flattening of the spectrum is further supported by considering the spectral index distribution across the galaxy. The nuclear area of IC342, where optically "fuzzy" emitting regions are seen, has the spectral index $\alpha = 0.6$. Further sources with flat spectra are found directly on the spiral arms where similar "fuzzy" regions are found. Presumably this is a mixture of the nonthermal emission with spectral index $\alpha = 1.2$ and thermal emission with $\alpha = 0.1$. Individual thermal sources account for some 20% of the emission of 4.8 GHz. From studies of the spectral index it can be shown that some 70% of the total flux is nonthermal so that 10% is due to either smaller obscured HII concentrations or distributed in the disk of the galaxy.

The thermal content of a normal galaxy was originally discussed by Segalovitz (1977, A.A. 61, 59), and was recently re-examined by van der Kruit et al. (1977, A.A. 55, 421) and van der Kruit (1977, A.A. 59, 359). The conclusions of the latter authors for NGC 6946 and M51 disagree with the results of Segalovitz and indicate the emergence of thermal emission at f > 3 GHz. Our results on IC342 support these conclusions. New observations in Effelsberg of normal galaxies at a number of frequencies above 5 GHz are in progress, specifically aimed at settling the issue beyond any doubt.

BYSTEDT: M31 has been observed with the Westerbork telescope at 49 cm (Israel, de Bruyn and Bystedt). The HPBW of the synthesized beam is 0!9 x 1!4 and the r.m.s. noise level at the map center about 0.8 mJy/beam. As van der Kruit mentioned, it has been noted by Wielebinski that there seems to be an excess of 5C3 sources approximately in the minor axis direction of M31. Our map gives the impression that the asymmetry is seen also among sources weaker than the 5C3 sources, and that the asymmetry is present even close to the central part of M31.

WIELEBINSKI: A radio telescope with a beam elongated in declination and the point source distribution near M31 gives a very realistic halo around M31. In addition, a galactic spur complicates the issue. I agree with Dr. van der Kruit that renewed investigation of the halo (if any) around M31 should be made, particularly at low frequencies.

PFLEIDERER: DECONVOLVED SINGLE DISH RADIO OBSERVATIONS OF NGC 6946

The problem of finding the true flux density distribution of extended objects if the measurements are smeared out by the finite beam of the observing instrument cannot be solved exactly (Bracewell and Roberts 1954, Austr. J. Phys. 7, 615). The main reason is that the instrument does not transmit higher Fourier components. However, by observing a large enough area it may be possible to make a deconvolution which is fairly unique. An extreme case is a single point source which is effectively deconvolved simply by determining its position and flux density. Högbom (1974, A.A. Suppl. 15, 417) argues on similar lines.

As a cooperation between Dr. Wielebinski's group in Bonn and the Institut für Astronomie, Innsbruck, we have developed a deconvolution procedure in which the true map is replaced by a model of point sources on a grid. The flux density of each point source is chosen such that the convolution with the beam (which must be known) reproduces the observed map within the noise level. The procedure is similar to the first part (deconvolution part) of the CLEAN procedure (with a small loop gain) as described by Högbom (1974). Some smoothing can be achieved by a superposition of the results for different point source grids.

The first completed example is a 6-cm map from Effelsberg (HPBW = 2.6) of NGC 6946 which has an optical extension of about 5' x 8'. Unfortunately, our radio data are markedly distorted by bad weather. Some details become apparent in the deconvolved map which are quite difficult to deduct directly from the observed map: the central source is little or not extended and has only about one quarter of the total flux density. It is surrounded by extended emission which is entirely restricted to the optically visible region of the galaxy. Some resemblance of this emission to the spiral structure is indicated. Both maps and more details will be published in Mitteil. Astron. Ges. No. 43 (1978).

The theory of our procedure is not yet completed. It should, however, be noted that the informational content of the observed and the deconvolved map is the same within the noise level. The question is therefore not so much whether or not the deconvolved map is correct but rather how to interpret it - in particular, how to avoid an overinterpretation. It should be possible to learn how to read these maps, as we have learned how to read observed maps.

VAN WOERDEN: I agree that deconvolution is possible, and may be necessary, provided one has a stable, accurately known instrumental profile and a good signal-to-noise ratio. How precisely is the antenna pattern of the 100-meter telescope known, and is it a function of time, temperature, altitude, azimuth, etc.?

PFLEIDERER: At 6.2 cm the sidelobes do not play an important role. The shape of the main lobe is important but that we know pretty well.

EKERS: I think the right question to ask is how you tell which of the infinite number of possible solutions is the correct one (see Bracewell and Roberts, Austr. J. Phys. ∾ 1962).

PFLEIDERER: It is a trial and error process; the structure must come out the same each time. The solution is unique for the main structure but not for the details.

THE LARGE SCALE DISTRIBUTION OF RADIO CONTINUUM IN E AND S0 GALAXIES

R.D. Ekers,
Kapteyn Astronomical Institute,
University of Groningen

If we look at the radio properties of the nearby ellipticals we find a situation considerably different from that just described by van der Kruit for the spiral galaxies. For example NGC 5128 (Cen A), the nearest giant elliptical galaxy, is a thousand times more powerful a radio source than the brightest spiral galaxies and furthermore its radio emission comes from a multiple lobed radio structure which bears no resemblance to the optical light distribution (e.g. Ekers, 1975). The other radio emitting elliptical galaxies in our neighbourhood, NGC 1316 (Fornax A), IC 4296 (1333-33), have similar morphology. A question which then arises is whether at lower levels we can detect radio emission coming from the optical image of the elliptical galaxies and which may be more closely related to the kind of emission seen in the spiral galaxies.

Since elliptical galaxies are less numerous than spiral galaxies we have to search out to the Virgo cluster to obtain a good sample. Some results from a Westerbork map of the central region of the Virgo cluster at 1.4 GHz (Kotanyi and Ekers, in preparation) is given in the Table.

Radio Emission from Galaxies in the core of the Virgo Cluster

Name NGC	Hubble Type	m_p	Flux density (10^{-29} W m^{-2} Hz^{-1})	
4374	E1	10.8	6200	3C 272.1
4388	Sc	12.2	140	
4402	Sd	13.6	60	
4406	E3	10.9	< 4	
4425	S0	13.3	< 4	
4435	S0	11.9	< 5	
4438	S pec	12.0	150	

This result is typical for spiral and elliptical galaxies and illustrates the different properties quite well. All the spiral galaxies in this field are easily detectable and have radio brightness similar to those discussed by van der Kruit. One of the ellipticals NGC 4374 (3C 272.1) is 100 times brighter than the spiral galaxies whereas the other, NGC 4406, which is optically one of the brightest Virgo cluster elliptical galaxies, is at least 10 times weaker than even the relatively small spiral galaxies in this field. Neither of the two S0 galaxies are detected.

If we compare the ratio of radio to optical luminosity for a larger sample of elliptical and spiral galaxies we find that the spiral galaxies have a scatter of about a factor of ten in this ratio but this ratio for the elliptical galaxies ranges from very much greater than the spiral galaxy values down to significantly less. This very broad range in radio power from elliptical galaxies compared with that from spiral galaxies is a consequence of the flatter radio luminosity function for elliptical galaxies (Ekers, 1976). The new result which I want to stress here is that there are also a number of elliptical galaxies with less radio emission than the spiral galaxies. A similar conclusion is reached by Dressel and Condon (preprint) from analysis of the new Arecibo Survey of 2000 galaxies from the Uppsala General Catalogue.

In another field of the Virgo cluster one of the brighter elliptical galaxies, NGC 4472, is detected at a brightness level comparable to that seen in the normal spiral galaxies, but Ekers and Kotanyi (1977) have shown that this still has a morphology typical of the radio galaxies although in this case it is smaller than the optical diameter of the galaxy.

Finally some comments on radio emission from S0 galaxies. In general they have radio properties similar to the elliptical galaxies; i.e. some are double sources, they have compact nuclear sources (see discussion on the radio nuclei) and have a wide range of luminosity. However, the total number of detected S0 galaxies is small and the sample of radio emitting S0's contains a number of peculiar objects whose classification is not entirely clear (e.g. Cen A, Fornax A, NGC 2911). Because of this classification difficulty in individual cases it is better to look at some general statistical results. Ekers and Ekers (1973) had suggested that the distribution of axial ratios for radio detected S0's was different from that of all S0's. Since this distribution is determined by the random projection angles (e.g. van den Bergh, 1977) it requires a rather implausible anisotropy in the radio emission to explain this result if these radio S0's are drawn randomly from the total S0 population. This effect persists in the new Westerbork data and is also seen in a larger and independent sample from Dressel and Condon (preprint). The major difference is an absence of radio detected S0 galaxies which are seen edge on. This could be explained if many of the radio emitting galaxies are E or D systems misclassified as S0. Alternatively, the radio emitting S0's may contain a relatively large bulge component which could increase the minor axis diameter of an edge-on S0.

To conclude, I would like to reemphasize that the continuum emission from elliptical and S0 galaxies is completely different from that for

the spiral galaxies. We have no cases where the radio emission from
ellipticals is anything like the distribution of light. We also have
found some elliptical galaxies where the radio emission is substantially
less than that from the spiral galaxies. The lack of an equivalent to
the type of radio emission seen in the spiral galaxies may result from
a difference in the source of relativistic electrons but is more likely
to be related to the lower gas densities and hence lower magnetic fields
in the body of the elliptical galaxies.

REFERENCES

van den Bergh, S. 1977, Observatory, 97, 81
Ekers, R.D. and Ekers, J.A. 1973, Astron. Astrophys. 24, 247
Ekers, R.D. 1975, in "Structure and Evolution of Galaxies", ed.
    G. Setti, p. 217
Ekers, R.D. 1976, in "The Physics of Non-Thermal Radio Sources",
    ed. G. Setti, p. 83

DISCUSSION FOLLOWING PAPER I.5 GIVEN BY R.D. EKERS

VAN WOERDEN: From Dr. Ekers' discussion it appears that the radio
continuum properties of spirals are predictable, while ellipticals show
great variety and lenticulars possibly as well. The situation with
neutral hydrogen appears similar: for spiral galaxies, the ratio M_H/L_B
of hydrogen mass to blue luminosity varies by roughly a factor 10
within one morphological (sub) type; for lenticulars (S0 galaxies) it
varies at least a factor 100, and similarly for ellipticals [see below].

VAN DER LAAN: In spirals the radio power is the cumulative result of
many stellar-scale events. In ellipticals the radio emission seems
always attributable to spectacular events in the nucleus. It is not
surprising then that the distribution of L_R/L_{opt} is much broader for
the ellipticals than for the spirals.

OORT: But why do ellipticals not have SN-events?

EKERS: I don't know, but even if there are as many relativistic
particles in ellipticals as in spirals they are poorly contained because there is no gas to hold the magnetic field. So the electrons can
escape without radiating in the disk of the galaxy.

GALLAGHER: Isn't it true that the ellipticals you detect are the ones
with interstellar matter? So I don't see how your previous point about
the absence of interstellar matter follows.

EKERS: The mass of interstellar matter seen in these elliptical
galaxies is still much less than that in the large spiral galaxies with
which I was making the comparison. However, I do agree that if you
want to argue that the interstellar matter is the critical parameter

then we would have to say that both too much <u>and</u> too little inhibit the formation of this type of radio continuum emission.

WIELEBINSKI: You used Centaurus A and 3C31 as examples of ellipticals. Both have unusually large radio continuum emission features; Centaurus is known to be over 1 Mpc in linear size. Our recent observations of 3C31 at 11 cm show emission over 45 arc min, i.e. $\sim$ 1.5 Mpc. Can those two unusual ellipticals be considered normal?

EKERS: They are not normal ellipticals. They come into the sample because they are close, not because they are large. Centaurus A and 3C31 are weak radio galaxies.

BALDWIN: In how many elliptical galaxies could the radio observations have detected emission of the same surface brightness as we see in spirals?

EKERS: The only quantitative answer I can easily give is based on the distribution of the ratio of total radio-to-optical luminosity which I showed. From this we see that there are about 30 elliptical or S0 galaxies with limits on this ratio which are less than the value for the majority of spiral galaxies. Since the diameters of these elliptical galaxies are usually less than those for the spiral galaxies, this result should also apply for surface brightness.

FREEMAN: How many <u>unambiguous</u> S0s have double radio sources?

EKERS: None! But how many unambiguous S0s are there?

TOOMRE: After your nice compact talk, I am left bothered that we don't seem to see any good in-between cases at all as far as the continuum radio pictures are concerned. We seem to find only the continuum disks or else the double radio sources, but not both in one and the same system.

VAN WOERDEN: Van den Bergh says that the S0s are <u>not</u> a transition between ellipticals and spirals; they are gasless spirals (or rather: gasless disks), forming a sequence parallel to those of gasrich spirals and gaspoor (anemic) spirals. And that's not true either: as I'll show you some S0s are gasrich!

VAN WOERDEN: THE GAS CONTENT OF LENTICULAR GALAXIES

With the 64-meter radiotelescope at Parkes, we have made a survey of neutral hydrogen in all southern ($\delta < -18°$) lenticular (S0 or S0/a) galaxies of diameter $D \geq 2$ arcmin in the Reference Catalogue. Among 55 objects observed, we have 16 strong detections, unconfused by other galaxies in the beam. Their (distance-independent) hydrogen-to-blue-luminosity ratios M_H/L_B range from 0.05 (and < 0.03) to 1.4. For comparison, the average values per morphological type found by Balkowski (1973, A.A. 29, 43) are: 0.1 for Sa, 0.3 for Sc, 0.9 for Im. Clearly, M_H/L_B varies widely among galaxies of quite "early" type (van Woerden

1977, "Topics in Interstellar Matter", ed. van Woerden, p. 261).
Inspection of deep Siding Spring Schmidt plates shows that some of the galaxies classified S0 or S0/a in the Reference Catalogue have well-developed spiral arms, though these are rarely bright; NGC 6902 is an outstanding example (van Woerden et al. 1976, P.A.S.A. 3, 68). Others, such as NGC 1533 and NGC 5102, while rich in gas, show no trace of spiral structure. We have looked for correlations of M_H/L_B with colour, bulge-to-disk ratio, and luminosity, but without success. So far, gas richness in lenticular galaxies appears unrelated to any other property.

Several lines of further investigation may be pointed out. As shown by Gallagher during this Symposium, deep large-scale photographs may throw light on the morphology and structure of these objects. Colorimetry would help to analyze the stellar composition. Photographs through Hα filters could locate HII regions and bring evidence of recent star formation. Spectroscopy could then reveal the chemical composition of the gas. Aperture-synthesis HI studies could provide the large-scale distribution and motions of gas in these systems; this would contribute to an understanding of their evolution. A vital question is why these gasrich systems have little or no spiral structure.

A detailed account goes to Astronomy and Astrophysics.

KERR: HI OBSERVATIONS OF A LARGE SAMPLE OF ELLIPTICAL GALAXIES

In cooperation with G.R. Knapp and B.A. Williams a very sensitive observational search was made for HI emission from 38 early-type galaxies, mostly ellipticals, using the Arecibo 305-meter telescope. Ellipticals are especially interesting because they contain very little interstellar matter, and it is useful to set better limits for their gas content.

Thirty-two of the galaxies were not detected: using estimates of the velocity width for each galaxy scaled to its luminosity, very low upper limits were set for the HI content in each case. These limits are inconsistent with the amounts predicted from stellar mass loss, and they are also inconsistent with continuing star formation in the galaxies. They support the suggestion that gas is continuously removed from these galaxies by a galactic wind mechanism.

Six of the observed galaxies were detected. Only one of these was a normal elliptical galaxy, namely NGC 4278, which has been previously detected in HI by Bottinelli and Gouguenheim (1977, A.A. 54, 641) and Gallagher et al. (1977, Ap.J. 215, 463). NGC 4278 was mapped, showing that the gas lies in a rotating disk which extends well beyond the visible body of the galaxy. The HI rotation curve for this galaxy is flat, and its M/L ratio is greater than 20. This observation shows that this elliptical, like many spiral galaxies, is embedded in a massive, low-luminosity halo.

No signal was apparent from NGC 5846, a detection of which was recently reported by Huchtmeier et al. (1977, A.A. 57, 313) with the Bonn 100-meter telescope.

BERTOLA: How do you reconcile the rotation curve of NGC 4278 derived from HI measurements with the velocity gradient observed along the major axis in the optical emission lines?

KERR: The optical and radio rotation curves seem to be independent of each other, as if the HI disk and the elliptical are semi-separate entities.

GALLAGHER: HI IN NGC 1052 AND NGC 4636

The elliptical galaxy NGC 1052 (E4) has been measured with the NRAO 300-foot telescope in 21-cm line emission. The HI mass is $(8 \pm 3) \times 10^8$ $M_\odot$ and $M_{HI}/L_{pg} = 0.06$. The properties of NGC 1052 are very similar to those of NGC 4278 in regard to nuclear activity, HI content, the absence of detectable star formation, and galaxy group characteristics. This suggests the two unique features of these galaxies, large HI content and nuclear activity, are physically related. The data and a more complete discussion are given in a paper by Knapp et al. (submitted to Ap.J.).

We have also obtained a more marginal measurement for HI in NGC 4636 (E0), which confirms the tentative detection of HI by Huchtmeier et al. (1975, A.A. 42, 205) and is in agreement with the recent results obtained by L. Bottinelli and L. Gouguenheim with the Nançay radiotelescope.

VAN WOERDEN: Can the analogy between NGC 1052 and 4278 be extended further? I believe that NGC 4278 too is a member of a pair.

GALLAGHER: Yes, NGC 4278 forms a pair with NGC 4274, an Sa with a fair amount of HI, about 20' north of NGC 4278.

KOTANYI: Would you say that, as previously suggested for NGC 4278, the HI profile in NGC 1052 is indicative of a turbulent disk?

GALLAGHER: I wouldn't; I don't know.

MEBOLD: We confirm the detection of the 21-cm emission line at the position of NGC 1052 (Fosbury et al. 1977). At a position 8' to the NE of NGC 1052, i.e. on the side away from NGC 1042, we detected a line with about the same parameters. As to the confusion with NGC 1042 we can add that we mapped the emission of NGC 1042. The result proves that the flux of NGC 1042 that is picked up by the telescope if pointed at the position of NGC 1052 is less than 10% of the line of NGC 1052.

GIOVANELLI: I would like to enter a note of caution. Observations of Haynes and myself have revealed a "cloud" of HI with similar properties to your profile of NGC 1052, a few diameters away from M51. Therefore the possible relation to NGC 1042 should not be underestimated. I shall describe this later [see Discussion V.2].

VAN WOERDEN: Another possible example of dust (and gas) being accreted may be the pure lenticular NGC 5102. The hydrogen in this system is not widespread (angular size < optical diameter). We found a pronounced dust lane reaching right into the nucleus. A point of difference with the ellipticals is that this S0 has blue colours indicating recent star formation.

CAPACCIOLI: THE DISCONTINUITY BETWEEN ELLIPTICAL AND DISK GALAXIES
Since a few years a number of rotation curves of elliptical and S0 galaxies were made available. For some of these galaxies the rotation curve extends far enough to give a reliable estimate of the turnover velocity (for references see Bertola and Capaccioli, in prep.). After proper corrections to the observed velocities, the main value of the turnover velocities for ellipticals turns out to be almost one third of that for S0s. On the other hand S0 galaxies exhibit maximum rotational velocities in the range of values typical of spirals of comparable masses. Therefore a strong dynamical discontinuity exists between elliptical galaxies on one side and S0s and spirals on the other, i.e. between galaxies having only the spheroidal component and those having, in addition, a disk. The presence of such a discontinuity is also indicated by other evidence concerning: M/L ratio, photometric properties, intrinsic flattening, HI content, HI parameters and morphological characteristics. In conclusion, several physical reasons suggest to group the morphological classes of the Hubble sequence according to the fact that the disk is present or not.

WIELEN: In measuring the rotation curve of elliptical galaxies, one is probably measuring essentially the "peak velocity", i.e. the velocity where the maximum in the velocity distribution occurs. It is possible to construct self-consistent models of ellipticals in which this peak velocity (corresponding to the center of the absorption line) differs significantly from the mean rotational velocity which determines the angular momentum. Hence, the possibility should be kept in mind that ellipticals may have a larger angular momentum than is derived from identifying the peak velocity with the mean rotational velocity. The problem may be settled by obtaining high-dispersion spectra in which the absorption line profiles should reflect the possible asymmetries in the velocity distribution.

"The peanut-shape that you see in these edge-on bulges has never been modelled successfully, as far as I know. People have tried. We have tried it, but the guy who was doing it decided that he didn't want to do astronomy after trying this."

K.C. Freeman in Discussion I.1

GLOBAL DYNAMICS OF THE INTERSTELLAR GAS, MAGNETIC FIELD, AND COSMIC RAYS

E. H. LEVY
Department of Planetary Sciences and Lunar and Planetary
Laboratory, University of Arizona, Tucson, Arizona, USA

I have been asked to review the physical principles which underlie the dynamical equilibrium and stability of a composite system of gas, magnetic field, and cosmic rays. What is of particular concern here are those aspects which control the distribution of magnetic field and cosmic rays, and thus influence the morphology of galaxies as seen in nonthermal radio emission.

The salient features of the nonthermal spiral galaxy radio emission which are relevant to this discussion can be summarized briefly. Several of the observational data are ambiguous because of the difficulty of uniquely disentangling the nonthermal emission from the thermal emission of hot regions. It appears possible that nonthermal disk emission has a brightness variation which follows the general distribution of galactic mass, suggesting that cosmic rays are generated by some process which occurs in the overall stellar population (Van der Kruit et al. 1977). Away from the plane, spiral galaxies show nonthermal radio emission from thick disks or limited halos extending of the order of 1 to about 10 kiloparsecs from the galactic plane. Spectral indices of the observed radiation generally fall in the range -0.7 to -0.9 which suggests energetic electron spectra similar to what is observed in our own galaxy.

For our present purposes we will assume that galactic cosmic rays are produced in and largely confined to galaxies. This conservative assumption is consistent with the known properties of cosmic rays. The alternate possibility, that cosmic rays fill large regions of the universe, has been discussed extensively elsewhere (Brecher and Burbidge 1972). (Particles having energies above about 10^{17} eV/nucleon are likely to fill large regions of the universe, as these cannot be confined to galaxies. These particles possess a negligible fraction of the total cosmic-ray energy density in galaxies, and thus have no effect on the dynamics of galactic material.) The cosmic rays can be thought to comprise a high temperature, low density gas. In a typical galaxy the thermal energy of the cosmic-ray gas is more than 10^7 times greater than its binding energy in the galaxy's gravitational field. Thus cosmic rays remain in galaxies only insofar as they are constrained by other agencies. It can be shown through use of the virial equation that the net magnetic field stress is purely expansive, no matter how complicated and contrived the field morphology may be (Parker 1954). Thus a galactic magnetic field will expand away to infinity unless anchored by some other force. By virtue of the gas' electrical conductivity and the cosmic rays' relatively small gyration radii, the cosmic rays, gas, and magnetic field are constrained to move together as a single, composite medium. Taking interstellar conditions in our own galaxy to be representative, the composite medium can be visualized as a single fluid for spatial scales greater than about one parsec and time scales longer than some 10^4 years. Such a composite interstellar medium is confined to a galaxy by the gravitational force which acts predominantly on the gas. In our own galaxy the electron component of the cosmic rays contains only a small fraction, about one percent, of the total particle energy. Thus while the electrons provide only a small part of the total interstellar pressure, they are tied to the composite medium and their distribution is controlled by the dynamics of the composite medium. Thus galactic nonthermal radio morphology traces the gross dynamics of the complete system of gas, field and particles.

The physical character of the equilibrium of an interstellar medium can be pictured by concentrating on the force balance in the z-direction, perpendicular to a galactic disk. For simplicity suppose that the z-component of the gravitational force (due almost entirely to the stars) is a given constant, $-g\underline{e}_z$, above the galactic plane. If $\underline{B} = (B_x, B_y, B_z)$ is the magnetic field, if P is the cosmic-ray pressure, and if p is the gas pressure, then the condition for hydrostatic equilibrium in the z-direction is

$$\frac{d}{dz}\left[P + p + \frac{B_x^2 + B_y^2 - B_z^2}{8\pi}\right] = -\rho g , \qquad (1)$$

where ρ is the gas density. First consider a stratified interstellar medium (Parker 1966) in which $B_z^2 \ll B_x^2 + B_y^2 \equiv B^2$ and in which the cosmic ray and magnetic field pressures are proportional to the gas pressure. Writing $P = \beta\rho u^2 = \beta p$ and $B^2/8\pi = \alpha\rho u^2$, the solution of equation (1) is

$$\rho(z) = \rho(0) e^{-\Lambda/2} , \qquad (2)$$

where Λ is the characteristic scale height of the gaseous disk and is defined by

$$\frac{\Lambda}{2} = \frac{u^2(1 + \alpha + \beta)}{g} = \frac{p + P + B^2/8\pi}{\langle\rho\rangle g} \qquad (3)$$

where $\langle\rho\rangle$ is the average gas density in the disk. Equation (3) then relates the gas density and pressure, cosmic ray pressure and magnetic field strength to the thickness of the gaseous disk. Equation (3) offers a consistent description of the known properties of the gaseous disk of our own galaxy. But, as we will see below, equation (3) only applies in a crude way because the stratification assumptions are inevitably unrealistic. Parker (1966) showed that such a stratified interstellar medium is dynamically unstable in a Rayleigh-Taylor sense. The equilibrium in which a heavy interstellar gas confines buoyant cosmic rays and magnetic field against their expansionary tendencies is similar to the equilibrium of a light fluid overlain by a heavy fluid in a gravitational field. Each system is unstable to sinking of the heavy fluid through the light.

In galaxies the instability evolves to the form shown in Figure 1. Large gas complexes accumulate at localized regions as material slides along the field lines to gather in troughs of the magnetic field. In the spaces between the gas accumulations the magnetic field is relieved of its overburden of gas. Thus freed of the confining gravitational stress, the combined pressures of cosmic rays and magnetic field inflate the field to produce a system of magnetic arches extending to large distances above and below the galactic plane. The modes of this instability have been investigated in some detail (Parker 1967a,b, 1968a,b; Lerche and Parker 1967, 1968; Shu 1974). The growth time for the instability is a few times 10^7 years and the characteristic scale lengths along the magnetic field are several hundred parsecs to a kiloparsec.

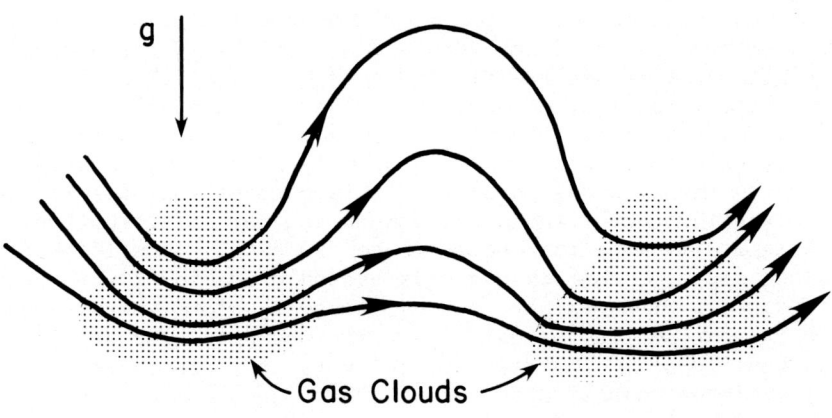

FIGURE 1

Dynamical instability of a composite interstellar medium produces large gas accumulations and an extensive system of arching magnetic loops.

Mouschovias (1974, 1975) has explored numerical models of the stationary equilibrium end states that result from the fully developed instability. He showed that in the absence of significant cosmic ray sources the stationary equilibrium has magnetic arches extending above and below the galactic plane to distances approximately equal to the separation of the gas complexes. Such an equilibrium would account for thick radio disks of nonthermal emission extending on the order of a half to one kiloparsec on either side of a galactic disk.

Now the equilibrium shown in Figure 1 is not stationary in real galaxies. Star formation, supernovae, galactic differential rotation, cloud motion, etc. act to continually disrupt the equilibrium states; at the same time the magnetic field is continually regenerated by dynamo action of the fluid motions in the differentially rotating gaseous disk (Parker 1971; Vainshtein and Ruzmaiken 1972). Furthermore cosmic rays are generated continually in galaxies, and this continual supply of cosmic ray energy has important dynamical consequences. The pressure of the cosmic ray gas inflates the magnetic field lines which protrude through the surface of the disk (Parker 1965). The field line inflation proceeds until the cosmic ray stresses completely overwhelm the field stresses in the inflated magnetic arch. At this point the cosmic rays escape the hold of the magnetic field. The further inflation of the field lines may produce a substantially more extensive halo than that which results from the equilibrium magnetic field arches. This inflation process constitutes, in essence, a pressure relief valve which enforces the approximate equality of the magnetic and cosmic ray energy densities. Such an approximate equipartition is observed in our own galaxy. Equipartition is generally presumed applicable to cosmic radio sources; field line inflation provides a dynamical basis for it.

The global dynamics of a system of magnetic field, gas, and cosmic rays which we have reviewed provides a dynamical basis for understanding the distribution of magnetic field and energetic particles which produces the radio morphology of galaxies. The following picture emerges from these considerations. Cosmic rays are generated broadly throughout the disks of galaxies. Since the nonthermal radio emission profile seems to match the disk mass profile, it is evident that cosmic rays are not well mixed throughout the disk, otherwise their distribution would smear substantially more than seems to be the case. Thus the weight of the evidence suggests that cosmic rays escape predominantly through the faces of galactic disks rather than by streaming along spiral magnetic fields to the peripheries of galaxies. The mechanism for particle escape perpendicular to the disk evidently is provided by the dynamical instability of the composite system of field, gas, and cosmic rays and by the further inflation of the field by cosmic ray pressure. These combine to produce an extended distribution of field and particles which may be observed in nonthermal emission as a thick disk or halo. As we will mention presently cosmic rays seem to pass freely between the disk and halo. However, the particles do not seem to mix freely throughout the halo as this also would smear the distribution more than is observed.

It is worth mentioning that several of the directly observed properties of cosmic rays in our own galaxy provide insight into these problems. It is well known on the one hand that the relative abundance of Li, Be, B, and C, N, O in Galactic cosmic rays implies that the average cosmic ray particle passes through 5 gm/cm^2 of matter during its life in the Galaxy. If the mean density of gas in the disk is one hydrogen atom per cm^3, then the disk residence time is about 3×10^6 years (cf. Meyer 1969). On the other hand the cosmic ray abundance of the radioactive isotope ^{10}Be, with a half-life of 1.5×10^6 years, provides a direct measure of the total cosmic ray residence time in the Galaxy. The most recent measurements (Garcia-Munoz et al. 1975; Webber et al. 1977) suggest total residence times of the order of 2×10^7 years. The most straightforward interpretation of the discrepancy is that these cosmic rays spend the largest part of their Galactic residence time in a halo where they encounter little matter and that the particles pass relatively freely between the disk and the halo. In the simplest case of a stationary halo, the above lifetimes suggest a halo volume perhaps five times as large as the disk volume. This would correspond to a "thick disk" extending about half a kiloparsec above and below the galactic plane.

A further point is that the small value of the observed cosmic-ray anisotropy in the Galaxy, combined with the short residence times, indicates that cosmic rays have only enough time to escape in the direction perpendicular to the galactic disk. This supports the idea that cosmic rays do not spread freely through large regions of the disk.

Note that consistent with the fabric of dynamical behavior which we have reviewed here are many possibilities for the detailed motion of the cosmic-ray particles. Each model has specific implications which can be compared with observations, but unique interpretations of the observations are not yet possible. A number of the specialized models of cosmic ray motion have been reviewed in detail recently by Ginzberg and Ptuskin (1976) and we refer the interested reader to their article.

The ideas described here lead naturally to the notion of expanding galactic halos (Ipavich 1975). Models of dynamical halos have been constructed recently by Owens and Jokipii (1977a,b) with which they have explored the consequences for radio emission from energetic electrons. The models are complicated by a number of free parameters and by the fact that the electrons suffer energy loss through adiabatic deceleration as well as through radiation. One significant point that emerges from their calculations is that electrons near the peripheries of the halo have on the average suffered the greatest energy loss. This results in a frequency dependence of the halo morphology which is equivalent to a steepening of the emission spectrum with increasing distance from the galaxy.

REFERENCES

Brecher, K. and Burbidge, G.R.: 1972, Astrophys. J. 174, 253.
Garcia-Munoz, M., Mason, G.M. and Simpson, J.A.: 1975, Astrophys. J. Letters 210, L141.
Ginzburg, V.L. and Ptuskin, V.S.: 1976, Rev. Mod. Phys. 48, 161.
Ipavich, F.M.: 1975, Astrophys. J. 196, 107.
Lerche, I. and Parker, E.N.: 1967, Astrophys. J. 149, 559.
Lerche, I. and Parker, E.N.: 1968, Astrophys. J. 154, 515.
Mouschovias, T.Ch.: 1974, Astrophys. J. 192, 37.
Mouschovias, T.Ch.: 1975, Astron. Astrophys. 40, 191.
Owens, A.J. and Jokipii, J.R.: 1977, Astrophys. J. 215, 677.
Owens, A.J. and Jokipii, J.R.: 1977, Astrophys. J. 215, 685.
Parker, E.N.: 1954, Phys. Rev. 96, 1686.
Parker, E.N.: 1965, Astrophys. J. 142, 584.
Parker, E.N.: 1966, Astrophys. J. 145, 811.
Parker, E.N.: 1967a, Astrophys. J. 149, 517.
Parker, E.N.: 1967b, Astrophys. J. 149, 535.
Parker, E.N.: 1968a, Astrophys. J. 154, 57.
Parker, E.N.: 1968b, Astrophys. J. 154, 875.
Parker, E.N.: 1971, Astrophys. J. 163, 255.
Shu, F.H.: 1974, Astron. Astrophys. 33, 55.
Vainshtein, S.I. and Ruzmaiken, A.A.: 1972, Sov. Astron. A.J. 15, 714.
Van der Kruit, P.C., Allen, R.J. and Rots, A.H.: 1977, Astron. Astrophys. 55, 421.
Webber, W.R., Lezniak, J.A., Kish, J.C. and Simpson, G.A.: 1977, Astrophys. Letters 18, 125.

DISCUSSION FOLLOWING REVIEW I.6 GIVEN BY E.H. LEVY

BURKE: What density did you assume for the local interstellar matter?

LEVY: About 0.1 to 0.3 cm^{-3}.

BURKE: Such a density is entirely acceptable for the sun's neighborhood, and then there is no discrepancy between the spallation lifetime and the Be^{10} lifetime for cosmic rays.

ALLEN: What is the variation of the magnetic field in Mouschovias calculations?

SHU: I don't remember the exact numbers, but the scale height for the magnetic field in the raised portions (where most of the cosmic rays are) is much larger than the scale height of the gas. In any case, I would stress again that the original stratified state is a physical artifact in that such an equilibrium, even if artificially created initially, cannot be stable.

ALLEN: COMMENTS ON THE RADIO CONTINUUM EMISSION FROM NORMAL DISK GALAXIES

The fine review by Dr. Levy on the theoretical situation has promted me to draw your attention to some features in the high-resolution radio aperture synthesis observations which are perhaps not easily explained by the existing models for the cosmic-ray origin and diffusion in galaxies. These features are:

(1) In normal flattened galaxies which show relatively bright radio disks ($<T_b> \gtrsim 1$ K at 21 cm) the radio and optical surface brightnesses correlate well over the main parts of the disks. This has been shown for the ring-averaged radial distributions in the face-on galaxies NGC 6946 (van der Kruit et al. 1977, A.A. 55, 421) and M51 (van der Kruit 1977, A.A. 59, 359), and may also occur in the z distribution of NGC 891 from about 20" to 60" above the plane (unpublished work by Allen, Spinrad and Bruzual).

(2) The variations in the radio spectral index are rather small over the main parts of the disks for both the face-on galaxies referenced above (after removal of the thermal component), and for the edge-on system NGC 891 (Allen et al. 1977, A.A., in press, Figure 5) up to a distance of about 2.5 kpc above the plane. The frequently-noted steepening of the spectral index in fact begins to be incontrovertible only at the extremities in r and z.

(3) The morphology of the HI distribution differs substantially from that of the radio continuum in these bright-radio-disk systems both in r and in z; the latter is dramatically illustrated for NGC 891 in Figs 1 and 2 of Sancisi et al. (1975, La Dynamique des Galaxies spirales, ed. L. Weliachew, p. 295). The radio emission can apparently be strong in places where the HI is faint (central regions of the disk in e.g. M51 and NGC 6946, and at high z distances from the plane in e.g. NGC 891) and vice versa (large radial distances from the center). Note that there are obviously other components which contribute to the nonthermal radio continuum emission such as supernova remnants and the possible compression in a density wave shock. These components appear superposed on the disks referred to above. They apparently dominate in galaxies like M81 and M31 which have only relatively faint radio disks. My concern here is with galaxies which have bright radio disks.

These features in the observations suggest at least two inferences:
A. Any model in which the synchrotron radio volume emissivity is a monotonic function of the local gas density would seem to be too simple.
B. There may be a more intimate relationship between the electron component of the cosmic rays and an older stellar population than has hitherto been assumed.

Point A above may of course be circumvented if the distribution of ionized gas is substantially different from that of the HI. Although this situation may not yet be entirely excluded by observations, another hypothesis which has the possible virtue of corroborating all the known features is the following: Energetic electrons are produced not only by explosions of type II SN near spiral arms, but also by some other sources which are distributed throughout a galaxy more-or-less as an older stellar population. The electrons lose essentially all of their relativistic energy by synchrotron emission within one kpc or so of their sources; this requires magnetic fields stronger than about 10^{-5} gauss for electrons radiating mainly at 1500 MHz. Under these assumptions it is clear that the total observed synchrotron radio emission integrated over frequency must be independent of the magnetic field strength. From the standard formulae in e.g. Moffet (1975, Galaxies and the Universe, ed. A. Sandage, M. Sandage, J. Kristian, p. 211) it is easy to show that the synchrotron volume emissivity goes like $\varepsilon_\nu \propto Q_0 \, B^{\alpha-1} \, \nu^{-\alpha}$, i.e. linear in the density of sources of relativistic electrons, and as $B^{-0.3}$ for $\alpha = 0.7$. The required value of $\geq 10^{-5}$ gauss is roughly that which is found from equipartition arguments in e.g. NGC 891 (Allen et al. 1977, A.A., in press). The sources could be things like pulsars, flare stars, magnetic white dwarfs, X and γ-ray bursters, Type I supernovae, etc. etc. Note that one Type I SN per 30 years in a typical disk galaxy would result in more than 100 sources in a 1 kpc^3 volume of the disk in the $\sim 10^7$ year lifetime of the relativistic electrons. The steepening of the radio spectral index at the extremities in r and z would then be associated with the rapid decrease of B in these regions and with the subsequent breakdown of the "total confinement" model sketched above.

WAXMAN: ON THE NATURE OF A GALACTIC BOUNDARY LAYER FLOW

In studying fluid dynamical interactions between a galactic disk and a surrounding gaseous halo, two physical conditions are necessary for the existence of a boundary layer circulation. The first concerns the presence of a vertical shear in the azimuthal flow of gas, and the localization of this shear to the vicinity of the disk. If one treats the disk as a density enhancement in the equatorial plane of the halo, then this shear is due to the inclination between the equidensity surfaces of this enhancement and the equipressure surfaces of the background halo gas. That is, the vertical shear is a manifestation of the baroclinicity inherent to disk-halo systems. The second condition is that there must be an effective viscous coupling of the disk to the halo, localized to the disk region as well. This viscous coupling is provided by the stochastic momentum transport of interstellar gas clouds colliding with each other and their interaction with the intercloud medium.

In this work the fluid dynamical equations, for the case of a density enhancement which is small compared to the gas density in the halo, are solved by the method of matched asymptotic expansions. The result is a circulation of gas in the meridional plane extending over several disk scale heights from the galactic plane. Radial flows, inward in the disk and outward in the halo, should be of the order of 20 km/s. The vertical flows, of the order of 0.5 km/s are upward within about 10 kpc of the rotation axis and downward beyond this point. It is tempting to identify this weak downflow of gas with the low velocity "galactic infall" observed in the solar neighborhood. In addition, this circulation implies a thorough mixing of the metal-enriched disk gas with the overlying halo gas on a timescale of about a billion years. Density enhancements which simulate a significant decrease in the gas density near the central region of the disk give rise to new features in the flow pattern which are suggestive of a "3-kpc arm". A thorough stability analysis of this boundary layer circulation is presently under way.

The details of this analysis are to appear in the Astrophysical Journal in the near future.

ELVIUS: OPTICAL POLARIZATION IN GALAXIES

During several visits to the Lowell Observatory, Flagstaff, Arizona, USA, I have made observations of the polarization of light in galaxies. The latest results were obtained in early 1977. A scanning dual-beam polarimeter was used together with a computerized data acquisition system. Some results will be mentioned here. Details will be published elsewhere.

In the Seyfert galaxy NGC 1068 polarization of light has now been found also in regions outside the nucleus. About 10 seconds of arc to the north-east of the nucleus 4 percent of polarization in P.A. $135°$ indicates the presence of large clouds of scattering particles.

Polarization effects analogous to the 'interstellar polarization' in the Milky Way have been found in dark regions of many spiral galaxies like NGC 3190, 3623, 4216, 4565, 4594, 4826 and 5005. These observations indicate large-scale magnetic fields along the spiral arms. Similar polarization effects along dark features were also found in the barred spiral NGC 5383, in the nucleus of the peculiar galaxy NGC 3718 and in the filaments seen in absorption against the main body of NGC 2685.

Polarization due to the scattering of light has previously been observed in M82 and was mentioned above for NGC 1068. It was also observed in filaments outside the main body of NGC 2685, where the direction of the electric vector indicates that the light from the nucleus of NGC 2685 is scattered by non-spherical particles aligned in a magnetic field along the filament.

If the polarization observed on the brighter side of some galaxies like NGC 3623, 4216 and 7331 is also interpreted as due to scattering of light from the nucleus in clouds of aligned particles in the spiral arms, we are led to the controversial conclusion that the brighter side of the galaxy is the nearer one and that the arms are leading.

II   SPIRAL STRUCTURE AND STAR FORMATION

"We do not pretend that we have a theory about every spiral galaxy."

C.C. Lin in Discussion II.4

THE EVOLUTION OF DISK GALAXIES

S. E. Strom and K. M. Strom
Kitt Peak National Observatory*

1. INTRODUCTION

The past decade has witnessed dramatic changes both in our conceptual models of disk-system formation and evolution and in the power of new observational techniques to confront, challenge, and redefine these models. In this contribution, we would like to review recent optical wavelength studies of spiral and S0 galaxies which appear to influence our understanding of disk-system evolution. Particular emphasis will be placed on the effects of environment on evolutionary processes, since it appears likely that the addition or removal of gas during the lifetime of a disk system may often be dominant in controlling its appearance.

2. THEORETICAL OVERVIEW

A typical disk system is composed of two morphologically distinct components: a spheroidal component, the bulge; and a flattened component, the disk. In some cases, the disk component is forming stars at the current epoch (in galaxies of type Sa-Sc and in some irregular systems), while in others (S0 galaxies and "smooth-arm" spirals) there is no evidence of star formation. The relative prominence of bulge and disk components, expressed as a bulge-to-disk ratio (B/D), appears to vary continuously among observable systems. Among relatively luminous galaxies, disk systems appear to be the dominant morphological type in the field and in low-density groups of galaxies. In the great clusters, the frequency with which star-forming disk galaxies appear decreases dramatically; it is also possible that the frequency of all disk systems is lower in such regions. Recent theoretical efforts have been directed first toward explaining the morphological appearance of disk systems, and next toward understanding their relative frequency in differing environments.

Most prominent among recent contributions to our understanding of disk-system formation have been those of Larson (1976) and Gott and
_____
*Operated under NSF contract No. AST 74-04129 with AURA, Inc.

Thuan (1976). Both sets of models presuppose the existence of a rotating protogalactic gas cloud in which star formation accompanies collapse. Collision between gas clouds in the collapsing protogalaxy leads to dissipation of energy in the gas (through cloud heating and subsequent radiative processes) and the eventual formation of a thin disk. The prominence of the disk and bulge components is determined by the relative efficiency of star formation during the galaxy-formation epoch. Those systems in which star formation is relatively efficient at early epochs form large spheroidal components; little gas remains to form a disk. Conversely, when few stars are formed initially and when dissipative processes in the gaseous component dominate the early evolution, the disk component is most prominent. The beliefs of a decade ago, which argued that the Hubble sequence from elliptical to spiral galaxies represented a sequence of increasing initial angular momentum, are not supported by current galaxy collapse models. Current speculation centers on the initial density of the protogalactic cloud as the primary determinant of the relative time scales for star formation and for collapse of gas to a disk and thereby the B/D ratio. If the star-formation rate (number of stars formed/volume/time) $\sim \rho^n$, then the time scale for star formation $\tau_s \sim \rho^{1-n}$. The time scale for collapse to a disk is on the order of the free-fall time scale $\tau_{ff} \sim \rho^{-\frac{1}{2}}$. Hence, $\tau_s/\tau_{ff} \sim \rho^{1.5-n}$. The estimates by Schmidt (1959) of the star-formation rate in our Galaxy, and the more recent theoretical estimates of Talbot and Arnett (1975), suggest $1.7 \lesssim n \lesssim 2$. In protogalactic condensations of high initial density, the star-formation rate is therefore expected to be relatively high and consequently not much gas may remain to form a disk. In regions of lower density, however, the time scale for star formation may be longer than the free-fall time scale in the protogalactic cloud, and these systems may be dominated by the disk component. Gott and Thuan (1976) have argued that more spheroidal galaxies may be formed in dense clusters of galaxies if the mean density of the protogalactic clouds is in some way related to cluster-formation conditions. They believe that the "seeds" for great clusters are found in regions of above-average density enhancements. In such regions, systems of high B/D are expected to predominate.

2.1. Evolution of the disk postformation: spiral galaxies

The current epoch appearance of disk galaxies depends on three factors: 1. the amount of gas remaining (to form stars) in the disk subsequent to its formation; 2. the rate at which gas is consumed in astration events; 3. the effects, if any, of mechanisms which add or remove gas from the disk (and thereby enhance or truncate star formation).

The collapse into disk form involves collisions of subcondensations within the protogalactic cloud; the relative velocities of these subcondensations are on the order of several hundred km sec^{-1}. Star formation may proceed vigorously in regions of high compression behind shocks induced in the supersonic cloud-cloud collisions. It is not yet clear, however, how much star formation takes place during these final

II.1 THE EVOLUTION OF DISK GALAXIES

disk-collapse phases, and what fraction of the "initial" disk is stellar or gaseous. Sandage et al. (1970) argue that the fraction of remaining gas is the dominant factor which determines the Hubble type of a galaxy. At the extremes, in their view, S0 galaxies represent systems in which little postformation gas remains, whereas Sc and irregular galaxies represent systems with initially gas-rich disks. While this suggestion may be correct, by itself it is insufficient to explain the detailed relationship between bulge prominence and arm appearance characteristic of the Hubble sequence.

Perhaps the greatest advances in understanding postformation disk evolution have come from recent theoretical studies of spiral galaxies. From inspection at optical wavelengths, the dominant features of these galaxies are regions of active star formation extending over scales of many kpc and arranged in a regular pattern of spiral arms. The regularity of the spiral patterns and their apparent persistence on time scales significant compared to a Hubble time led to the hypothesis that the arms represent a quasi-permanent, spiral wave pattern in the density distribution of the underlying old disk stars (Lindblad 1960; Lin and Shu 1964). A theory describing these density waves has been extensively developed by Lin and his collaborators over the past 15 years. In this theory, the wave pattern, which is characterized by an angular velocity of rotation, Ω_p (the pattern speed), results from a self-sustaining departure from the axisymmetric gravitational field of the disk system. The importance of star formation in spiral arms is believed to result from interaction of any remaining disk gas with the spiral wave pattern sustained by the underlying disk stars.

At present, considerable controversy surrounds discussion of the physical processes by which spiral-density waves are initially induced, and the processes which amplify and damp the waves and thereby determine the wave lifetime. However, much progress has been made both in understanding (a) the dependence of the wave pattern on galaxy mass size, and the distribution of mass within a galaxy, and (b) the role of gas-density wave interactions in triggering star formation.

The first impression of the spiral-arm pattern in a galaxy is derived from the "openness" of the pattern; this quality can be expressed in terms of the "pitch angle" i. The quantity i (the angle between the spiral wave, at any radial distance r from the center and a circle of radius r centered on the galactic nucleus) is primarily related to the degree of central concentration in the galaxy (Roberts et al. 1975). Galaxies exhibiting high central concentration or large bulges support wave patterns which have small values of the pitch angle (tightly-wound arms); open wave patterns are most easily supported in galaxies of low central mass concentration.

Another feature which directly affects the visual perception of the spiral pattern is the relative prominence and distribution of the recently-formed stellar population in the arms. Galaxy-wide shocks induced by interaction of disk gas with the density wave appear to provide a most promising mechanism for driving star-forming events in spiral arms. In this picture, gas at a given radial distance r moving at an angular speed $\Omega(r)$ encounters the density wave with an unperturbed (by the gravitational field of the arms) velocity perpendicular to the arms given by $w_{\perp_0} = (\Omega - \Omega_p) r \sin i$. For a typical massive spiral galaxy, the maximum circular velocity is on the order of 250 km sec^{-1}; the pitch angle i is on the order of 5-15 degrees. Hence, $w_{\perp_0}$ is on the order of 25-60 km sec^{-1}. For an idealized two-component (cloud-intercloud) model of the interstellar gas (Field et al. 1969), the value of $w_{\perp_0}$ exceeds the expected sound speed (~ 8 km sec^{-1}) in the intercloud gas ($T \sim 10\,000^\circ$K). Hence, as the gas encounters the spiral-wave crest supersonically, a shock wave is formed. For a given wave amplitude, the strength of the shock and the compression are proportional to $(w_{\perp_0}/a)^2$, where a is the effective accoustic speed in the gas. For large values of $w_{\perp_0}$, the shocks are strong and regions of higher compression are narrow. For small $w_{\perp_0}$, shocks are weak and the region of compression is broad. Even for $w_{\perp_0} < a$, some of the gas can nevertheless be accelerated by the spiral gravitational field near the wave crest to transonic values and produce a shock wave, if the wave amplitude is sufficiently large.

One effect of compressing the intercloud gas in shock regions is to force some intercloud material into the cold-cloud phase. The greater ambient pressure in the intercloud medium in regions of high compression may trigger the collapse of both ambient and newly-formed cold clouds. The contraction of these clouds is presumed to result in star formation. [Woodward (1976) has attempted some more quantitative studies of shock-driven implosion of cold clouds located in compressed intercloud material. His results suggest in greater detail how star formation may proceed.] In galaxies characterized by high values of $w_{\perp_0}$, newly forming stars are thought to be confined to the narrow, post-shock, high-compression regions. In galaxies where $w_{\perp_0}$ is generally small, new stars may be formed in relatively broad regions of weak compression.

The hypothesis that star formation is triggered by galactic shocks is very attractive because it provides a natural explanation for the predominance of star formation in spiral arms. No other proposed mechanism can account readily for the coherence of star-forming episodes on scales of many kpc.

If the picture of galactic shock-induced star formation is correct, it suggests that the star-formation rate (and the rate of gas depletion) depends on the frequency with which disk gas encounters the density wave, $(\Omega - \Omega_p)$. Furthermore, it also seems natural to suggest that the efficiency with which stars are produced at each encounter depends on the degree of compression (greater star-formation efficiency in high-compression regions), although no direct theoretical justification for this statement is available. These beliefs have important implications for understanding the evolution of disks or galaxies of different masses, sizes, and degrees of central concentration. For example, we expect that $(\Omega - \Omega_p)$ will be largest in galaxies in which the ratio M_{galaxy}/R_{galaxy} is large since $\Omega \sim \sqrt{GM/R}/R$. Furthermore, we expect that the values of $w_{\perp_0}$ are largest in such galaxies and in those for which the pitch angle is relatively large, since $w_{\perp_0} \sim \sqrt{GM/R}$ f(central concentration). Finally, we expect that the degree of compression is highest not only when $w_{\perp_0}$ is high, but when the wave amplitude is high as well. In Figure 1, we show a typical run of the quantity $(\Omega - \Omega_p)$

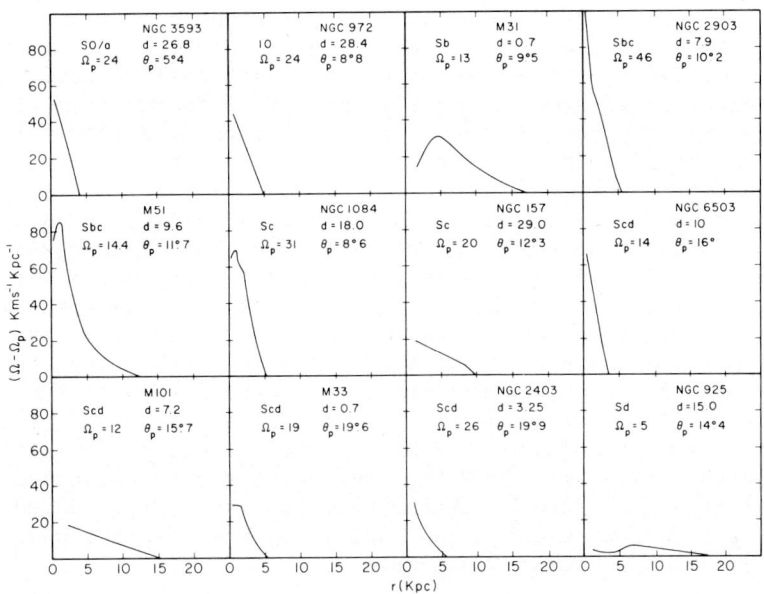

Figure 1. A plot of $(\Omega - \Omega_p)$ against galactocentric distance r for 12 spiral galaxies.

against r for a series of galaxies of differing morphological type. We expect that the formation of stars, the depletion of gas, and the chemical enrichment in the remaining gas within a given galaxy will be highest in regions of high $\Omega - \Omega_p$, $w_{\perp_0}$, and wave amplitude. From Figure 1, we deduce that the gas will be depleted first in the inner regions of the galaxy and last in the outer regions. Furthermore, we expect that the chemical enrichment will be highest in the inner regions (where star formation, element production, and gas depletion rates are high) and lowest in the outer regions. In galaxies having low values of M/R (low values of $w_{\perp_0}$ and weak compression), the star-formation efficiency is low and less gas is processed, and the rate of element formation is low.

In the above discussion, we assume that the evolution of the disk proceeds in isolation. However, Larson (1972a,b) has suggested the possible evolutionary significance of external gaseous material added to galactic disks over a Hubble time. The addition of such material might result from the infall of gas bound to the galaxy but located in an extensive halo several hundred kpc in size, or by direct accretion as the galaxy moves subsonically through relatively dense pockets of intergalactic material. Assuming that the high-velocity clouds observed in our Galaxy (Oort 1970) represent infalling material, Larson finds that the infall rate is sufficient to account for a large fraction of the current disk-gas content and for the average observed star-formation rate. He argues that the Hubble sequence might conceivably be understood in terms of the fraction of gas available for infall after the initial collapse and the time since the last infall episode. In this rather extreme picture, he regards the spiral features as transient, material arms produced by the combined effects of differential rotation and star formation in gas recently introduced into the disk.

The chemical evolution of the disk is also influenced by the infall of gas. If the time scale for conversion of infalling gas-to-stars is τ, then for infalling gas comprised of pure hydrogen $Z = \alpha + (Z_o - \alpha) e^{-t/\tau}$, where α is the yield (the fraction of material going into star formation and re-ejected in the form of heavy elements), and Z_o the initial metal abundance of the disk. Larson further argues that the evolution of disks may be different in clusters than in the field. Disk systems in clusters may be formed from denser protogalactic condensations (see Gott and Thuan 1976). These dense condensations will have shorter free-fall times and consequently less extensive halos at the current epoch. Hence, little halo gas may be introduced into the disk at the current epoch, thus explaining the relative absence of spiral galaxies in great clusters.

Removal of disk gas will also play a major role in affecting the evolution of disk galaxies. If the gas is removed from the disk, star formation will cease unless gas can be replenished. Several mechanisms for the removal of disk gas have been proposed.

2.2. Galaxy-galaxy collisions

In this process first discussed by Baade and Spitzer (1951), galaxies are assumed to collide in regions of high galaxy density. In such collisions, the stellar subsystems are relatively unperturbed, whereas the disk gas is removed from the system both (a) because the gas is heated to temperatures which exceed the effective escape temperature from the combined colliding systems, and (b) because the velocity of the remaining gas is low relative to the center of gravity of the two stellar subsystems; hence this gas is left behind as the systems move in opposite directions at velocities greater than the escape velocity from either system.

2.3. Stripping by intergalactic material

Gunn and Gott (1972) propose that as spiral galaxies move through the intergalactic medium known to pervade some rich clusters of galaxies, the disk gas can be stripped by ablation, if

II.1 THE EVOLUTION OF DISK GALAXIES

$$\rho_{IGM} \, V^2_{galaxy} > 2 \pi G \, \sigma_{stars} \, \sigma_{gas} \qquad (1)$$

ρ_{IGM} is the density of intergalactic gas; V_{galaxy} is the velocity of the galaxy relative to the gas; σ_{gas} and σ_{stars} are, respectively, the surface densities of the gas and stars in the disk of the spiral. For the Coma cluster, the density of the intergalactic medium can be estimated from X-ray observations, and V_{galaxy} from the observed velocity dispersion in the cluster; for reasonable estimates of σ_{stars} and σ_{gas}, Gunn and Gott argue that no gas-bearing spirals can survive in the center of Coma. We should note that the processes which strip disk gas in clusters similar to Coma can also strip the galaxy of any halo gas thereby eliminating the evolutionary consequences of gas infall. Once a galaxy's motion carries it into environment of lower intergalactic density in the outer regions of the cluster, ablative stripping becomes unimportant. The mass loss from disk stars is expected to replenish the interstellar medium in the galaxy at a rate of $\sim 1 \, M_\odot \, yr^{-1}$; hence in $\sim 10^9$ years, the disk gas may comprise several percent of the total mass unless other gas-removal mechanisms are important.

2.4. Removal of gas by galactic winds

Mathews and Baker (1971) and Faber and Gallagher (1976) have proposed that gas may be removed from disk galaxies by the action of galactic winds. These winds are generated in the nuclear bulge of the disk system, driven by two heating mechanisms: 1. supernova heating; 2. heating by collisions (at velocities determined by the velocity dispersion of the stars in the nuclear bulge) between shells of gas ejected by dying stars. If the heating due to these effects is sufficient, the equilibrium temperature of the gas is so high that the gas is no longer bound to the bulge. In disk systems, winds generated in the central bulge may be sufficient to remove gas not only from the bulge region but from the inner parts of the disk as well. A recent calculation by Bregman (1976) suggests that over a wide range of B/D ratios, once a galaxy is stripped by mechanism 2.3 it remains stripped by the action of intergalactic winds.

2.5. The role of galactic halos

Ostriker and Peebles (1973) suggest that cold disk systems (whether comprised of gas or stars) are subject to large-amplitude, irreversible, bar-like instabilities. These authors propose that extended halos with $M_{halo}/M_{disk} \gtrsim 1$ represent plausible entities for stabilizing the disk. Such halos might be expected to have a major influence on the chemical evolution of the disk as well (Ostriker and Thuan 1975). Furthermore, energy and angular momentum exchange between spiral-density waves and the halo may significantly affect the amplification of these waves (Mark 1976).

Ostriker and Peebles suggest that the constituents of such putative halos must have large mass-to-light ratios, since halos of the proposed size and mass composed of the usual nuclear bulge population mix would not have escaped detection. Late-type M dwarfs have been put forth as plausible candidates for the dominant halo constituents. The successful detection of massive halos would be significant not only because of the implications for the structure and evolution of disks, but because the mass contained in such halos might represent the majority of the mass in the universe.

The above discussion suggests that the evolution of disk galaxies depends both on normal astration processes driven primarily by galactic shocks and on interactions with the environment.

We would like to explore now the observational evidence bearing on the evolution of disk systems. Because of the possible importance of environmental effects, we will consider separately relatively isolated, "normal" spiral galaxies and cluster disk galaxies. We shall first explore the extent to which the morphology and evolution of spiral galaxies can be understood in terms of interaction of gas with the density-wave pattern. Next, we shall discuss the nature of disk systems in which there is no evidence of recent star formation. We will focus here primarily on systems located in clusters of galaxies where environmental factors may predominate. Finally, we will discuss a class of relatively nearby spiral galaxies in which the gas content may be quite large and from which we may possibly hope to deduce the characteristics of normal spiral galaxies at much earlier evolutionary phases.

3. RECENT OPTICAL OBSERVATIONS

3.1. Spiral galaxies

The Hubble sequence. The main Hubble classification criteria for spiral galaxies are: 1. the prominence of the bulge relative to the disk; 2. the openness of the spiral arms. Galaxies of type Sa have tightly-wound arms (small pitch angle) and relatively large nuclear bulge regions, while those of type Sc have the most open-arm patterns (large pitch angle) and smallest bulges. Roberts et al. (1975) have shown that the computed pitch angle of the spiral arms is greatest for galaxy mass distributions which have a low degree of central concentration, whereas wave patterns computed for models with high central concentration are tightly wound. Hence, the relationship between bulge prominence and arm openness implicit in the Hubble classification scheme seems well understood on the basis of the wave patterns permissible for given galaxy mass distributions.

Luminosity class. van den Bergh (1960a,b) has shown that the luminosity of a spiral galaxy is related to the qualitative appearance of the spiral arms. Galaxies with prominent, narrow spiral arms are intrinsically the most luminous, while galaxies exhibiting patchy, broad arms have the lowest intrinsic brightness. Roberts et al. (1975) have argued that the width and prominence of the arms are directly related to the strength of the galactic shock induced by interaction of disk gas with the density-wave pattern. Where $w_{\perp_0} \gg a$, the degree of compression in the shock region is large and the width of the region of high compression is small. If star-forming efficiency is related to the compression suffered by the gas, and if the width of the spiral arm (as measured by the angular extent of recently-formed stars) is related to the width of the region of high compression, then one expects those galaxies characterized by large values of $w_{\perp_0}$ to have the narrowest, most prominent arms. Because $w_{\perp_0}$ is related to galactic mass ($w_{\perp_0} \sim M^2$), both this quantity and the arm appearance are expected to be correlated with intrinsic galactic luminosity as well (if $M/L \sim$ constant). This prediction has been borne out by a comparison of the luminosity classes assigned by van den Bergh with the mean value of $w_{\perp_0}$ derived from observed galaxy rotation curves (Roberts et al. 1975).

Choice of pattern speed, Ω_p. The pattern speed Ω_p cannot at present be predicted directly from density-wave theory. Therefore when comparing computed and observed wave patterns, Ω_p is treated as a free parameter. Roberts et al. (1975) have argued that an approximate value of the pattern speed can be estimated from the location of the outermost H II region in the spiral galaxy. They reason that this region indicates the approximate radius beyond which star formation cannot be initiated by galactic shocks. If we associate the outermost H II region with the "corotation radius" [at which $(\Omega - \Omega_p)$ and hence $w_{\perp_0} = 0$], we can derive Ω_p from the observed angular velocity of this region Ω. This choice of Ω_p leads to quite satisfactory fits to the wave patterns of 24 spiral galaxies for which rotation curves provide an estimate of an appropriate mass model (Roberts et al. 1975).

An independent check on the choice of pattern speed may be provided if the inner or outer Lindblad resonance in a galaxy can be located. At the inner resonance, $\Omega - \Omega_p = \kappa/2$. Here, κ is the free oscillation frequency of the stars (which can be computed directly from the observed rotation curve). A possible observational consequence of the inner Lindblad resonance in spiral galaxies is the presence of bright rings of young stars and H II regions, recently formed as disk gas encounters the tightly-wound, high-amplitude wave pattern predicted for the region just outside the resonance (Mark 1975). The galaxy NGC 5364 (Figure 2) is an excellent example of a galaxy exhibiting a prominent ring of H II regions. A rotation curve for this galaxy was derived by Goad et al. (1975). These authors conclude that the inner Lindblad resonance is located ~ 1 kpc inward of the ring of H II regions if a value of Ω_p equal to the observed angular speed of the outermost H II region in NGC 5364 is selected. This result provides encouraging support to the Roberts et al. criterion for selecting pattern speeds.

Wave amplitudes. Schweizer's (1976) observation of a wave pattern in the old disk population of several prominent spiral galaxies provided the first direct evidence of stellar density-wave arms. The amplitudes of the waves observed by Schweizer varied from ±5 percent to ±30 percent of the background disk-surface brightness. These values are somewhat larger than the amplitudes which have been inferred from

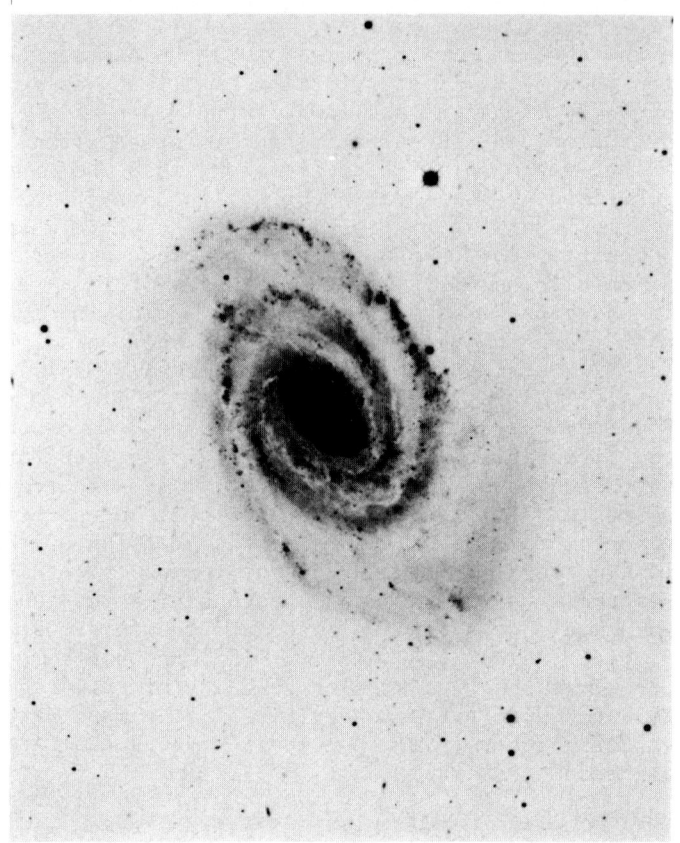

Figure 2. A blue-light photograph (GG 385 + IIIa-J) of NGC 5364 taken at the prime focus of the Mayall 4-m telescope by C. R. Lynds. Note the prominent central ring of H II regions. North is at the top and east at the left.

analysis of stellar orbits in our own galaxy, and which have been commonly adopted in most models of density-wave-driven star formation. Currently, the amplitude of the wave pattern in a given galaxy cannot be predicted directly from density-wave theory. Yet, the departures from the axisymmetric gravitational field produced by the wave play an important role in determining the degree of compression in galactic shocks. As a consequence, it is of some importance to determine the range of wave amplitudes characterizing spiral galaxies of differing morphological type. Eric Jensen of Rice University has undertaken such a study. In order to emphasize the contribution of the underlying disk population to the observed wave amplitude, he has chosen to observe his sample of galaxies at wavelengths of 8500 Å and 1 μ. At these wavelengths, the red K giant population of the old disk dominates the young stars in the spiral arms. Thus far, he has completed an analysis of two galaxies, M51 and M101. His results suggest that as the galactocentric distance increases the wave amplitudes increase from values of ±5 percent to ±40 percent of the background-disk surface brightness, thus confirming Schweizer's conclusions (derived from observations at shorter wavelengths).

<u>Luminosity and color evolution across spiral arms</u>. If star formation is triggered by passage of disk gas through the density-wave pattern, then the spiral-arm regions should exhibit the following evolutionary pattern: 1. near the concave, inner edge of the spiral arm, evidence of recent compression in the form of dust lanes in dark clouds; 2. in an intermediate zone, OB associations in H II regions formed from gas compressed at an earlier epoch; 3. on the outermost (convex) edge of the arm, aging clusters and associations (evolved from OB associations formed at an earlier epoch). B. Lynds (1970) has presented strong evidence which confirms that dust lanes are confined to the inner edge of spiral arms. Both Schweizer (1976) and Dixon et al. (1972) have computed the luminosity and color profiles across spiral arms expected from density-wave-driven star formation. The angular drift ϕ of the newly-formed stars relative to the "edge of the arm" as defined by the dust lanes is given by $d\phi(r)/dt = \Omega(r) - \Omega_p$. Schweizer's observed luminosity profiles provide some evidence in favor of "drift," although his results are not conclusive. An attempt to derive color changes indicative of an age sequence of the type described above has been made by Talbot et al. (1977) for M83. Thus far, their analysis of observed colors provides no definitive evidence which suggests aging across spiral arms in this galaxy. It should be noted that analysis of color and luminosity profiles across spiral arms is considerably complicated by the presence of dust and uncertainty in the time between compression and the appearance of observable (at visible wavelengths) young stars. Moreover, in order to estimate ages of the newly-formed stellar population, one must accurately subtract the contribution of the underlying density wave. More accurate photometric studies may eventually provide evidence of the expected age drift across spiral arms. It would be embarrassing to the shock-induced star-formation picture if such changes were not observed.

<u>Chemical enrichment</u>. If density-wave-driven star formation dominates postformation disk evolution, one expects that the frequency of star-forming events will depend on $(\Omega - \Omega_p)$, while the efficiency of star formation will be related to the compression suffered by disk gas as it passes through the density-wave crest. Radial changes in chemical composition can be related directly to the star-forming frequency. Regions where stars form most frequently should be those (a) where the disk gas is consumed most rapidly, and (b) in which the chemical composition of the remaining gas is high (since the ejecta from previous generations of stars easily contaminate the remaining material). In a recent study, Jensen et al. (1976) observed several abundance-sensitive emission-line ratios in H II regions located in the disks of 12 spiral galaxies. They attempted to correlate the inferred chemical composition with the star-formation frequency and efficiency inferred from

density-wave models. In agreement with the predictions of galactic shock models, they find that (a) the metal abundance is highest in regions of high $\Omega - \Omega_p$, and (b) galaxies characterized by high mean values of w_{1_o} show significantly higher mean abundances. While other, more *ad hoc* models might explain the results, Jensen et al. believe that their data provide a strong consistency check on the predictions of the density-wave model.

3.2. S0 galaxies and the effects of environment on evolution of disk systems

Galaxies classified as S0 are systems having featureless disks which exhibit no evidence of spiral density waves or recent episodes of star formation. They have long been viewed as "transition" objects between the elliptical and spiral sequences. Their true evolutionary status is at present not clear. It is possible that S0 galaxies represent (a) systems in which the amount of disk gas remaining after formation was small and in which star formation consequently ceased soon thereafter, (b) former spiral galaxies in which the evolutionary processes described in the previous section have exhausted the disk gas in the relatively recent past, (c) former spiral galaxies in which disk gas has been somehow removed either by interaction with the intergalactic environment or by other processes, or (d) some combination of the above. Freeman (1970) carried out a pioneering quantitative study of the characteristics of S0 galaxies in an attempt to compare them with actively star-forming systems. He concluded that the disk light exhibits an exponential light profile of the form $I = I(o)\ e^{-\alpha r}$ (see also de Vaucouleurs 1959). The exponential scale length for S0 galaxies is similar to those derived for galaxies of types Sa-Sbc ($2 \lesssim \alpha^{-1} \lesssim 10$ kpc), although different (for his sample) from the α^{-1} values characterizing Hubble types Sc and Scd (2-5 kpc).

Furthermore, his study shows that the projected central surface brightnesses derived for S0 disks are identical to those found for spiral galaxies. Freeman also notes that the bulge/disk ratios for "field" S0 galaxies are not discernably different from those of later Hubble types. Sandage et al. (1970) use these data to argue that because the intrinsic "bulk" properties of S0 galaxies are no different from spirals, it is illogical to assume that the astration rates (and hence gas depletion rates) differ between the disks of spirals and S0 galaxies. Because the fractional gas content is zero or nearly so in most S0 galaxies, while that of later-type galaxies is significantly greater, they conclude that the basic difference between S0s and spirals results from a difference in the amount of gas remaining in the disk subsequent to disk formation. However, arguments based on more recent observational studies may obviate this conclusion. First, Kormendy (1977) has demonstrated that the apparent exponential light distribution in disk systems is in part an observational artifact arising from the combined contributions of the bulge and disk regions to the observed surface brightness distributions. Several of the disk light distributions derived by Kormendy by careful subtraction of the bulge component show a distinctly non-exponential character. Moreover, the projected central surface brightnesses of the disk light distribution do not appear to have a "universal" value. Hence, it is no longer clear that (a) the bulk characteristics of spiral and S0 systems are indeed identical, and that (b) as a consequence, the rates of star formation and gas depletion in the disks must be identical as well.

It is also necessary for Sandage et al. to demand a difference in disk-system formation conditions between rich clusters and the field, since rich clusters contain a much smaller fraction of spiral galaxies (compared to S0s); disk systems in clusters must somehow be formed in a manner such that the amount of gas remaining in the disks is much lower. While this possibility is by no means ruled out, other mechanisms have been invoked which appear to offer a far less *ad hoc* explanation of the absence of spiral galaxies in rich clusters. In this section, we shall explore in some detail recent studies of the effects of environment on disk-system evolution. However, we must bear in mind that not all S0 galaxies are found in rich clusters, and that effects other than environmental influences may well be important in accounting for the simultaneous presence of spiral and S0 galaxies in the field.

Morphology of disk systems as a function of environment. Recently, Oemler (1974) investigated the frequency distributions of ellipticals, S0s, and spiral galaxies in clusters of galaxies differing in structure and appearance. He was able to discern three types of clusters: 1. cD; 2. spiral-poor; 3. spiral-rich. Spiral-rich clusters have a mixture of galaxy types most similar to the field (dominated by spirals and S0s and poor in E-type systems). They are irregular in appearance, have a low mean density of galaxies, and no tendency toward central concentration. Spiral-poor and cD clusters, on the other hand, are deficient in spiral galaxies and, according to Oemler, exhibit a much higher percentage of elliptical galaxies. cD clusters are dominated by central supergiant galaxies and tend to be dense, centrally concentrated, and spherical. Spiral galaxies are virtually absent in the cores of these clusters. Spiral-poor clusters represent cases intermediate in character between the cD and spiral-rich clusters; they are not quite as regular, compact, or centrally concentrated as the extreme cD clusters. Oemler suggests that to a large extent the difference in type results from dynamical evolution of the clusters. The high-density, short-collapse-time, spiral-poor and cD clusters are presumed to be the most dynamically evolved. Both their smooth mass distributions and high central concentration suggest a considerable period during which two-body relaxation processes have been operative. Conversely, the low density of spiral-rich clusters implies long cluster collapse times. Furthermore, the lack of central concentration and the irregular mass distribution of these clusters indicate a lack of any significant relaxation.

The predominance of S0 galaxies in cD and spiral-poor clusters is supposed to result from transmutation of spirals to S0 galaxies as a consequence of ablative stripping in the dense cores of these cluster types. Observations of the X-ray luminosity and velocity dispersion for clusters representative of these types suggest that, at least near the cluster center, the intergalactic gas density is sufficient to remove disk gas from most spiral galaxies. X-ray observations of spiral-rich clusters suggest the absence of intergalactic gas at densities sufficient to effect stripping. Evidently, dense intergalactic media can exist only in clusters already collapsed and dynamically relaxed (and possibly in those currently undergoing collapse).

It is not clear, however, that all differences in the distribution of morphological types can be attributed solely to environmental effects. If the ratio of ellipticals/(spirals + S0 galaxies) is truly different between spiral-rich and spiral-poor and cD clusters (Oemler 1974), one must accept either (a) that S0 galaxies can be transmuted to ellipticals (see Richstone 1976; Marchant and Shapiro 1977), or (b) a difference in the initial distribution of galaxy-bulge/disk ratios which depends on conditions in the protocluster environment. It is essential to determine the true fraction of ellipticals, S0s, and spiral galaxies based on quantitative analysis of galaxy profiles. It would also be important to determine the difference, if any, between S0 galaxies located in cD and spiral-poor clusters and those located either in irregular, spiral-rich clusters or in the field. Presumably, S0 galaxies in the field have completed their "normal" evolutionary development. If, for example, field S0s represent galaxies that consume their gas most rapidly (presumably those with highest $\Omega - \Omega_p$ and w_{t_o}), then the distribution of B/D ratios and M/R ratios for these galaxies might differ significantly from those characterizing S0 galaxies in rich clusters where normal evolutionary development has been truncated by stripping.

<u>The fraction of spiral galaxies as a function of X-ray luminosity and cluster position.</u> Galaxies presently in the center of clusters similar to Coma are moving through intracluster gas of a density apparently sufficient to effect stripping by ablation. Hence, all spiral galaxies in the cluster cores probably have been stripped. However, there is reason to suppose that the fate of spirals located in the outer regions of the cluster is not as certain. Because the density of intergalactic gas appears to decrease outward from the cluster center, the gas density in the outer regions appears too low to result in spiral stripping. Nevertheless, it is possible that some galaxies now on the outside of the cluster have passed through the cluster center in the past. However, since the typical crossing time for a rich cluster of galaxies similar to Coma is on the order of a few billion years, it seems likely that galaxies presently in the outer regions of such a cluster will not have passed through the cluster center many times during a Hubble time even if (a) their orbits are all radial or (b) the cluster relaxed at a very early epoch. Furthermore, it is not clear at what epoch the intergalactic medium achieved a density sufficiently high to strip spirals. If the medium is of relatively recent origin, or if galaxy orbits are not primarily radial, systems located in the outer regions of the cluster may remain unaffected by stripping.

II.1 THE EVOLUTION OF DISK GALAXIES

Melnick and Sargent (1977) have classified galaxies in a number of clusters known to exhibit X-ray emission. Their data suggest that the fraction of spiral galaxies increases from near zero in the cluster cores to values typical of the field in the outer cluster regions. Moreover, they find that the fraction of spiral galaxies is largest in those galaxies with weakest X-ray emission and smallest in those galaxies where X-ray emission is stronger (see also Bahcall 1977). These authors regard this evidence as strongly favoring stripping caused by the ram pressure of intergalactic gas.

Although their results strongly support the idea of transmutation from spiral to S0 galaxy types, it is also conceivable that formation rather than environmental conditions might also account for the observations. For example, the conditions under which galaxies are formed in clusters may favor the production of elliptical and rapidly evolving spiral galaxies (those with relatively large bulge-to-disk ratios, high values of $\Omega - \Omega_p$ and high values of w_{1_o}) in cluster cores. In order to test the importance of stripping, it would be of value to: 1. identify recently stripped galaxies; 2. search for (a) the presence of a greater fraction of (unstripped) spiral galaxies at earlier epochs (greater look-back times) in clusters similar in morphological appearance to nearby spiral-poor and cD clusters, or (b) changes with radial distance from the cluster center, indicating not only a change in the spiral fraction but in the properties of presumably stripped S0 systems as well.

<u>Smooth-arm spiral galaxies.</u> A possible time sequence for spiral galaxy evolution "post-stripping" might be (a) loss of Population I tracers (dust, OB associations, and H II regions), (b) weakening of the density wave in the disk population, and (c) the final S0 state in which no density waves are discernible. Because the integrated color of the disk should become redder as the mean age of the stellar population in the disk increases, recently stripped spiral galaxies should show the bluest disk colors and the strongest relic arms, while the oldest stripped systems should exhibit red disks and no arms (S0s). A few years ago, Strom et al. (1976a) identified two smooth-arm galaxies in which no evidence of Population I tracers was found. In both cases, the galaxies were located in known X-ray clusters and appear to be ideal candidates for identification as stripped spirals. The absence of a Population I component in the arms was demonstrated not only from the appearance of the galaxy, but from a quantitative comparison of arm and disk colors. No difference between arm and disk $(U - R)$ colors was observable. More recently, Wilkerson et al. (1977) have identified nearly 25 smooth-arm spirals located in clusters known to be X-ray sources. In Figure 3, we present photographs of three smooth-arm spiral systems representative of the range in observed arm amplitudes for systems of this class. The results of a preliminary study of the 25-galaxy sample suggest that the spiral waves of largest amplitude are found in disks having the bluest colors, as might be expected if these systems were stripped most recently. In systems with weak arm amplitudes, the disk colors appear reddest, a result consistent with the belief that systems stripped relatively long ago should show lower arm amplitudes.

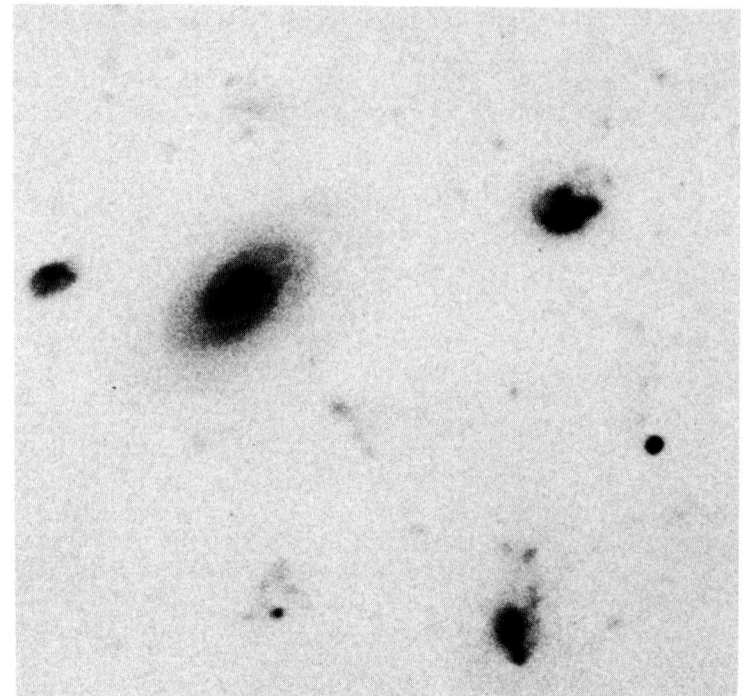

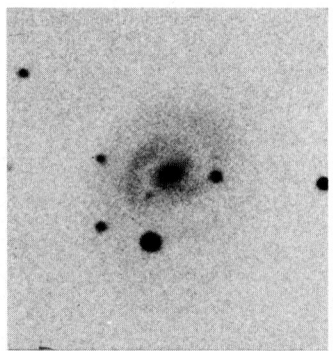
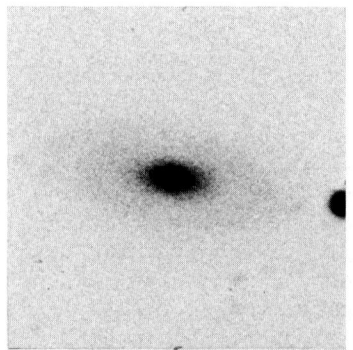

Figure 3. Ultraviolet(U)-band (UG 2 + IIIa-J) photographs of three smooth-arm spiral galaxies taken with the Mayall 4-m telescope. Top (a): NGC 3860 located in the cluster Abell 1367; bottom left (b): NGC 1268, Perseus cluster; bottom right (c): IC 2951, Abell 1367. Note the progressive decrease in wave amplitude from galaxy (a) to galaxy (c). The orientations of individual galaxies are arbitrary.

A particularly intriguing case of a smooth-arm spiral was found in the cluster Abell 1367 and is illustrated in Figure 3 (a). For this galaxy (NGC 3860), not only do we observe a high amplitude, smooth-arm spiral pattern, but also a number of irregular "blobs" scattered about the galaxy. It is conceivable that these blobs might represent shreds of material stripped from the galaxy and in which star formation was recently induced. Further spectroscopic study is required in order to confirm this speculation.

To further test the belief that smooth-arm spirals have been stripped, it would be of great importance to determine the neutral hydrogen content for galaxies of this class (since we presume that hydrogen has been removed from the disk!). Without confirmation of low hydrogen content, it is not possible to dismiss the possibility that the lack of Population I constituents in smooth-arm galaxies results not from the absence of disk gas but from physical conditions in the gas (high temperature, large velocity dispersion) which preclude (temporarily?) star formation at the current epoch (see Strom et al. 1976a; Scott et al. 1977).

If we are correct in believing that smooth-arm spirals represent the initial stages in the transmutation of a spiral to an S0 galaxy, it is important to note that we demand a decrease in the wave amplitude when the gas is removed and star-forming events in the disk cease. The physical cause for the decay of density waves under such conditions is at present not understood and merits careful theoretical treatment.

It is possible that smooth-arm spiral galaxies might also be found in the field. In such cases, one might speculate that a galactic wind or normal evolution has significantly reduced the gas content in these galaxies. In Figure 4, we present a photograph of a system (NGC 4622) which appears to have smooth arms in the inner regions and clumpy complexes of OB stars and H II regions in its outer parts. This system might well be one in which gas has been exhausted through evolutionary processes or removed by galactic winds in the inner parts of the disk and might therefore be intermediate in character between normal and smooth-arm spiral systems. Some of the galaxies classified as "anemic" by van den Bergh (1976) may be similar representatives of galaxies in transition between the normal spiral and smooth-arm evolutionary stages.

Disk colors as a function of cluster position. Strom and Strom (1977) have recently studied the disk $U - R$ colors of edge-on S0 galaxies in the Coma cluster. In Figure 5, we present histograms which depict the color distribution among (a) disks located within 18 arcmin (0.7 Mpc) of the cluster center, and (b) those located outside 18 arcmin. We deduce from Figure 5 that many more blue S0 disks are found in the outer region of the cluster. This result can be interpreted by assuming (a) that a larger fraction of the outer region S0 disks have been stripped more recently than those located in the inner parts of the cluster or (b) that the outer-region S0 galaxies have completed their disk evolution relatively unaffected by environmental effects;

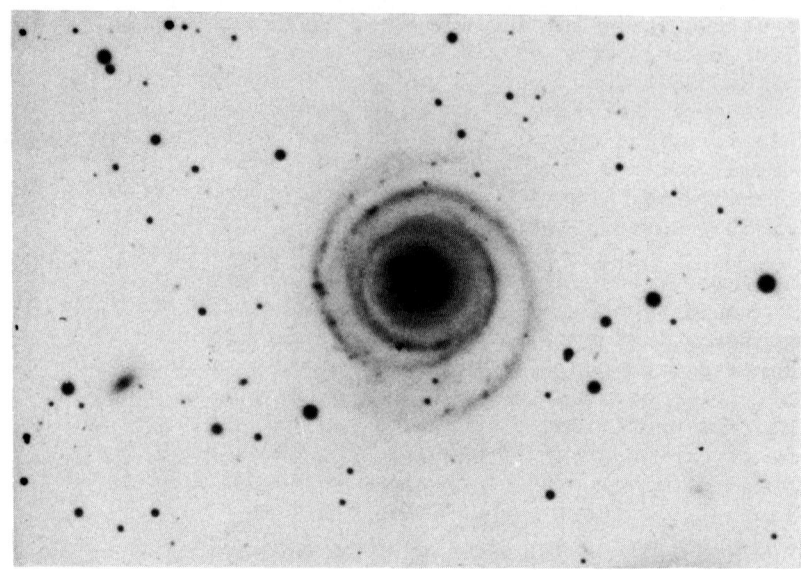

Figure 4. A blue-light photograph (GG 385 + IIIa-J) of NGC 4622 (Centaurus cluster) taken with the CTIO 4-m telescope.

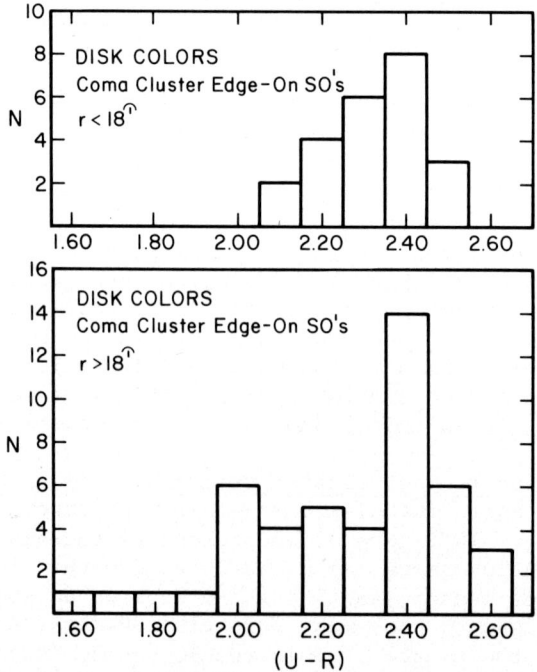

Figure 5. Histograms depicting the distributions of disk colors [$(U-R)$] for (top) disk systems located within 18 arcmin of the Coma cluster core and for (bottom) those located in the outer regions ($r > 18'$) of Coma.

hence, the observed color distribution reflects a (variable) range of times between the present and the last episode of disk-star formation. Hypothesis (b) can be checked by comparing the disk color distribution observed in the outer parts of Coma with that of field S0s. In either case, these data provide an important confirmation of the Melnick and Sargent (1977) hypothesis that stripping rather than initial conditions account for the increase in spiral fraction with increasing distance from the cluster center.

Disk colors in distant clusters of galaxies. Butcher and Oemler (1977) have studied the distribution of galaxy colors in two distant, rich, and centrally condensed clusters (3C 295, $z = 0.46$; Cl 0024 +1654, $z = 0.39$) similar in structure to the Coma cluster. They conclude that in distinct contrast to nearby clusters of this type (which contain galaxies primarily of the E and S0 type) between one-third and one-half of the galaxies in the distant clusters have blue colors similar to those characterizing spiral galaxies. Moreover, the fraction of blue galaxies increases with increasing distance from the center of these clusters. Butcher and Oemler's data support the belief that a smaller fraction of spirals has been stripped at the epoch corresponding to $z = 0.4$ as compared to the present epoch, and that the fraction of stripped spirals decreases with increasing distance from the cluster center.

Disk sizes. Larson (1972a,b) has argued that infall from low-density, gaseous halos (possibly remaining from protogalactic condensations several hundred kpc in size) significantly affects disk-system evolution over a large fraction of a Hubble time. If such extensive halos are common, it is conceivable that the outer regions of disk systems were formed during the last several billion years. A possible indirect indication of the importance of relatively recent disk-star formation resulting from the collapse of the outer regions of extensive halos might be afforded by the examination of disk sizes as a function of radial position in rich clusters. Ablative (or possibly tidal and collisional stripping) should remove halos surrounding disk galaxies located in the central regions of the clusters; some disk systems in the outer-cluster region might be unaffected owing to the lower density of intergalactic gas and of other galaxies. Strom and Strom (1977) have begun to examine disk-galaxy sizes for edge-on S0 galaxies in the Coma cluster. Sizes of S0 disks (as measured to an isophote corresponding approximately to $\mu_R = 25$ mag per square arcsec) were measured for a sample of nearly 70 galaxies in Coma. A significant increase in the fraction of "large" disk systems was noted for those disk galaxies located beyond 18 arcmin from the cluster center. While this result is preliminary at present (and subject to analysis of such selection effects as (a) systematic differences in the true orientation of the galaxies and (b) the relative contribution of disk and bulge to the total observed surface brightness), it suggests that formation of the outer regions of disk systems may be truncated in the dense cores of rich clusters. It is, of course, also possible that initial conditions favor the formation of galaxies having high bulge-to-disk ratios in the central cluster regions. It would be of considerable interest to know how disk sizes and B/D ratios vary from spiral-rich to spiral-poor clusters in order to assess the relative effects of environmental and formation conditions on the bulge-to-disk ratio.

4. LOW SURFACE BRIGHTNESS SPIRAL GALAXIES: EARLY STAGES OF SPIRAL EVOLUTION?

Examination of 48-inch Schmidt and 4-m plate material has revealed a class of spiral galaxies characterized by apparently low values of disk surface brightness μ_d. In Figure 6, we present U and R photographs of two such systems, NGC 4411 (a) and (b).

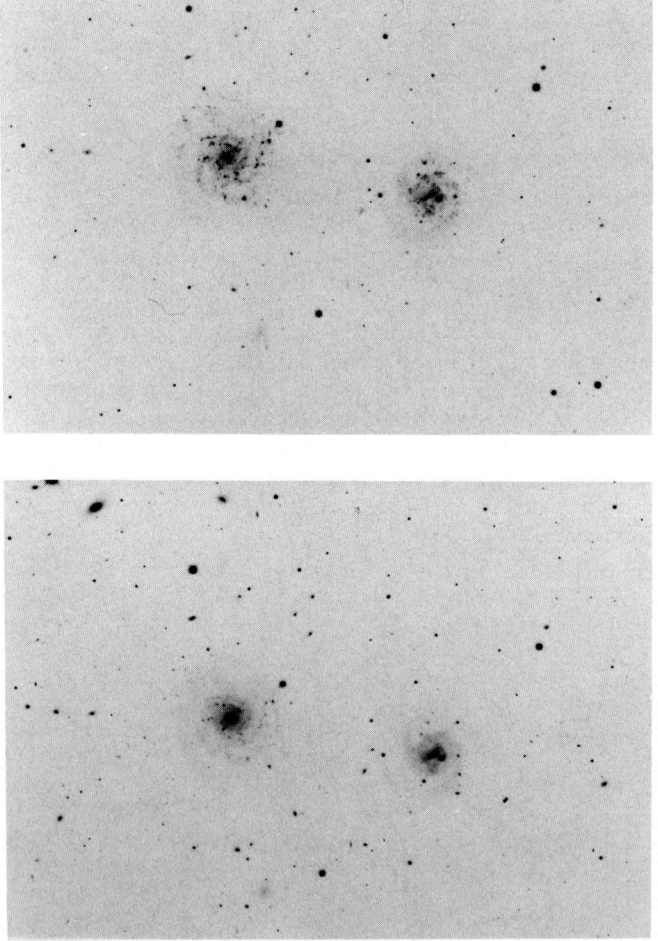

Figure 6. Mayall 4-m telescope prime focus, ultraviolet (UG 2 + IIIa-J; top) and red (RG 610 + 127-04; bottom) photographs of the low surface brightness spiral galaxies NGC 4411 (a) (right) and NGC 4411 (b) (left). North is at the top and east is at the left.

Low values of μ_d suggest that the total number of stars formed in the disk over a Hubble time has been small. We felt that an examination of such systems might be of considerable importance since they may represent (a) galaxies in which stellar formation is either inefficient, or (b) "young" galaxies in which the bulk of star formation has taken place only during the last few billion years. Romanishin et al. (1977) have undertaken an optical study of 12 such galaxies. Our major results to date suggest that (a) disk surface brightnesses in these systems are approximately 2-5 times smaller than those typical of bright, prominent spirals such as M81, M51, or M101; (b) the disks are unusually blue; and (c) the surface brightness μ_d is correlated with the disk color index. Those galaxies of lowest μ_d are also the bluest galaxies.

The blue disk colors may result from the dominant contribution of stars formed at a comparatively recent epoch. If so, either the average disk stars are relatively young or the initial mass function of older stellar generations is such that these stars have few descendents currently detectable at optical wavelengths. Alternatively, blue disk colors could result from the dominance of older, metal-poor disk populations.

Whether low surface brightness spiral galaxies are young, metal-poor, or both, it seems reasonable to assume that the fraction of gas converted to stars and heavy elements might be low. As a consequence, the neutral hydrogen content of the disk should be large. In collaboration with N. Krumm and E. Salpeter, we have obtained preliminary values of the ratio of total hydrogen mass M_H to photographic luminosity L_{pg} for three low surface brightness spirals; M_H/L_{pg} appears to lie in the range 0.5 to >2. These values are larger than those characteristic of normal disk galaxies and suggest that a smaller fraction of the gas in these low surface brightness systems may have been processed in star-forming episodes.

We are currently seeking indirect evidence bearing on the possible age of these systems. In collaboration with G. Knapp, Strom, Strom, and Romanishin are attempting to search for extensive neutral hydrogen clouds surrounding these systems. If such clouds are discovered, their presence might suggest that star formation on a galactic scale was initiated in these clouds at a relatively recent epoch.

If the galaxies are not intrinsically young, why is the disk gas content relatively high? One hint may come from the M/R ratio as estimated from the observed system luminosities and Holmberg radii for these galaxies. These ratios are smaller by factors of between 3 and 10 as compared to those characteristic of normal spiral systems. As discussed in Sec. 2.1, galaxies showing small values of M/R should be characterized by low values of $\Omega - \Omega_p$ and $w_{1\,0}$. Consequently, the star-formation rate in low surface brightness systems, as compared to prominent, "normal" spiral systems, may be significantly lower. Hence, the number of disk stars formed over a Hubble time in such galaxies will be smaller, the disk surface brightness lower, and the amount remaining neutral hydrogen larger. To test the "low star-formation rate" hypothesis, we have compared the amplitude of the spiral waves as derived from ultraviolet (U) plates of several low surface brightness galaxies with the U amplitude for normal spiral systems. The measured U amplitude should provide a crude index of the current star-formation rate in the arms, since the OB associations and H II regions contribute predominantly to the total light observed at this wavelength. Our results show that the low surface brightness galaxies exhibit U amplitudes 2-3 times smaller than those which appear to characterize "normal" spirals, therefore suggesting a lower rate of star formation at the present epoch.

Does the existence of very low amplitude arms in these systems suggest an M/R below which no spiral structure is possible? Is the dominance of irregular systems among low luminosity a manifestation of the inability of low-mass disk systems to support spiral waves?

Whether low surface brightness spiral galaxies are young or systems in which star formation has proceeded slowly over a Hubble time, their relatively high hydrogen content suggests that they might give some hint of the initial gas distribution characteristic of normal spirals at earlier stages in their evolutionary history. High resolution studies at 21 cm therefore seem merited for a few examples of this class. Moreover, if few stars are forming in their disks at present, observations of low surface brightness galaxies may provide crude limits on the fraction of hydrogen converted to stars during the first burst of star formation following the collapse of a protogalactic cloud to disk form.

5. OPTICAL AND INFRARED SEARCHES FOR MASSIVE HALOS

Attempts to detect at optical wavelengths a massive halo component in external galaxies have thus far proven unsuccessful (Davis 1975; Freeman et al. 1975). Perhaps the most sensitive test thus far reported is that of Gallagher and Hudson (1976). They attempted to observe the halo of edge-on spiral galaxy IC 2233. This system has a thin disk which exhibits no discernable bar or other instabilities. By using the chopping secondary of the University of Minnesota-University of California (San Diego) telescope, they were able to switch rapidly between halo and sky locations thereby cancelling short-time-scale variations in sky brightness. No halo component, to the level of 1-5 percent of the central disk surface brightness, was detected at wavelengths between 0.4 to 0.8 μ. They conclude that the M/L ratio for the halo component must therefore exceed 100.

Strom et al. (1978) have obtained scans out to galactocentric distances of ~ 5 kpc of the bulge components of NGC 3115 (E7/S0) and NGC 2768 (E6) at a wavelength of 2.2 μ. These observations should be extremely sensitive to any increase of the M dwarf population in the halo regions of these galaxies; if such a change in population mix were present, a color index such as $(V - K)$ should become redder at increasing galactocentric distances. Instead, Strom et al.'s data suggest that the $(V - K)$ color index grows monotonically bluer outward from the galaxy centers. The observed $(V - K)$ color index at the outermost points excludes a halo M/L greater than 50 if the bulk of the halo mass is contained in stars of type M8 V (see also Strom et al. 1976b).

Indirect optical evidence in support of massive halos can be found in Schweizer's (1977) study of the rotation curve and light distribution in the disk of the Sombrero galaxy. His data suggest a monotonic increase in M/L from the center to the edge of the observable disk. A similar result has been obtained in a study of NGC 4378 by Rubin et al. (1977).

We wish to thank Dr. E. Jensen, W. Rice, W. Romanishin, and M. S. Wilkerson, who have collaborated with us on a number of the research programs reported in this review. Their insight and diligence as well as their tolerance of our demanding personalities are acknowledged with gratitude. We also wish to note the many hours of stimulating and at times critical discussions with a number of our colleagues: Drs. H. Butcher, S. Faber, J. Gallagher, J. Goad, J. R. Gott, G. Illingworth, R. Larson, G. Oemler, L. Thompson, and B. Tinsley. Finally, we thank Dr. Don Wells of KPNO for his contributions to the development of the basic interactive picture processing system (IPPS). Without his work many of the new data discussed here could not have been reduced and analyzed.

REFERENCES

Baade, W. and Spitzer, L.: 1951, *Astrophys. J.* **113**, 413.
Bahcall, N.: 1977, *Astrophys. J. Letters* (in press).
Bregman, J.: 1976, *Bull. Am. Astron. Soc.* **8**, 539.
Butcher, H. and Oemler, A.: 1977, *Astrophys. J.* (in press).
Davis, M.: 1975, *Astron. J.* **80**, 188.
de Vaucouleurs, G.: 1959, *Handbuch der Physik* **53**, 311.
Dixon, M. E., Ford, V. L., and Robertson, J. W.: 1972, *Astrophys. J.* **174**, 17.
Faber, S. M. and Gallagher, J. S.: 1976, *Astrophys. J.* **204**, 365.
Field, G. B., Goldsmith, D. W., and Habing, H. J.: 1969, *Astrophys. J. Letters* **155**, 149.
Freeman, K. C.: 1970, *Astrophys. J.* **160**, 811.
Freeman, K. C., Carrick, D. W., and Craft, J. L.: 1975, *Astrophys. J. Letters* **198**, 93.
Gallagher, J. S. and Hudson, H. S.: 1976, *Astrophys. J.* **209**, 389.
Goad, J. W., Strom, S. E., and Goad, L. E.: 1975, *Bull. Am. Astron. Soc.* **7**, 395.
Gott, J. R., III and Thuan, T. X.: 1976, *Astrophys. J.* **204**, 649.
Gunn, J. E. and Gott, J. R., III: 1972, *Astrophys. J.* **176**, 1.
Jensen, E. B., Strom, K. M., and Strom, S. E.: 1976, *Astrophys. J.* **209**, 748.
Kormendy, J.: 1977, *Astrophys. J.* (in press).

Larson, R. B.: 1972a, *Nature* **236**, 21.
Larson, R. B.: 1972b, *Nature* **236**, 7.
Larson, R. B.: 1976, *Monthly Notices Roy. Astron. Soc.* **176**, 31.
Lin, C. C. and Shu, F. H.: 1964, *Astrophys. J.* **140**, 646.
Lindblad, P. O.: 1960, *Stockholm Obs. Ann.* **21**, 4.
Lynds, B. T.: 1970, in *Proc. IAU Symposium No. 38*, Reidel, Dordrecht, p. 26.
Marchant, A. B. and Shapiro, S. L.: 1977, *Astrophys. J.* **215**, 1.
Mark, J. W-K.: 1975 (private communication).
Mark, J. W-K.: 1976, *Astrophys. J.* **206**, 418.
Mathews, W. G. and Baker, J. C.: 1971, *Astrophys. J.* **170**, 241.
Melnick, K. and Sargent, W. L. W.: 1977, *Astrophys. J.* **215**, 401.
Oemler, A.: 1974, *Astrophys. J.* **194**, 1.
Oort, J. H.: 1970, *Astron. Astrophys.* **7**, 381.
Ostriker, J. P. and Peebles, P. J. E.: 1973, *Astrophys. J.* **186**, 467.
Ostriker, J. P. and Thuan, T. X.: 1975, *Astrophys. J.* **202**, 353.
Richstone, D. O.: 1976, *Astrophys. J.* **204**, 642.
Roberts, W. W., Jr., Roberts, M. S., and Shu, F. H.: 1975, *Astrophys. J.* **196**, 381.
Romanishin, W., Strom, K. M., and Strom, S. E.: 1977 (in preparation).
Rubin, V. C., Ford, W. K., Strom, K. M., Strom, S. E., and Romanishin, W.: 1977 (in preparation).
Sandage, A. R., Freeman, K. C., and Stokes, N. R.: 1970, *Astrophys. J.* **160**, 831.
Schmidt, M.: 1959, *Astrophys. J.* **129**, 243.
Schweizer, F.: 1976, *Astrophys. J. Suppl.* **31**, 313.
Schweizer, F.: 1977, *Astrophys. J.* (in press).
Scott, J., Jensen, E. B., and Roberts, W. W.: 1977, *Nature* (in press).
Strom, K. M., Strom, S. E., Wells, D. C., and Romanishin, W.: 1978, *Astrophys. J.* (in press).
Strom, S. E., Jensen, E. B., and Strom, K. M.: 1976a, *Astrophys. J. Letters* **206**, 11.
Strom, S. E. and Strom, K. M.: 1977 (in preparation).
Strom, S. E., Strom, K. M., Goad, J. W., Vrba, F. J., and Rice, W.: 1976b, *Astrophys. J.* **204**, 684.
Talbot, R. J., Jr. and Arnett, W. D.: 1975, *Astrophys. J.* **197**, 551.
Talbot, R. J., Dufour, R., and Jensen, E. B.: 1977, *Evolution of Galaxies and Stellar Populations*, Yale University Obs., New Haven, CT.
van den Bergh, S.: 1960a, *Astrophys. J.* **131**, 215.
van den Bergh, S.: 1960b, *Astrophys. J.* **131**, 558.
van den Bergh, S.: 1976, *Astrophys. J.* **206**, 883.
Wilkerson, M. S., Strom, K. M., and Strom, S. E.: 1977 (in preparation).
Woodward, P. R.: 1976, *Astrophys. J.* **207**, 484.

DISCUSSION FOLLOWING REVIEW II.1 GIVEN BY S.E. STROM

WAXMAN: Shouldn't we expect star formation to be occurring between the co-rotation point and outer Lindblad resonance on the leading edge of the wave in trailing wave systems?

STROM: Yes, there do exist galaxies which have a prominent spiral pattern on the inside, a gap which you might associate with co-rotation, and a spiral pattern on the outside. I have no quantitative data on such galaxies.

SHU: I find the discovery of the smooth-arm spirals to be extremely intriguing. It would be very important to determine exactly how much gas is in such systems to settle the old question of whether gas is a sine qua non for spiral structure. What are the future observing plans in this regard?

STROM: Currently observations are going on at Arecibo in order to look for HI gas in low surface brightness systems and in some of the smooth arm systems. Unfortunately, many of the systems we have identified are located outside the Arecibo declination range. I hope some people will be interested in observing these systems elsewhere.

MEBOLD: Together with Australian collaborators we have made a 21-cm line survey of the low surface brightness galaxies that have been detected with the Schmidt telescope at Siding Spring. Out of a total of about 160 galaxies approximately 70% have been detected in HI. The ratio M_H/L_B of these galaxies is exceedingly high. We have mapped several of the sufficiently large systems, and we find that in these systems HI is generally more centrally peaked than is usual for spiral galaxies.

WIELEN: DRIFT AND BROADENING OF AGEING SPIRAL ARMS

In order to study the ageing of spiral arms, H. Schwerdtfeger and I have calculated the drift and broadening of ageing spiral arms in the frame of density-wave theory. An aged spiral arm is defined by the positions of stars of a common age which have been migrated away from the zero-age spiral arm. We assume that the stars are formed at the global spiral shock front. For the initial systematic velocities of

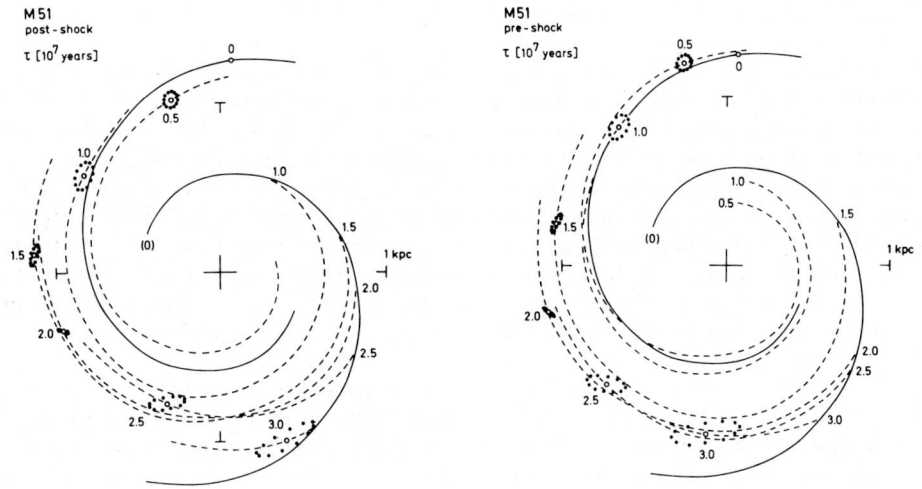

the stars, we consider two cases: In the "post-shock case", the stars reflect the motion of the gas immediately after the shock. In the "pre-shock case", the average initial velocity is the gas velocity before the shock. Orbit calculations for many test stars, including a velocity dispersion at birth (10 km/s in the figures), then provide the desired evolution of ageing spiral arms. As examples we show the results for the inner part of M51. The drift and broadening are very complicated, neither linear nor monotonic with age τ. In the post-shock case, the stars even move back to the inner side of the shock front. The pre-shock case seems to agree better with observations. In both cases (and in other galaxies with strong shocks), the newly born stars move essentially along the spiral shock front for a rather long period (about 40% of the circular rotation period at half the corotation radius), thus minimizing the drift at the beginning.

GROSBOL: I would like to report on the calculation of birthplaces of 100 - 200 B5-A0 stars in different density-wave potentials with pattern speeds ranging from 10 to 37 km s^{-1} kpc^{-1}. It shows that for two pattern speeds, namely $\sim$ 14 km s^{-1} kpc^{-1} and $\sim$ 32 km s^{-1} kpc^{-1} more birthplaces are located in the arms than expected in a random model.

GALLAGHER: COMMENTS ON NGC 3312, NGC 1291 AND NGC 1079

In the preceeding paper by Dr. Strom, a scenario for stripping of cluster galaxies by the intracluster medium (ICM) was presented. I would like to suggest that NGC 3312 in the Hydra I cluster of galaxies (Abell 1060) is an example of a galaxy that is presently experiencing ram-pressure-stripping. On a limiting IIIaJ exposure obtained by Drs. M. Smith and D. Weedman, a series of faint, filamentary extensions are seen to the southeast of the disk of NGC 3312. These seem to be material which has been removed from the galaxy. Tidal collisions, internal activity, a low velocity interaction with an intergalactic cloud, or ram pressure stripping by an ICM are possible interpretations. However, since Hydra I is a known X-ray cluster, the ablation hypothesis appears consistent with both the properties of the cluster and the morphology of the disturbance in NGC 3312, although other models at present cannot be rigorously excluded.

I will also briefly comment on spiral-like structures that are found in a limited sample of galaxies which have been classified as S0 or S0/a and have been detected in the HI 21-cm line. NGC 1291 and NGC 1326 are usually classified as RSBC/a and are both good examples of "θ" galaxies. As G. de Vaucouleurs has emphasized, NGC 1291 is probably the best example of such galaxies. Based on a blue IIIaJ photograph obtained with the CTIO 4-m telescope, the inner region which contains the lens and bar, has a rather smooth light distribution. This and the colors are in agreement with the presence of an old stellar population. The outer ring, however, shows many condensations which are very similar to the knotty structures found in spiral arms of Sb or Sa systems. In fact the entire outer ring structure appears to primarily result from the overlap of 2 spiral arms with very small pitch angle and low surface brightness. Thus the global characteristics of NGC 1291 may not be so different than those of normal Sb spirals which have similar hydrogen mass to luminosity ratios. It also seems possible that the dominant bar has produced significant changes in star formation patterns as compared to normal galaxies.

Another interesting case is NGC 1079, which has a relative HI content appropriate to a typical Sc. A CTIO 4-m plate shows a moderately high surface brightness "S" shaped bar imbedded in a very low surface brightness disk. There is little evidence for star formation activity beyond some condensations in very faint outer arms. Thus this galaxy may be related to the low surface brightness spirals that were discussed by Strom. Like these spirals NGC 1079 has abnormal disk properties. The Kitt Peak Interactive Picture Processing System has been used to produce a mean surface brightness profile and there is no exponential disk to a level of 4 magnitudes below the blue sky brightness.

SHOSTAK: Could you describe the HI profiles for the two "revealed spirals" you have shown?

GALLAGHER: No, we were unable to use the HI profile to classify NGC 1291 as the profile is very sharp due to the face-on orientation of the galaxy.

Anonymous: *"You apply the universal law of linearity ..."*

W.W. Shane: *"But that's what everybody does!"*

Discussion II.2

SOME COMMENTS ON RADIO OBSERVATIONS OF SPIRAL ARMS

W.W. Shane and J. Bystedt
Leiden Observatory

If we view spiral structure in galaxies as a manifestation of some dynamical process which we wish to understand, then the observer can distinguish two facets of the problems. In the first place he must consider the observable consequences suggested by theoretical developments. In the second place he must endeavour to interpret the available observations in terms of realistic physical models, where such models may require detailed information beyond that required for an understanding of spiral structure as a dynamical phenomenon. We shall consider these two aspects in turn.

1. OBSERVATIONS AND THEORY

Looking back about three years, as we often do in reviewing developments in a field such as this, some of us recall the conference on spiral galaxies held in Bures-sur-Yvette (Weliachew, 1975). The conference was timely and the general tone optimistic. The questions which had been raised about the relevance of the density-wave mechanism to spiral structure (e.g. Piddington, 1973) had been answered adequately (Bok and Bok, 1974) and the required next steps seemed to be clear enough. It is true that most of the problems that are troubling us now were already with us then, but their significance was probably not clear to most observers. At the time we were mainly concerned with identifying the density wave (for example, through matching the velocity and density patterns in a specific model, e.g. Visser, 1975) and with determining the density-wave parameters which best fit the observations of a specific galaxy (e.g. Rots, 1975; Guibert, 1975). The first attempts to do this had been reasonably promising (Shu et al., 1971) and we seemed to need only observations with a slightly more favorable combination of resolution and sensitivity to do the job properly.

It now appears that this view was far too simple. The theoretical difficulties have remained with us, despite many valuable suggestions for their solution, and have become more pressing and more evident to the observers. At the same time we saw that the observational data required for an attack upon these problems were much less easily accessable than we might have wished. We can list some of the major questions, without attempting to discuss their theoretical implications.

1. Stability problems. In order that the disk be protected against large-scale instabilities, Ostriker and Peebles (1973) have postulated a massive halo whose existance cannot easily be confirmed through direct observations. On the other hand, in order that the spiral structure may survive, the disk must be unstable with respect to certain spiral modes, implying an upper limit to the admissible velocity disperisons of the disk stars (Lau and Bertin, this symposium). Observational determinations of these velocity dispersions are, except in the Galaxy, beyond our current capabilities.

2. Propagation, maintenance and regeneration of spiral waves. There has been no lack of suggestions as to the means by which the spiral pattern can be maintained or regenerated. Both internal and external processes may play a role. The reflection and amplification mechanism suggested by Lin and his coworkers (e.g. Lin, 1975) appears promising but a detailed study of the resonance regions is required, whereas one of the long-standing observational difficulties has been the very identification of resonance regions in galaxies (Oort, 1975). The suggestion of tidal excitation of spiral waves has acquired new meaning through the realization that, aside from the perturbing effects of companions in some systems so convincingly illustrated by Toomre and Toomre (1972), many galaxies may be disturbed by massive infalling intergalactic gas clouds whose individual identities are hard to establish (Gunn, 1977).

3. Simultaneous presence of several spiral modes. Although this also is not a new suggestion (Lin, 1967), it has become increasingly clear that spiral patterns in galaxies can seldom be represented by a single spiral mode (Bertin et al., 1977), nor is there any theoretical objection to the simultaneous presence of several modes with comparable amplitude. Unfortunately this possibility introduces so many free parameters into any proposed model that fitting such a model to the observations becomes a rather unrewarding exercise in which it becomes possible to fit any conceivable set of observations through a suitable choice of parameters. In the absence of additional observational constraints it is not entirely clear how we should proceed further along this line.

Meanwhile improved instruments have not rewarded the observers, as some of us had hoped, with a clearer picture of spiral structure. Rather they have revealed a baffling complexity in the structure of galaxies which threatens to frustrate any attempt at a simple interpretation. Many of these complexities, in fact, have long been evident. Three examples will illustrate some of the problems.

M51 is a beautiful spiral (the first nebula whose spiral morphology was recognized) and has been studied in great detail. The arms are well defined in H I, H II, dust and non-thermal continuum and their relative positions have been analyzed in terms of a density wave with gratifying results. But who, in view of the analysis by Toomre and Toomre (1972) believes that the galaxy would look the same in the absence of its companion? And why is it that, despite the apparent symmetry, the optical major axis departs by at least 45° from the kinematically determined line of nodes?

M101, on the other hand, is strongly asymmetric in appearance without there being any prominent companion which might help to account for this. It has been studied in detail in the 21-cm line (Allen et al. 1973) and radio continuum (Israel et al., 1975) and, although very important results have been derived, it remains an unattractive object for the testing of dynamical theories.

M31 appears to possess well-defined spiral arms, although the high inclination angle obscures the pattern. Nevertheless, the arms in the inner region, where most spiral galaxies exhibit the simplest structure, show large radial motions (Shane, this symposium) reminiscent of those detected in the Milky Way. No simple density-wave model can account for these motions.

One nearby galaxy does seem to behave as one might expect from a simple density-wave model, and this is M81. Perhaps this should surprise us since M81 is surrounded by companions and intergalactic gas clouds and even has a peculiar nucleus. Nevertheless, with all of the reserve that is appropriate in considering a unique case, this seems to offer us our best opportunity to test a density-wave model. Herman Visser of the Kapteyn Laboratory in Groningen has done this in a study which represents the state of the art in fitting density-wave models to galaxies (Visser, this symposium). Perhaps the most difficult decision to be made is the degree of complexity of the model to be examined. On the one hand, it must be realistic, within the limits of the observations, and on the other, it must not possess so may degrees of freedom as to make a fit meaningless. Visser's results suggest that he has reached a suitable compromise in this respect.

It would be wrong to conclude even a brief discussion of model fitting without mentioning the particular problems of barred spirals. Here the theoretical analysis (Sanders and Huntley, 1976) has not progressed far beyond calculations of the response of a gas to an imposed gravitational perturbation. A self-consistent solution has not yet appeared. The calculations predict pronounced gas streaming parallel to the bar with shocks approximately in the position of the dust lanes. Direct observational confirmation is difficult because the most conspicuous barred spirals will have the bar close to the plane of the sky and the streaming velocities will be unobservable. In such systems a less direct check is possible by examining the behavior of the apparent rotation velocity inside and outside the barred region. The elongated gas orbits in the inner region should result in a measured rotational velocity in the inner (but not the outer) part substantially below that required for circular rotation. Work in progress on NGC 1300 and 4236 does not seem to indicate this but the results are very preliminary. Much more suggestive are the radio (Sancisi, private communication) and optical observations of NGC 5383 (Peterson et al., 1977), a barred spiral whose bar may well be quite inclined with respect to the plane of the sky. The observations here are mutually consistent and, although the interpretation is still open to lively discussion, some evidence for streaming parallel to the bar, as predicted by the calculations, seems to be present. It appears probable that we can expect important developments in the study of barred spirals in the near future.

2. INTERPRETATION OF THE OBSERVATIONS

If we are to make proper use of the available observational material we must look carefully at the interpretation. In this section we will discuss some of the more urgent problems.

Van der Kruit (this Symposium) has already discussed in detail the problem of separation of thermal and non-thermal contributions. Attempts to estimate the thermal contribution from optical data are fraught with difficulties. Once the individual color excesses of the H II regions have been determined, no mean task in itself, and converted into extinctions, the problem remains of accounting for the non-uniformity of the extinction over (and within) the H II region. Application of standard formulae will always lead to an underestimate of the total extinction. Van der Kruit's discussion of M51 (v.d. Kruit, 1977) suggests that this may amount to about 0.5 mag. in that particular galaxy, and selection effects (heavily obscured H II regions will be optically too faint to be included) may increase this. In view of these and other difficulties we heartily subscribe to van der Laan's comment (this Symposium) to the effect that there is no satisfactory alternative to observations at several short wavelengths (< 10 cm) as a means of isolating the thermal contribution.

Understanding the non-thermal radio radiation also requires a knowledge of the physical conditions beyond our capability of direct observation. The well-known expression for volume emissivity is $\varepsilon \propto \rho_e B^{1-\alpha}$. Since both the density of relativistic electrons, ρ_e, and the magnetic field strength, B, are (in some models) proportional to the gas density ρ, we often write $\varepsilon \propto \rho^{2-\alpha}$ (Mathewson et al., 1972) and, replacing both ε and ρ by their volume averages, derive the useful, if incorrect, relation, $<\varepsilon> \propto <\rho>^{2-\alpha}$(!). The situation is further complicated when we realize that the mixture of physical conditions in regions of different average density will be quite dissimilar, as will be the degree of inhomogeneity. It would seem that further progress in interpreting the non-thermal continuum radiation from galaxies will depend upon a better understanding of the local physical conditions.

Even interpretation of the line radiation is subject to uncertainties. It is customary to accept the integrated brightness temperature as a measure of the column density of gas. Two considerations throw doubt upon the validity of this assumption. In the first place, although the measured brightness temperature is never large enough to suggest

II.2 SOME COMMENTS ON RADIO OBSERVATIONS OF SPIRAL ARMS

significant self-absorption, beam dilution makes it quite possible that it is important on a small scale so that considerable H I may be hiding in or behind dense clouds (e.g. Shane, 1971). In the second place it is not clear that H I need be conserved. Molecule formation may play a significant role in the spiral arms as may ionization in large regions of low density between the arms. Happily, Visser has shown that the observations of M81 can be interpreted on the basis of the simplest assumptions without introducing serious inconsistencies, but we should remain on guard.

One of the often quoted quantities in discussions of spiral galaxies is the arm to inter-arm ratio of continuum radiation intensity (e.g. Segalovitz, 1976). It is taken as an indication of the degree of compression of the gas in the spiral arms and is required for the separation of the spiral arm contributions from the base disk (Mathewson et al. 1972). But the quantity determined from the observations does not always represent what one would wish. The peak represents an average taken over a resolution element and would be higher if the resolution was better, and it also varies eratically along the arm, whereas the minimum, or even the average, between arms is usually close to or below the sensitivity limit of the instrument. Thus what is often quoted is the ratio of peak to the mean over a circle in the galaxy. This is quite a different quantity and has more to do with the width and profile of the arm than with the interarm values. This should be kept in mind when interpreting these results.

One of the conclusions which has been drawn from analysis of this kind of data is that the enhancement of continuum radiation in the arm does not exceed that of line radiation by as much as one would be lead to believe from the volume emissivity relation quoted above (Shane, 1975), at least when one considers a simple two-dimensional model (Mathewson et al. 1972). The solution has been sought in the escape of relativistic plasma from the arms through Parker instabilities (Mouschovias et al. 1974) and it is clear that this offers a promising means of maintaining a halo or thick disk in the non-thermal continuum (e.g. Levy, this symposium) which, among other things, will reduce the arm to inter-arm ratio. Less clear is it that the Parker instability can account for the "beads on a string" appearance of spiral arms. Given the usual length (1 kpc) and time (10^8 years) scales, bulk gas velocities approaching 10 km/s are required if the beads are to be formed before passing out of the string. Such velocities should be observable using new instruments. What is not evident is how the magnetic field, conspiring with the gravitational potential of the disk, can provide the required acceleration on such a scale. More detailed calculations are required. Meanwhile the observations, which first seemed to indicate the expected periodic clumping in the H I (Oort, 1974), are, with improved resolution, showing a more chaotic picture in which the beads are a good deal less apparent (Rots and Shane, 1975).

Finally, two relatively new sorts of observations should be mentioned, whose interpretation ought to lead to more insights into the nature of spiral arms. After the detection of CO in external galaxies (Rickard et al. 1977) we have entered the mapping stage, and the first results (Combes et al. 1977a, b) are as encouraging as they are impressive. If we may adopt the suggestion (Bash and Peters, 1976) that CO clouds represent a population with a maximum age of about 3×10^7 years, then the rules of star migration (e.g. Wielen, this Symposium), will apply equally to CO clouds, while we will enjoy the additional advantage of good velocity data. Thus radio astronomers will enter a field which has thus far been virtually the monopoly of our optical colleagues.

The first radio polarization map of a spiral galaxy has been produced (Segalovitz et al. 1976) and, although the galaxy observed, M51, is in many ways atypical, a preliminary attempt at interpretation is possible. The observations were discussed in terms of a model in which a fraction f of the volume of the galaxy is occupied by a uniform magnetic field (with azimuthal orientation) and the remainder by a randomly oriented field. Taking account of a substantial thermal contribution (which was not done in the original discussion), we can estimate $f=0.2$. In a quiescent galaxy the magnetic field will be drawn out into uniform rings by differential rotation. The time required for this might be estimated as about $3A \simeq 5 \times 10^7$ years (adopting galactic values for convenience). We call this the combing time, t_c, following a suggestion by Toomre. But from time to time the magnetic field is randomized by a mixing event. A simple calculation serves to specify the ratio of this mixing time, t_m, to t_c in terms of f, and we find that $f = 0.2$ requires $t_m = 0.4\ t_c = 2 \times 10^7$ years. Supposing that these mixing events are supernova explosions and that supernovae occur once per 50 years (Katgert and Oort, 1967) and are distributed over a volume of 200 kpc^3 (again adopting galactic estimates) we may deduce as characteristic radius of action of a single supernova explosion about 50 pc, a not unreasonable value. But clearly a more realistic model is required, taking into account the very different conditions in and between the arms, before serious calculations can be considered. Nevertheless, we hope this small example will serve to emphasize the need for more polarization observations at wavelengths of 6 cm or less.

REFERENCES

Allen, R.J., Goss, W.M., and van Woerden, H.: 1973, *Astron. Astrophys.* 29, 447.
Bash, F.N. and Peters, W.L.: 1976, *Astrophys. J.* 205, 786.
Bertin, G., Lau, Y.Y., Lin, C.C., Mark, J.W-K., and Sugiyama, L.: 1977, submitted to *Astrophys. J. Letters*.
Bok, B.J. and Bok, P.F.: 1974, *The Milky Way*, Harvard Univ. Press, 4th ed., p. 228.
Combes, F., Encrenaz, P.J., Lucas, R., and Weliachew, L.: 1977a, *Astron. Astrophys.* 55, 311.
Combes, F., Encrenaz, P.J., Lucas, R., and Weliachew, L.: 1977b, submitted to *Astron. Astrophys.*

Guibert, J.: 1975, in Weliachew 1975, p. 263.
Gunn, J.E.: 1977, in B.M. Tinsley and R.B. Larson (eds.), *The Evolution of Galaxies and Stellar Populations*, Yale Univ. Observatory, p.445.
Israel, F.P., Goss, W.M. and Allen, R.J.: 1975, *Astron. Astrophys.* 40, 421.
Katgert, P. and Oort, J.H.: 1967, *Bull. Astr. Inst. Netherl.* 19, 239.
Kruit, P.C. van der: 1977, *Astron. Astrophys.* 59, 359.
Lin, C.C.: 1967, *I.A.U. Symposium* No. 31, 313.
Lin, C.C.: 1975, in Weliachew 1975, p. 493.
Mathewson, D.S., Kruit, P.C. van der and Brouw, W.N.: 1972, *Astron. Astrophys.* 17, 468.
Mouschovias, T.Ch., Shu, F.H. and Woodward, P.: 1974, *Astron. Astrophys.* 33, 73.
Oort, J.H.: 1974, in *Galaxies and Relativistic Astrophysics* (B. Barbanis and J.D. Hadjidemetriou, eds.), Springer Verlag, p. 1.
Oort, J.H.: 1975, in Weliachew 1975, p. 533.
Ostriker, J.P. and Peebles, P.J.E.: 1973, *Astrophys. J.* 186, 467.
Peterson, C.J., Rubin, V.C., Ford, W.K.Jr. and Thonnard, N.: 1977, *Bull. Am. Astron. Soc.* 9, 336.
Piddington, J.H.: 1973, *Astrophys. J.* 179, 755.
Rickard, L.J., Palmer, P., Morris, M., Zuckerman, B. and Turner, B.E.: 1975, *Astrophys. J. Letters* 199, 75.
Rots, A.H.: 1975, in Weliachew 1975, p. 201.
Rots, A.H. and Shane, W.W.: 1975, *Astron. Astrophys.* 45, 25.
Sanders, R.H. and Huntley, J.M.: 1976, *Astrophys. J.* 209, 53.
Segalovitz, A.: 1976, *Astron. Astrophys.* 52, 167.
Segalovitz, A., Shane, W.W. and de Bruyn, A.G.: 1976, *Nature* 264, 222.
Shane, W.W.: 1971, *Astron. Astrophys. Suppl.* 4, 315.
Shane, W.W.: 1975, in Weliachew 1975, p. 217.
Shu, F.H., Stachnik, R.V. and Yost, J.C.: 1971, *Astrophys. J.* 166, 465.
Toomre, A. and Toomre, J.: 1972, *Astrophys. J.* 178, 623.
Visser, H.C.D.: 1975, in Weliachew 1975, p. 211.
Weliachew, L. (ed.): 1975, *La Dynamique des Galaxies Spirales*, CNRS, Paris.

DISCUSSION FOLLOWING REVIEW II.2 GIVEN BY W.W. SHANE

LEVY: I would like to comment on some questions raised by Dr. Shane concerning the Parker instabilities in the disk.
(1) Observability of streaming motions of the gas: The present state of the gas in our galaxy or in other galaxies can best be described as a state that exists after the full development of the instability. So while one might occasionally expect to see some streaming motion, in gas accumulations that have existed for a sufficiently long period of time one wouldn't expect to see much gas streaming anymore.
(2) Availability of energy to drive the instability: The gravitational energy of the stellar disk, which predominates, and the magnetic energy are sufficient. The time scale for the growth of the instability is the free-fall time for the material in the rotational field, so that there is automatically enough force to drive the instability.

(3) Material is never lifted! The instability is the result of material sliding down the magnetic field lines, not of material being lifted. So you need not account for lifting material up any height at all.

THE DYNAMICS OF THE SPIRAL GALAXY M81

H.C.D. Visser
Kapteyn Astronomical Institute
University of Groningen, The Netherlands

The spiral galaxy M81 is a challenging object for testing a density-wave theory. It is of large angular size and favourably inclined to the line of sight; radio observations of neutral atomic hydrogen showing well defined spiral arms with noncircular motions are available (Rots and Shane 1975), and there is evidence from surface photometry for density waves in the old disk population (Schweizer 1976). Can these observations be unified in one density-wave model?

The first step is the construction of an axisymmetric mass model on the basis of a rotation curve. As will be discussed below (see Figure 6), the observed rotation curve derived from the HI measurements is distorted by density-wave motions. After correction for this effect a mass model can be constructed consisting of two spheroids in the central regions ($R \leq 3$ kpc), representing the nucleus and the bulge, and a Toomre disk.

The theoretical spiral pattern of the potential perturbation has been computed with linear stellar density-wave theory (Lin and Shu 1971), including the effects of the finite thickness of the stellar layer and neglecting the gas. A pattern speed Ω_p of 18 km s^{-1} kpc^{-1} was adopted; this value gives the best model with corotation radius 11.3 kpc and inner Lindblad resonance (ILR) at 2.5 kpc. For lower pattern speeds the ILR is so far from the centre that the gas perturbations observed at 3 kpc cannot be ascribed to a density wave. A higher pattern speed restricts too strongly the region where the gas flow can be computed. The velocity dispersions of the stars play an important role. For a disk which is just marginally stable against axisymmetric instabilities, no reasonable values for the pitch angle of the spiral can be achieved. If in the central regions the random motions of the stars are higher by about 70% with a marginally stable disk around corotation, a spiral pattern is obtained which can serve as a basis for the model (Figure 1).

The amplitude of the wave can be determined theoretically (Shu 1970) with a second-order analysis. However, we prefer a more direct method in which the azimuthal brightness variations determined from surface photo-

Figure 1. The theoretical spiral pattern of the potential perturbation (dashed line) and the shock front in the gas (full line) superimposed on a 200 inch photograph (Photo by Hale Observatories).

metry (Schweizer 1976) are transformed into amplitudes of the potential field perturbations, neglecting the self-gravity of the gas. This transformation can be done on the basis of the mass model, the scale heights of the stellar and gaseous disk, and the pitch angle of the spiral pattern.

The theoretical gas flow has been computed with nonlinear density-wave theory for a one-component gas (Roberts 1969, Shu et al. 1973). Since the amplitude of the wave is higher than the critical forcing, we can expect galactic shocks over the whole region computed in the model (3-8.7 kpc). The shock front has been indicated in Figure 1. There is a small displacement of this front with respect to the minimum of the potential well; the pitch angles are nearly the same except in the outer regions. At the shock front we may expect to find the principal dust lanes, and the agreement with the observed dust lanes appears acceptable. The discontinuity in radial velocity owing to the shock is of the order of 30 km/s on the minor axis. To compare the theoretical density and velocity fields with the hydrogen observations the model fields must be smoothed to the beam of the radio telescope. A special technique called "phase smoothing" (making use of the asymptotic approximation of density-wave theory) has been developed to reduce the two-dimensional smoothing to one dimension. Also the smoothing effects of integration over a gaseous layer of finite thickness along the line of sight can be treated in this way. For M81 the smoothing by a beam with a full width at half power (FWHP) of 25" is more important than the line-of-sight smoothing; however, at resolutions better than about 15" the line-of-sight smoothing may dominate. The smoothed fields are only weakly dependent on the velocity dispersion of the gas and the parameters used for the calculation of the stellar wave except Ω_p.

The smoothed theoretical velocity fields at 25" and 50" resolution are shown in Figure 2 and 3 together with the observations of Rots and Shane (1975). For the observed velocity fields part of the reduction has been redone to improve the velocity determination (see Bosma 1978, Visser 1978). The underlying picture represents the HI distribution at 25" resolution (Rots 1974). In the 25" velocity field the kinks in consecutive contours at the inside edge of the spiral arms are an indication of the shock front. The next important things to note are the secondary turnover points at 3.3 kpc from the centre along the major axis, in the observations as well as in the model. These secondary turnovers are entirely due to the density wave, because the unperturbed velocity field shows only one turnover point at 4.8 kpc on either side of the galaxy. At 50" resolution (corresponding to $\simeq 1/3$ of the arm spacing of the theoretical spiral pattern on the minor axis) there are no indications of the shock front; the velocity contours have a nearly sinusoidal shape. The agreement of the model with the observations is good at both resolutions, although the western arm outside 7 kpc shows some deviations. It is not yet clear if this asymmetry in the outer regions can be attributed to tidal interaction with companion galaxies. Also there is the possibility that in M81 more than one density wave is acting at the same time; the coexistence of more than one wave has been postulated on theoretical grounds (Mark 1977). Note that the azimuth of the underlying spiral of the potential wave has been chosen in such a way that the agreement between the observed and theoretical velocity fields is as good as possible, since the phase of the velocity perturbations is important for the correction of the observed rotation curve. The velocity perturbations were also analyzed along the minor and major axes at 25" and 50" resolution (Figure 4). The observational data on either side of the centre have been averaged together taking due account of the antisymmetry. The model velocities are in good general agreement with the observations although there are indeed differences in detail, such as the location of the minimum near 8 kpc along the minor axis. Note the difference in slope at the two resolutions near 7 kpc along the minor axis.

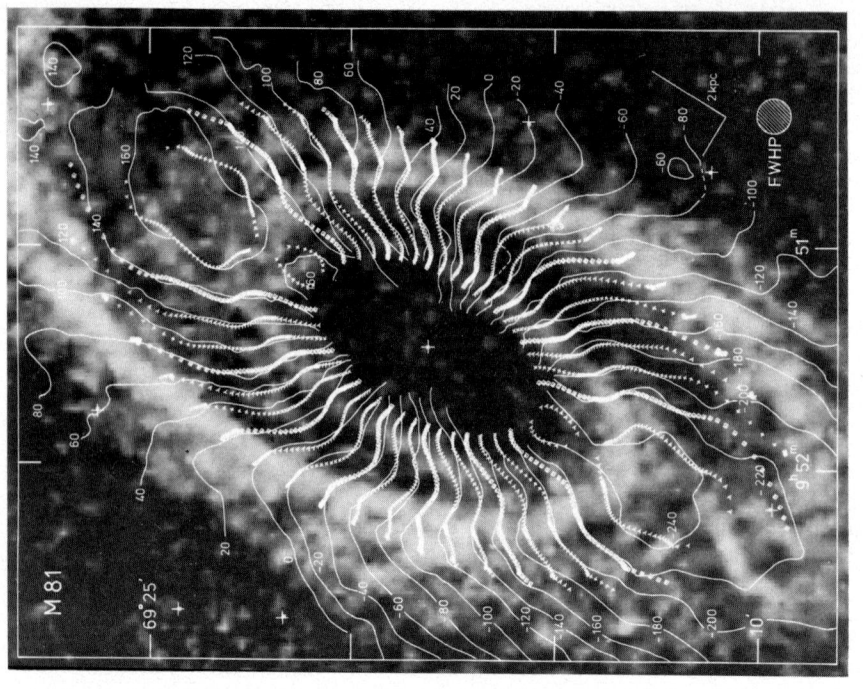

Figure 2

Theoretical (symbols) and observed (full and dashed lines) radial velocity fields at 25" (Figure 2) and 50" resolution (Figure 3) superimposed on a radiograph of the density distribution at 25" resolution (Photo courtesy of E.B. Jenkins). Also the beam and linear scales in the plane of the galaxy are shown. Dashed lines do not represent actually measured velocities, but indicate a possible continuity.

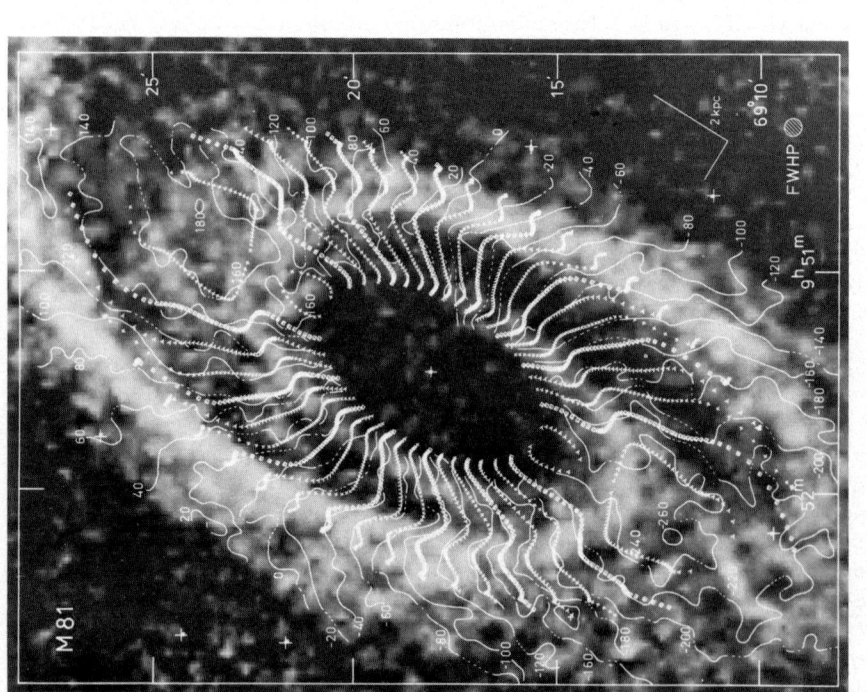

Figure 3

Figure 5 shows the arm profiles (HI-surface density as function of spiral phase) at 25" resolution for different ranges in radial distance for model and observations (Rots 1975). For $8 < R < 10$ kpc the model curve is less certain since this is in the neighbourhood of the second harmonic resonance. The region of the higher harmonic resonances occurs beyond 8.7 kpc; no solutions could be found here. Hence the observations have here been averaged over a wider range in R than the numerical data of the model. The secondary compression at phase $\simeq 60°$ may be an indication of the second harmonic resonance. To obtain the best fit of the density profiles with the model phase shifts of $-15°$, $-18°$ and $+7°$ have been applied to the theoretical profile in the upper, middle, and lower panel, respectively. We have no satisfactory explanation for this phase shift, and this appears to be a drawback of the model. When the amplitudes (i.e. the maxima) of the observed profiles of the two arms are averaged together, the agreement with the model appears reasonable. The asymmetry of the profiles of the eastern arm (phase $0°$) for $R > 6$ kpc is well represented. The general agreement for the region $4 < R < 6$ kpc is less satisfactory. This may be due to a poorer definition of spiral structure in the inner regions.

Since we constructed a symmetrical density-wave model, we cannot account for the observed asymmetries in the HI-density and velocity fields. In general the eastern arm shows a better agreement with the model.

Finally we shall discuss the effects of the density wave on the observed rotation curve. A rotation curve determined from the velocity fields in Figures 2 and 3 in some standard way (e.g. Rots 1974), turns out to be quite different from the rotation curve used for the axisymmetric mass model. This is demonstrated in Figure 6 which shows the rotation curve representing the axisymmetric mass distribution, the observed rotation curve at 25" resolution and the model curve. In practice we have determined the "axisymmetric" rotation curve from the observations by iteration.

The results may be summarized as follows:
i) A self-consistent model for the spiral galaxy M81 can be constructed inside $\simeq 9$ kpc on the basis of density-wave theory as formulated by Lin, Shu and Roberts;
ii) For an acceptable spiral pattern the random motions of the stars must be higher than those required for a marginally stable disk in the more central regions, assuming that the present choice of the pattern speed is justified;
iii) The amplitude of the stellar density wave as determined from surface photometry is consistent with the amplitudes of the observed density and velocity perturbations of the gas;
iv) Indications for the existence of shocks can be recognized as strong gradients in the velocity field if the beam of the radio telescope is of the order of 1/6 of the arm spacing on the minor axis;
v) Density-wave motions can be clearly detected if the beam is 1/3 of the arm spacing;
vi) The arm profiles are more or less consistent with the nonlinear density-wave theory after considering the smoothing effects of the

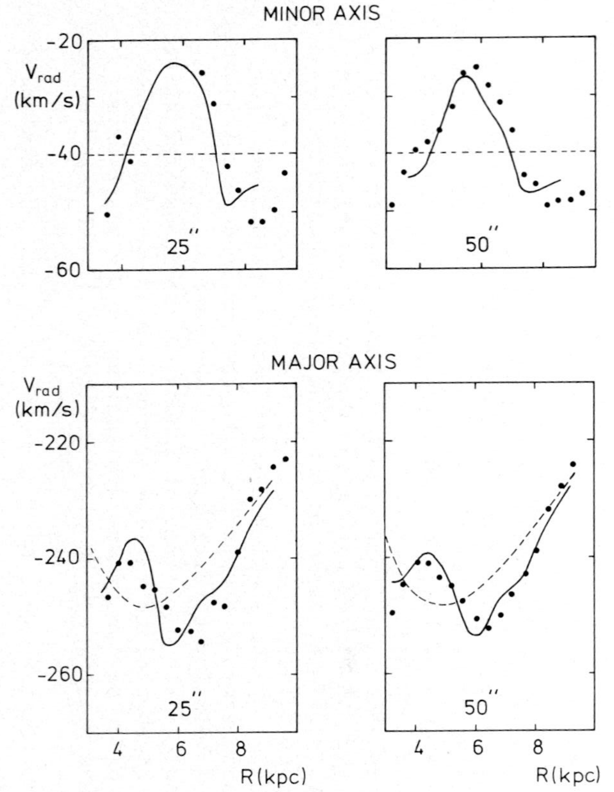

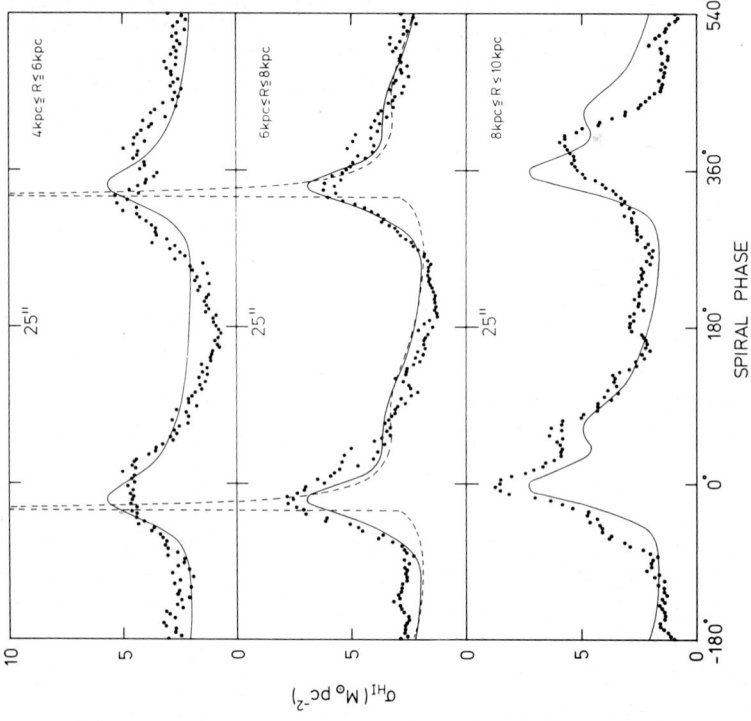

Figure 5. HI-surface densities as function of spiral phase for the observations (dots) and the model (full lines). The dashed line in the middle panel represents the unsmoothed shock profile.

Figure 4. Radial velocities along the minor and major axes at 25" and 50" resolution for the model (full lines) and the observations (dots). The radial velocities of the axisymmetric model are represented by a dashed line.

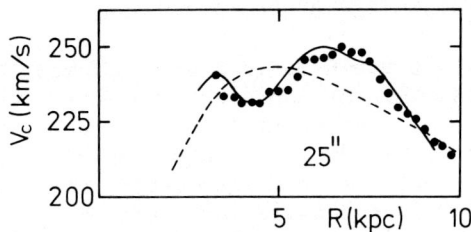

Figure 6. Distortion of the rotation curve by the density wave. The full line is the model curve, observations are represented by dots, and the "axisymmetric" rotation curve is indicated by a dashed line.

      beam. A phase shift was needed to get the maxima of the profiles at the right place;
vii) Smoothed density and velocity fields are weakly dependent on the parameters used for the calculation of the stellar wave except Ω_p;
viii) Density-wave motions can cause considerable distortions of the rotation curve; correction for this effect is then required for the construction of mass models from observed rotation curves.

A more detailed discussion will appear in a future paper (Visser 1978).

Acknowledgements

I thank T.S. van Albada, R.J. Allen and H. van Woerden for their interest, encouragement and discussions. F.H. Shu for providing several computer programs and assistance during the first stages of the work, W.W. Roberts for discussions and A. Toomre for valuable advice concerning the wave amplitude. I am indebted to A.H. Rots for provision of the observational material. I thank A. Bosma for his cooperation in the reduction of the velocity fields and J.M. van der Hulst for discussions. I am grateful to the Netherlands Organization for the Advancement of Pure Research (Z.W.O.) and the University of Groningen for financial support. The Westerbork Radio Observatory is operated by the Netherlands Foundation for Radio Astronomy with financial support of Z.W.O.

References

Bosma, A.: 1978, in preparation
Lin, C.C. and Shu, F.H.: 1971, in M.Chrétien, S.Deser and J.Goldstein(eds.), Astrophysics and General Relativity 2, Gordon and Breach, New York, p. 235
Mark, J.W-K.: 1977, Astrophys. J. 212, 645
Roberts, W.W.: 1969, Astrophys. J. 158, 123
Rots, A.H.: 1974, Distribution and kinematics of neutral hydrogen in the spiral galaxy M81, University of Groningen (Ph.D. Thesis)
Rots, A.H.: 1975, Astron. Astrophys. 45, 43
Rots, A.H. and Shane, W.W.: 1975, Astron. Astrophys. 45, 25
Schweizer, F.: 1976, Astrophys. J. Suppl. 31, 313
Shu, F.H.: 1970, Astrophys. J. 160, 99
Shu, F.H., Milione, V. and Roberts, W.W.: 1973, Astrophys. J. 183, 819
Visser, H.C.D.: 1978, The dynamics of the spiral galaxy M81 (in preparation)

DISCUSSION FOLLOWING PAPER II.3 GIVEN BY H.C.D. VISSER

MARK: Visser and I felt that it would be good to complement his talk with some preliminary results where we compare the observed wave amplitude of Schweizer with the relative amplitude as suggested by the theoretical calculation of spiral modes. For a reasonably realistic model of M81 which includes about 25% of bulge matter, we find that the amplitudes as observed by Schweizer compare very well with the amplitude distributions in the two dominant spiral modes.

TOOMRE: How did you get the amplitude?

MARK: Our relative amplitudes as stated give the values of the amplitude at one radius relative to that at other radii. The absolute value is not predicted by present theory but one chosen for the best fit of the observations. This run of amplitudes is already very suggestive.

VAN DEN BERGH: Recently a number of people have suggested that the optical peculiarities of M82 might be due to a recent encounter with M81. Do you see any evidence for such an encounter in the velocity field of M81?

VISSER: The major axis of the velocity field is not a straight line, but is curved in the outer regions. This indicates that in the outer regions the disk may be warped, which might be due to an encounter with M82.

"I think you should include numerical simulation in the category of observation too."

A.J. Kalnajs in Discussion II.4

A CONFRONTATION OF DENSITY WAVE THEORIES WITH OBSERVATIONS

Agris J. Kalnajs
Mount Stromlo and Siding Spring Observatory, Research School
of Physical Sciences, Australian National University

ABSTRACT

It would be a mistake to think that the density wave theories of spiral structure have reached the maturity where they can make unconditional predictions which can be tested. On the contrary, they are still very dependent on observations for help and guidance.

1. INTRODUCTION

I have been taught to believe that a theory is a theory only if it is capable of being falsified. A good theory should propose observations which can unequivocally contradict it, and this element of risk is essential in order to have a confrontation. A sifting of the observational data in search of features which can be explained by a theory, and the subsequent tabulation in two columns, separated by a line, one labelled "observation", the other "theoretical interpretation", may be a good advertisement for it, but it does not constitute a confrontation.

When I turn to what is commonly referred to as the density wave theory, I find it very difficult to think of an observation which could contradict it. There seems to be an overabundance of uncertain parameters, and "as yet imperfectly understood mechanisms", which allow plenty of room to manoeuver around any tight spot posed by observations. Under these circumstances we are not dealing so much with a confrontation between theory and observation, as with a reconciliation.

In this paper I will discuss my perception of the status of the theory, and speculate about the direction in which I expect it to progress. Observations can and will play an important role in determining future developments, but the sort of data that are of the most importance pertain to the structure of the disk which is supposed to support the density waves, and are extremely hard to obtain. One actually has to turn the problem around and ask what spiral structure tells us about the disks, and in order to be able to do this, more attention will have to be devoted firming up the dynamical foundations of the theory itself, which are not as firm as one would like to think.

2. DENSITY WAVES AND WKBJ DENSITY WAVES

There are two distinct approaches to the subject of density waves, and in order to understand how this has come about and what the distinctions between the two are, I must sketch a brief history. It will be very rough, and if you wish to find out who did what and when, I recommend the forthcoming review article by Toomre (1977).

It is safe to say that all parties in the field agree or at least hope that the underlying stellar disk is sufficiently close to an axisymmetric equilibrium, so that any deviations can be treated as small perturbations of it. It is also agreed that Newtonian gravitation is the dominant force.

From the assumed symmetry of the equilibrium it follows that the perturbations will take the form of density waves. Each wave is sinusoidal in the azimuthal direction, rotates without shear at a fixed angular rate which is known as the pattern speed, and it may also grow or decay in time. The radial structure, pattern speed and growth rate of each wave or mode is determined by an integral equation which depends on the equilibrium structure. The growing modes of finite disks are discrete, can be singly excited, and one rejects all equilibria as unrealistic, unless the largest growth rates are sufficiently low. The amplitudes of the modes are not specified, except to the extent that they should not be too large. Roughly speaking, this implies that the largest acceptable amplitude for the density perturbation should decrease with the length scale of the mode, which makes the largest scale modes the most interesting ones since their effects should be more easily observed.

It is also agreed that one should at first try to describe the effects of the density wave in the stars on the gas in an iterative fashion: compute the non-linear response of the gas due to the stellar force field, and then consider, at most, only the effects of the induced density Fourier component which has the same angular periodicity as the stellar forcing field. The out-of-phase component is the more interesting one of the two, since it describes the energy and angular momentum transfers between the two subsystems, which relate to the question of growth and decay of the waves.

A difference arose on the question of how one should tackle the rather complicated integral equation which determines the modes. At a rather early stage C.C. Lin scented the possibility of developing an analytical theory of density waves, provided that the underlying perturbed potential was in the form of a tightly wrapped spiral. In this case one could make use of the rapidly varying phase to develop asymptotic or WKBJ solutions and neatly sidestep the complications of the integral equation. It was also necessary to have analytical expressions for the equilibrium orbits, which in general are too complicated to be of much use, but could be approximated by epicycles provided the eccentricities are sufficiently small.

The WKBJ theory, worked out by Lin and Shu, is sometimes referred to as the "local theory", for in the lowest order of approximation one pretends that the galaxy is uniform over the distance of a wavelength or so, and therefore requires that the wave be self-supporting at each radius. It is then left to the higher order terms to piece these locally constructed waves together to form the grand design, calculate group velocities, pattern speeds, etc.

The tightly wrapped potential hypothesis allowed W.W. Roberts (1969) to work out a tidy asymptotic theory for shock formation in the gaseous component, which has stimulated theories of star formation and given a fairly concrete picture of what a spiral arm should in this case look like.

In making this great leap forward, Lin and his co-workers are not now only facing the observers in the front, but have exposed their rear to the sniping of those they left behind. This of course was anticipated (Lin 1967). But what was not, was the possibility of simulating disk galaxies in numerical experiments on large computers. These simulations showed that disk galaxies did not behave in a manner which could be described in the framework of WKBJ theory. Self-gravitating disks do not form long-lived spiral structure, instead they are prone to large-scale two-armed instabilities which heat up the disk. The end product of this evolutionary phase is usually a finite amplitude wave in the form of a bar.

The numerical simulations certainly convinced me to give up my original plan of pursuing mode calculations with an epicycle version of the integral equation and epicyclic equilibrium distributions, since epicycles became less and less adequate as one approached the center. In order to keep the orbit approximation one had to exclude the center, but this is where the instabilities in the simulations seemed to occur!

The use of finite eccentricity orbits has made the equilibrium and mode calculations quite lengthy, and while I have not yet succeeded in constructing stable disks, it has been possible to identify the causes of the fastest growing instabilities and to substantially dampen them by simply including retrograde orbits in the central parts. The cure for the instabilities points to models with fairly hot centers, where pressure plays as significant a role as rotation in opposing the gravitational field.

It should be recalled that the possibility of disks being able to support tightly wrapped spiral waves is only a working hypothesis (Lin 1967), albeit a very reasonable one. But even if it turns out to be correct, the spiral modes will not be of much interest if the disk has faster growing instabilities which will dominate the evolution.

After all, many of the numerical simulations started with the very same disks used to illustrate WKBJ waves. My contribution to the sniping stems in part from doubts about the chances of being able to suppress the large scale instabilities and at the same time keep the tightly wound ones. If the explanation for the latter is the outward transfer of angular momentum as suggested by Lynden-Bell and Kalnajs (1972), then as they also pointed out, whatever a spiral wave can do, a more open one can do it better. If one has to increase the pressure in order to stabilize the bar type instabilities, it does not appear likely that one is going to decrease the scale of the dominant mode.

Further sniping stems from the role of the gas in the WKBJ theory, which is, as Oort (1962) has stressed, the key element for the presence of spiral structure. For very tight spirals, gas is a liability, since it damps the waves. However it can be made to work for you, if you have stable negative energy modes, such as bars, which are allowed to interact with it (Lynden-Bell 1975; Kalnajs 1973). This has led me to my own working hypothesis, namely to assume that a stable bar mode will eventuate when one manages to produce a stable disk, and to see how it interacts with the gas. While this does not encompass the whole range of observed spiral structure, it seems to be a direction in which some progress can be made.

3. WHY GAS?

Gas differs from stars in one unique respect: it can dissipate energy. If you believe that spiral structure is a long lived phenomenon, and hope to explain it by stars alone, you will have to rely on a relatively few so called resonant stars to keep it going, since spirality implies a redistribution of angular momentum, and the resonant stars act as sources and sinks (Lynden-Bell and Kalnajs, 1972). The trouble is that these sources and sinks have a finite capacity, and if they absorb too much they start giving it back, and vice versa. Gas does not have this problem, and will shock easily, particularly so near the Lindblad resonances. The reason for this is that in a steady field it tries to settle down in periodic orbits, but since these orbits intersect (in space) around the resonances, it cannot avoid running into itself.

This verbal theory is best illustrated by a simulation of forced gas response, presently pursued by P. Schwarz at Mt. Stromlo. The gas is assumed to be in clouds, which collide inelastically, loosing a good fraction of their relative motion in each encounter. The forcing is due to a rotating bar, reminiscent of the sort of thing one obtains in numerical simulations. The bar is turned on slowly over a period of two bar revolutions, and the distribution shown in figure 1 at the end of the third revolution persists for a further 4 to 7 revolutions before it dissolves. The reason for its demise is that the torque exerted on the gas tends to depopulate the intersecting periodic orbits by pushing the gas outwards past the outer Lindblad resonance. The end state is a nearly closed ellipse with its major axis along the bar. The locus of the shock, or rather of the collisions, coincides with the density maxima.

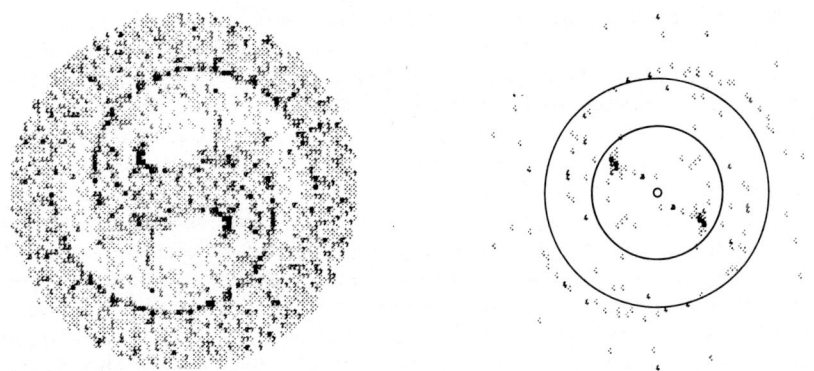

Figure 1. Quasi-stationary distribution of gas clouds driven by a bar (left), and the locus of the cloud collisions (right). The major axis of the bar is in the horizontal direction and the circles denote corotation and outer Lindblad radii.

While this is not the first demonstration that bars can force spiral patterns in dissipative gas, it is the one I understand best. The problem of making the pattern last is not unique to bar-type forcing, it occurs also when the field is spiral.[1] There are still a number of avenues to be explored before one is forced to the conclusion that the gas in the disk has to be replenished on a time scale of a billion years if spiral structure is to persist longer than that. But this is a distinct possibility.

4. THE PROBLEM OF Q'S, HOT DISKS, BULGES AND HALOES. IS Q=1?

I was instructed by the organizing committee to "concentrate on observable effects and avoid controversy of interest only to theoreticians". There is one issue that may arguably fall outside these guide-lines, but since it has far-reaching consequences, particularly for the WKBJ theory, it must be mentioned. It stems from our inability to construct cool self-gravitating equilibrium disks, which Toomre (1974) termed as a "near scandal" when he first reviewed it. The problem is still with us, but only in a slightly different form.

For nearly a decade since Toomre's WKBJ type analysis of the axisymmetric stability of disks (Toomre 1964), it was thought that once the fast-growing axisymmetric instabilities were taken care of by a sufficiently high radial velocity dispersion, the disks would be at most mildly unstable to non-axisymmetric perturbations. The ratio of the magnitude of the actual velocity dispersion to the minimum needed for

---

[1] For comparable azimuthal field strengths, the spiral field is at least as efficient in pushing the gas out past the Lindblad resonances as the bar.

stability Toomre denoted by Q. He also remarked that the observed parameters near the sun seemed to indicate that Q is close to 1, which no doubt helped to fuel the myth that "realistic" disks ought to be just barely stable in an axisymmetric sense. Thus it was somewhat of a surprise when the early numerical simulations by Miller, Prendergast, and Quirk (1970) showed up bar-like instabilities which proceeded to heat up the disk substantially above the Q=1 level. At that time the heating was not taken too seriously in view of the seemingly rough nature of the numerical work. The subsequent calculations by Hohl (1971, 1973), which were impeccable in this respect, confirmed the bar-instabilities and their associated heating. It was Ostriker and Peebles (1973) who first collected all the then available evidence and pointed out the shocking fact that the kinetic energy in the form of random motions has to exceed the rotational part by about a factor of 2.6 in order that a self-gravitating disk be stable. The typical values of Q associated with such largely pressure supported disks were of order of 4, with the lowest values around 2, and the tendency was for Q to increase with radius. Such values correspond to a velocity dispersion at least double of that observed near the sun, which led Ostriker and Peebles to propose the existence of haloes. A halo, which would be called a bulge if its radius was smaller than some characteristic disk scale length, can and will reduce the absolute level of random velocities needed to make the disk stable. This is almost self-evident, for by postulating that only a fraction of the equilibrium force field in the plane of the galaxy is provided by the disk, with the remainder coming from the rigid halo, you end up with a lower surface density and hence can afford to decrease the velocity dispersion and still remain stable. However if we insist on using Toomre's Q as a thermometer, its scale shrinks in proportion with the surface density, and there is no evidence that a halo reduces Q. On the contrary, I found that Q actually is increased by the presence of a stabilizing halo (Kalnajs 1972). The far more realistic simulations by Hohl (1976) designed specifically to study the stabilizing effects of a halo could not reduce Q below 2 around the solar radius.

The value of Q is quite important for WKBJ waves, and is implicitly assumed to be 1.0. If Q rises above 1.0, there is a region around corotation where the WKBJ waves cannot propagate, and as Toomre (1974, 1977) has stressed, that region spreads rapidly with Q, so that by the time Q reaches 2 the waves are squeezed out of existence. The reason why one wants propagating waves near corotation is that the proposed amplification mechanism which is necessary to keep the waves going, depends on the communication of angular momentum across this region (Mark 1976). The best available estimate of Q from observations in the solar neighbourhood puts the value at 1.5, with an uncertainty in the range 1.2 to 2.0 (Toomre 1974): While one might just manage to keep the WKBJ waves going with Q=1.2 by marshalling all the higher order corrections, Q=1.5 is a very serious problem for WKBJ waves, and with Q=2.0 they are ruled out.

II.4 A CONFRONTATION OF DENSITY WAVE THEORIES WITH OBSERVATIONS

The precise value of Q is not of great concern to the more general type density waves. The numerical simulations show that bar like instabilities are still present when Q is in the range 2 to 4. My own analytical studies, which agree well with the early stages of the simulations, confirm the persistent nature of the bar modes, although for historical reasons I have confined myself mainly to selfconsistent disk models with Q's in the range 1 to 1.5. Figure 2 illustrates the

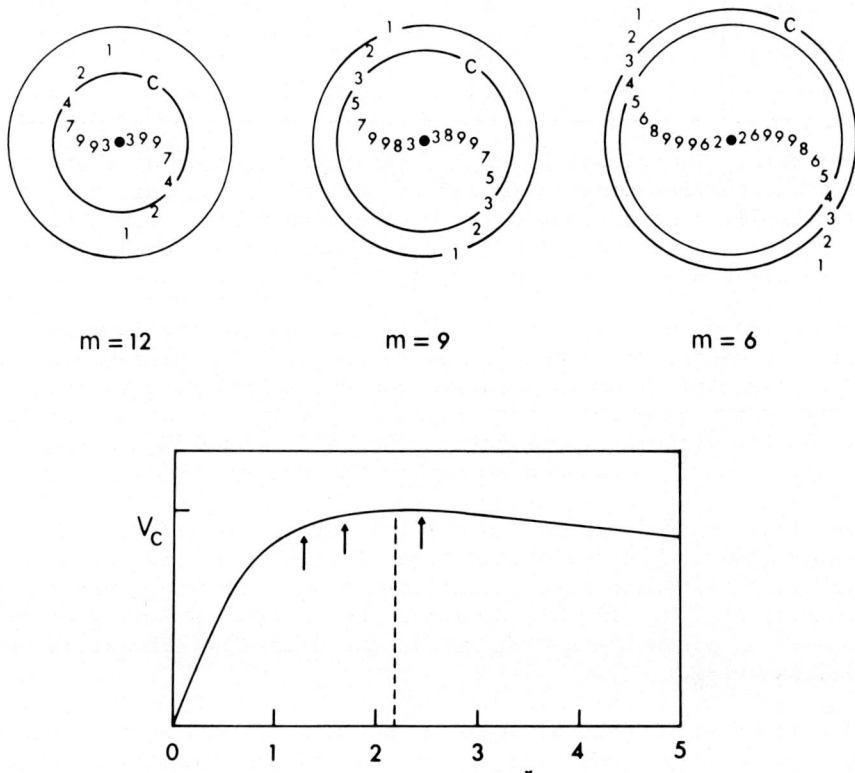

Figure 2. The lines of constant phase of the dominant mode potentials of three isochrone/m models (above), and the rotation curve of model (below). The circles labelled by C are the corotation radii, which are indicated by the arrows on the rotation curve. The unlabelled circles indicate the location of the maximum of the rotation curve.

appearance of the dominant two-armed modes of the isochrone/m models[2]. The relevant parameters of the three modes are found in Table 1. The pattern speeds and growth rates are expressed in units of the central

---

[2] The distribution functions (Kalnajs 1976) include a retrograde component, which is a function of $J_1 + |J_2|$, and matches up continuously with the direct part at $J_2 = 0$.

Table 1

	m = 12	m = 9	m = 6	m = 8	m = 8/1.5
Q_{center}	1.10	1.26	1.50	1.32	1.98
$Q_{corot.}$	1.00	1.14	1.46	1.22	1.83
Q_{gl}	.99	1.16	1.39	1.23	1.84
Pattern Speed	.59	.47	.34	.43	.26
Growth rate	.42	.29	.15	.25	.04
Corotation	1.32	1.73	2.45	1.90	3.03

rotation rate. The global Q, or Q_{gl} can be thought of as that factor by which the surface density should be multiplied in order to make the disk marginally stable to axisymmetric perturbations. The stellar disk instabilities are remarkably similar to those discovered by Erickson (1974) in his study of "softened" gravity disks.

As in the case of the simulations, the instability is mainly a central phenomenon; the potential peaks roughly at a distance of three mean epicycle radii from the center, and the spiral form is due to the inability of the outermost stars to keep up with the forcing field because of its sizable growth rate. The pattern speeds are fast in the sense that corotation falls well within the galaxy.

The last two columns in Table 1 illustrate the effects of a "halo" on the m=8 model. The "halo" takes the form of allowing only 2/3 of the disk to participate in the oscillation, which corresponds to a Q in the vicinity of 1.9. In this case the instability is fairly mild: if one were to place the sun at r=3.7, the e-folding time would be 200 million years.

One might argue that an apparent success of WKBJ waves would prove that Q must be very close to 1.0, at least in the vicinity of corotation. If so, then why? The best argument advanced so far why this might be the case is that corotation occurs in a region where the gas density dominates over stars, and because gas can cool, Q could hover just above 1.0 (Lin 1970). This would mean that corotation should be near the outer edge of the galaxy. Thus a determination of the pattern speed is of considerable interest.

If the interaction between the bar and gas is to grow spontaneously, a substantial part of the spiral structure must lie outside corotation. For this to be the case, corotation must be in the middle of the galaxy. The best chance of determining the pattern speed is from studies of the observed HI motions in external galaxies.

5. MOTIONS IN EXTERNAL GALAXIES

The most important observational consequence of a density wave is the density fluctuation and associated velocity field induced in the gas. The WKBJ theory does make a very strong prediction about the tangential velocity and the degree of compression of the gas. It stems from the fact that the azimuthal component of the spiral field can be neglected and therefore angular momentum is conserved. Consider a circular equilibrium flow, as illustrated by frame 1:1 in Figure 3.

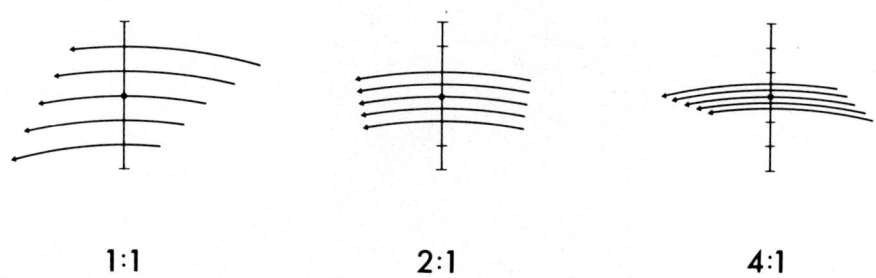

Figure 3. The effect of a radial compression on a sector of gas located at the radius where the rotation curve has its maximum. The uncompressed flow (1:1) shows differential rotation, which is arrested when the compression ratio is 2:1, and reversed when the ratio becomes 4:1.

Because of differential rotation, the gas that at some time in the past was contained inside a sector, shears out into a trailing feature. If for some reason the same patch of gas were to experience a uniform compression by a factor $(2\Omega/\kappa)^2$ (which at the maximum of the rotation curve is 2), the conservation of angular momentum would speed up the outer streamline with respect to the inner one in such a manner that the sector would appear to rotate at a uniform angular rate, as shown by frame 2:1. If the compression ratio were to be 4:1, the sense of shear would be reversed (and one might start to worry about the stability of the flow!). Thus it should be clear that as long as the forces are predominantly in the radial direction, one expects to see streaming motions on either side of an arm, irrespective whether it is leading, trailing, on the inside, or on the outside of corotation. The result is also independent on the rate at which the gas was compressed. The streaming motions will translate into wiggles in the 21cm velocity maps, and will be correlated with the position of the arm. The information about the WKBJ pattern speed is contained in the radial motion, in the sense that the motion in the arm is toward the center inside corotation, and outward outside, if the arm is trailing. In the leading case the radial component is reversed. After both components are combined and the resultant velocity field is convolved with a finite beam, the result is a sinusoidal wiggle. It is the phase of the wiggle with respect to the arm or shockfront which tells us where corotation is.

The same sort of relation between the tangential velocity and compression appears to hold, at least qualitatively, in the case when the arm is forced by a bar. This can be seen in Figure 4 which shows

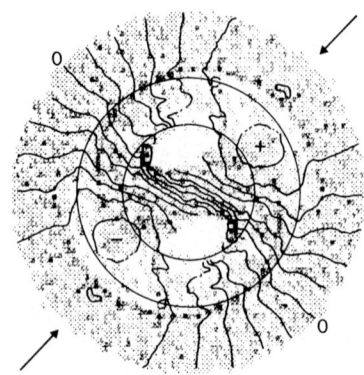

Figure 4. The velocity map of the gas cloud simulation shown in Figure 1. The true major axis is at 45 degrees to bar and indicated by the arrows. Also shown are the corotation and outer Lindblad resonance radii.

the velocity map superimposed on Schwarz's gas clouds. (Some of the smallest scale wiggles are due to the discrete nature of the clouds - the total number is 4000 and the mean per beamwidth is 3). In this case even the radial motion is in the same sense as predicted by the WKBJ theory.

Not too long ago hopes were high that one could, with the help of density waves, unravel the complexities of the 21 cm line profiles from the Galaxy, and learn something about the waves in return. These hopes visibly subsided once the magnitude of the task became apparent, and with the advent of aperture synthesis telescopes, external galaxies have replaced our own as the test bed.

Of the half-dozen nearest galaxies which have been observed with sufficient resolution to be possible candidates for a confrontation, M 33 and M 81 are the likeliest. The others can be excused for a variety of reasons, such as having too many arms, bad orientations, suffering from tidal effects, etc. The first observations of M 33 at Cambridge did not reveal any likely WKBJ pattern (Wright et al. 1972; Warner et al. 1973). The galaxy has been reobserved with a tenfold increase in sensitivity, but the data are yet to be reduced.

II.4 A CONFRONTATION OF DENSITY WAVE THEORIES WITH OBSERVATIONS

Visser has just given a detailed report on his model fitting of WKBJ patterns to the 21 cm observations of M 81 carried out by Rots (1975). I am not trying to belittle the effort that has gone into this comparison, when I say that you did not witness a confrontation. It is a convincing demonstration that we are dealing with a density wave of some sort, but if one were to argue that the good fit on the east side indicates that the corotation must be at 11 kpc, the discrepancy on the west side suggests that perhaps 5 kpc is the correct place, and vice versa. The galaxy is clearly asymmetric as a result of the tidal effects of the nearby companions as shown by van der Hulst, and I do not know which side is least affected, or how to go about correcting for such effects.

If I were to make a similar study, I would plot the theoretical convolved density and velocity fields on top of each other, and search for qualitative features which distinguish between being inside or outside corotation, and then try to discern these in the observations. As noted above, the distinction between the two cases depends on the relation between the shock and the phase of the wiggle. That relation is easily lost if one compares a theoretical velocity field with an observed density, since it is most unlikely that the latter will coincide with the theoretical one over the length of the spiral arm. Figure 4 illustrates how sensitive the phase relation can be.

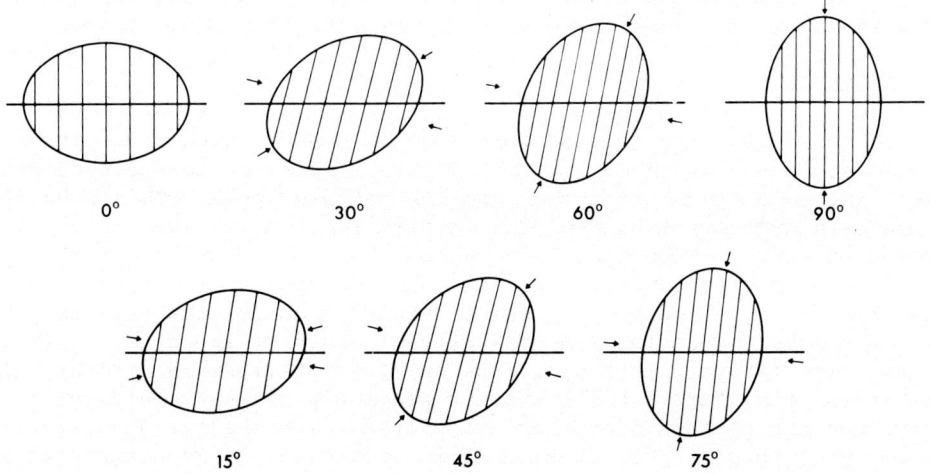

Figure 5. The velocity maps of the gas flow induced by a stable bar mode in a uniformly rotating stellar disk, shown for seven orientations of the bar.

One feature that is not very sensitive to detail is the tilt of the lines of constant radial velocity inside corotation, produced by a bar. This can be seen in Figure 4. The reason why it appears to be incorruptible is due to its large scale, and the fact that the streamlines are elongated in the same direction as the bar. The simplest illustration of this effect is provided by a stable bar mode of a uniformly rotating stellar disk. Figure 5 shows the velocity maps due to the motion of cold gas for various orientations of the bar. The bar is shown as it would appear viewed face on, and the major axis is in the horizontal direction. The arrows indicate the bar and kinematical major axes. There is a simple formula in this case which relates the angle v between the normal to the velocity contours and the true major axis, with the angle b between the bar major axis and true major axis,

$$\tan v = - a \sin 2b / (1 - a \cos 2b)$$

and

$$a = 0.75 \, e / \omega \, .$$

Here e is the ratio of the bar minor to major axes, and ω is the pattern speed expressed in units of the disk rotation rate. The amplitude of the mode has been chosen to make $e = 1/6$, and the pattern speed is $1/2$, hence $a = 1/4$.

Figure 5 shows that the only way a bar can go unnoticed is if $b = 0$, since in this orientation it can be confused with an inclined disk. It is also worth noting that the true major axis lies always somewhere between the kinematical and bar axes, and therefore a compromise between the two will be closer to the truth than the choice of one or the other. The 21cm observations of NGC 4151 (Bosma *et al.* 1977a) and NGC 4736 (Bosma *et al.* 1977b) show the sort of behaviour illustrated above.

6. PHOTOMETRY

The naive approach to star formation in a WKBJ shock would predict an increase in the width of a spiral arm in proportion to the difference between the angular rotation rate and the pattern speed. Thus Schweizer's measurements of the arm half-widths in different colours and as a function of radius (Schweizer 1976) should have been sufficient to determine the pattern speed. But they show no systematic trend. Wielen has just pointed out a possible explanation for the failure. There is another parameter, the velocity at birth of the stars which can make arms fat, and thus spoil the expected progression of dust, gas, young stars, old stars. While another seemingly simple method for determining the pattern speed has evaporated, nevertheless quantitative photometry of this kind is invaluable in providing the magnitudes of the spiral force fields. It would be nice if one could obtain a more direct determination of the stellar contribution to the arm light.

7. CONCLUSION

Implicit in the above ramblings is a warning, directed particularly to observers, not to spend too much time interpreting their results in terms of this density wave model or that. It is almost certain that the models will be obsolete next year, and the time spent in forcing the least-squares solutions is far better spent in gathering more data, and searching for features that appear to be sufficiently universal to require an explanation. To give an example: now that the WKBJ people are beginning to think in terms of modes, you can look forward to cooler and hence less massive disks, and proportionately more massive haloes. The reason for this is that whereas before one would fit just the short-wave branch of the dispersion relation to the data, now the longwaves must also be added. When the two are combined, the pattern becomes more open, and in order to restore the fit, the disk must be cooled.

REFERENCES

Bosma, A., Ekers, R.D., and Lequeux, J.: 1977a, *Astron. Astrophys.* 57, 97.
Bosma, A., van der Hulst, J.M., and Sullivan, W.T. III: 1977b, *Astron. Astrophys.* 57, 373.
Erickson, S.A.: 1974, Ph.D. Thesis, MIT.
Hohl, F.: 1971, *Astrophys. J.* 168, 343.
Hohl, F.: 1973, *Astrophys. J.* 184, 353.
Hohl, F.: 1976, *Astron. J.* 81, 30.
Kalnajs, A.J.: 1972, *Astrophys. J.* 175, 63.
Kalnajs, A.J.: 1973, *Proc. Astron. Soc. Aust.* 2, 174.
Kalnajs, A.J.: 1976, *Astrophys. J.* 205, 751.
Lin, C.C.: 1967, *Ann. Rev. Astron. Astrophys.* 5, 453.
Lin, C.C.: 1970, *Proc. IAU Symp. 38*, 377.
Lynden-Bell, D. and Kalnajs, A.J.: 1972, *Monthly Notices Roy. Astron. Soc.* 157, 1.
Lynden-Bell, D.: 1975, *C.N.R.S. Internat. Colloquium No. 241*, p. 91, (Paris: CNRS).
Mark, J.W.-K.: 1976, *Astrophys. J.* 205, 363.
Miller, R.H., Prendergast, K.H., and Quirk, W.J.: 1970, *Astrophys. J.* 161, 903.
Oort, J.H.: 1962, In *Interstellar Matter in Galaxies*, ed. L. Woltjer, p. 234 (New York: Benjamin).
Ostriker, J.P., and Peebles, P.J.E.: 1973, *Astrophys. J.* 186, 467.
Roberts, W.W.: 1969, *Astrophys. J.* 158, 123.
Rots, A., and Shane, W.W.: 1975, *Astron. Astrophys.* 45, 25.
Schweizer, F.: 1976, *Astrophys. J. Suppl.* 31, 313.
Toomre, A.: 1964, *Astrophys. J.* 139, 1217.
Toomre, A.: 1974, In *Highlights of Astronomy*, ed. G. Contopoulos, p. 457, (Dordrecht: Reidel).
Toomre, A.: 1977, *Ann. Rev. Astron. Astrophys.* 15, 437.
Wright, M.C.H., Warner, P.J., and Baldwin, J.E.: 1972, *Monthly Notices Roy. Astron. Soc.* 155, 337.
Warner, P.J., Wright, M.C.H., and Baldwin, J.E.: 1973, *Monthly Notices Roy. Astron. Soc.* 163, 163.

DISCUSSION FOLLOWING REVIEW II.4 GIVEN BY A.J. KALNAJS

LIN: There are (roughly speaking) two classes of mass models for galaxies: (1) those with a spheroidal component and a disk with a "hole" in the middle (cf. Einasto), and (2) those without a spheroidal component. Dr. Kalnajs has been stressing the second class of models. In this case, the system is likely to develop into an open spiral, or even a bar. The parameter Q is likely to be high. In the first case, the asymptotic theory can be applied to yield spiral patterns of relatively small pitch angles [see Mark's comment below].

MILLER: You mentioned that the dispersion relation, especially near corotation, is affected if $Q > 1$. You also mentioned that Q is redefined in the presence of a halo. Is the Lin-Shu dispersion relation affected by a halo in any way beyond that implied in the definition of Q?

KALNAJS: No, the dispersion relation is local and depends only on local parameters. It knows nothing about the global distribution of material which determines some of these parameters.

STROM: To what extent does the observation of wave patterns in disks (from near IR observations of the underlying disk stars) allow you to check the wave generation mechanism? Do you believe that wave patterns observed in Schweizer's galaxies represent density wave patterns? Is there any case where you see a bar-like instability in his galaxies?

KALNAJS: I don't know how to distinguish between a self-supporting wave in the sense of the WKBJ theory, and a forced spiral due to a bar near the center.
    Yes, I do believe that there are density waves in Schweizer's galaxies, and that there could be a bar in the center of M81.

WAXMAN: What effect does spiral structure excited by a bar have on the bar itself; and therefore, how long can we expect the bar to maintain the spiral structure?

KALNAJS: It can either amplify or damp it. In the situation I illustrated, the bar is a negative angular momentum mode in the stellar component, and because of the energy loss the spiral structure in the gas orients itself so as to pull the bar backwards and thus amplifies it. The spiral structure decays in this case because the colliding gas gains angular momentum and moves out.

OORT: Referring to your remark about Visser's confrontation of the M81 observations with spiral-wave theory I felt that you were asking too much. I think that it is already an important accomplishment that in one case a good representation has been obtained of the various observed characteristics with a theory of spiral structure.

TOOMRE: What is it you still question about the wave in M81 (whose existence I would have thought Visser, Rots, Shane and others have now demonstrated beyond all reasonable doubt)?

KALNAJS: I do not question the presence of some sort of a density wave in M81. It is just not clear to me whether it is a WKBJ type wave or the effect of a large-scale global mode in the stars.

VISSER: First I want to say that my model of M81 does not prove that other models are impossible and that density-wave theory has to be correct in all its consequences because of this model. On the other hand, I have to point out that the model is not so perfect as it seems. If the velocity perturbations of the model are as best as possible in phase with the observed wiggle, there is a phase shift of the observed spiral arms with respect to the model arms. The phase shift can be as large as $22°$ ($11°$ in azimuth).

SCHWEIZER: How massive must the bars which you use be to produce the spiral patterns you are interested in? Do you need bar-disk contrasts of factors 3, or can you do with 10% to 20% contrasts? In M81, there just isn't much room for a massive bar, if the light distribution is at all indicative of the mass distribution.

KALNAJS: One needs a very modest field in the vicinity of the outer Lindblad resonance to produce a spiral shock. For the example I just cited, the total field strength is just a bit over 3% of the axisymmetric component. Such a field can be produced by a small large-contrast bar, or a large small-contrast one. We chose the former, without having any particular galaxy in mind.

FREEMAN: Is there a problem understanding the spiral structure in systems like M33 where the disk obviously has no hole at the center, there is no bar and the bulge is quite insignificant?

MONNET: In the center of M33 there is indeed a bulge with a mass of about 2% of the total mass of the galaxy.

LIN: I would just like to make one simple general statement, namely we do not pretend that we have a theory about every spiral galaxy.

FREEMAN: But M33 is the simplest galaxy you'll possibly ever see up close!

LIN: Is it? No, it has quite a complicated disk structure.

MARK: DISCRETE SPIRAL MODES IN DISK GALAXIES
Discrete spiral modes are calculated based on stellar dynamics, according to the most recent form of the asymptotic theory of density waves. These self-excited modes are maintained wave systems which no longer propagate in the radial direction. They are expected to grow to observable amplitudes in a few billion years, and several of them may coexist in a given galaxy. A quasi-stationary spiral structure may be found in many galaxies due to a balance between wave amplification and certain dissipative processes, such as shocks. The theory can be applied to realistic galaxy models. Shown in the left figure below are the rotation curve and the mass distribution of a galaxy model which

resembles M31 in many respects. Shown in the right figure are the perturbation density contours for the lowest two-armed spiral mode. For each individual spiral mode, their radial density distribution in the stellar arms is qualitatively similar to that observed in nearby galaxies (for example by F. Schweizer). Details will appear in a forthcoming publication (Bertin et al., Proc. Nat. Acad. Sci., USA).

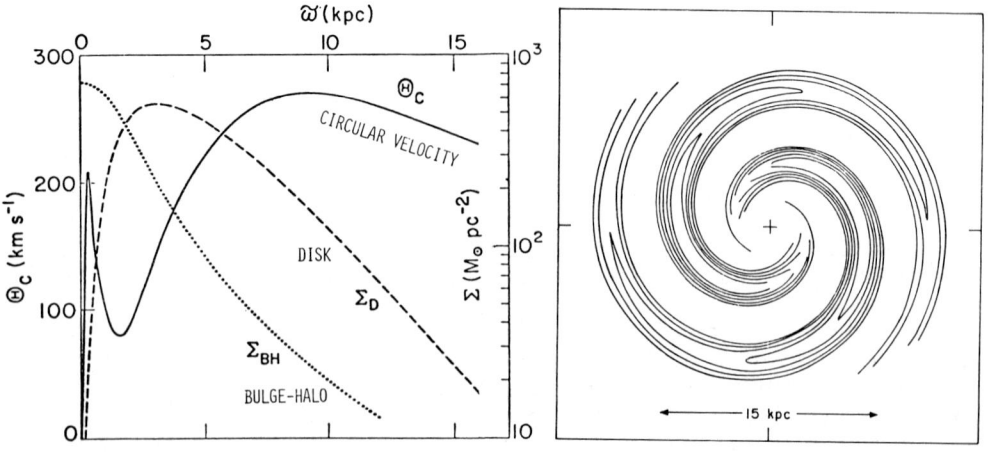

BERTIN: GROWTH OF SPIRAL WAVES IN DISK GALAXIES

A local dispersion relation for spiral waves in a gaseous disk is derived. In addition to the well-known parameter $Q = \kappa a/\pi G \sigma_0$, another dimensionless parameter $J = 2m(\pi G \sigma_0/\kappa^2 r)/\sqrt{(1/s - 1/2)}$ is found to be important in characterizing the growth of spiral waves. Here a is the acoustic speed, σ_0 is the surface density, m is the number of spiral arms, κ and Ω are respectively the epicyclic and circular frequency, r is the radial distance from the galactic center, and $s = -d\ln\Omega/d\ln r$.

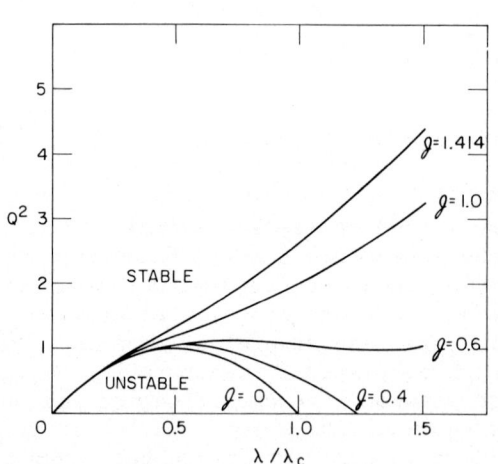

Specifically, for stability against disturbances of local wavelength λ, $Q^2 > 4[\lambda/\lambda_c - (\lambda/\lambda_c)^2/(1 + J^2 \lambda^2/\lambda_c^2)]$ where $\lambda_c = 4\pi^2 G\sigma_0/\kappa^2$. The marginal stability curves are shown on the Figure for various values of J. For axisymmetric disturbances, $J = 0$ and the curve resembles that of Toomre (1964, Ap.J. 139, 1217). The diagram indicates that non-axisymmetric waves are more difficult to suppress than the axisymmetric ones, in general agreement with the work by Goldreich and Lynden-Bell (1965, M.N.R.A.S. 130, 125) and by Julian and Toomre (1966, Ap.J. 146, 810).

The details of this work, together with the role of J in the growth of spiral modes, will be given in a forthcoming paper (Lau and Bertin, in prep.).

LIN: You might have added that for large values of J, one needs a higher value of Q for stability.

W.W. ROBERTS: GALACTIC SHOCKS IN OPEN-ARMED NORMAL SPIRALS AND BARRED SPIRALS

In the steady state gas dynamical studies of the late 1960s (Fujimoto, 1968, I.A.U. Symp. No. 25, 435; Roberts, 1969, Ap.J. 158, 123) the dark dust lanes, which are observed along the inside edges of spiral arms, were identified as tracers of large-scale galactic shock waves which form in the gas. Thought also to be tracers of the shock phenomenon were the strikingly narrow and rather straight dust lanes often observed along the leading edges of the bar structures in barred spirals (see Lin, 1970, I.A.U. Symp. No. 38, 377). More recent two-dimensional, time dependent, numerical hydrodynamical calculations have been carried out by Sorensen et al. (1976, Ap.Sp.Sci. 43, 491), Sanders and Huntley (1976, Ap.J. 209, 53), and Huntley et al. (1977, in prep.) for the response of rotating disks of gas to bar-like perturbations in galactic gravitational fields. All of these time-evolutionary calculations evolve through a state in which the viscous, differentially-rotating disk of gas forms a central gas bar with two trailing spiral waves and exhibits features resembling shocks.

Here, in cooperation with J.M. Huntley and C.C. Lin, we discuss an approach in which we have been able to generalize the steady state gas dynamical studies for tightly-wound normal spirals to include normal spirals with open spiral arms and barred spirals with prominent bar structures in the inner parts. The response of the gas to a barred spiral-like perturbation, for example, is calculated by means of an analysis which enables the two-dimensional flow to be broken up into two physical regimes. In regime I where the gas flow is highly supersonic, the flow is determined through an asymptotic approximation that neglects secondary terms proportional to the square of the dispersion speed, such as the transverse gradient of pressure. In Regime II near and within the bar (and spiral arms) the flow is determined through an asymptotic approximation that neglects the small variation of the velocity, density, and pressure along a shock with respect to their variation normal to the shock. The composite picture for the steady state flow of gas is constructed by joining the two regimes of flow in the transition layer between regimes.

This composite picture is illustrated for one case, model A, in Figure 1. Arrows on two of the gas streamlines (left panel) indicate the clockwise sense of gas circulation about the disk and through the shocks (----) which form near the potential minimum (——) of the barred spiral perturbation. A photographic simulation (right panel) of the gas density distribution illustrates the strong compression of the shock on the inside edge of each gas arm.

Figure 2 illustrates the corresponding results for a second case, model B, in which the flow equations contain a "friction type" term

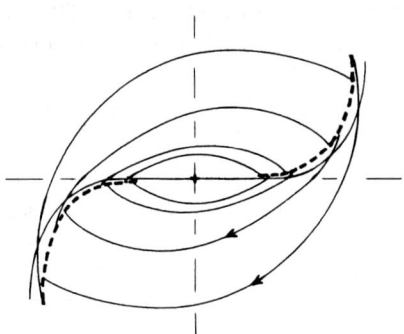 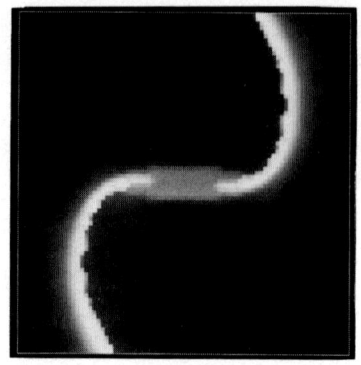

Figure 1. Gas streamlines and gas density distribution for model A.

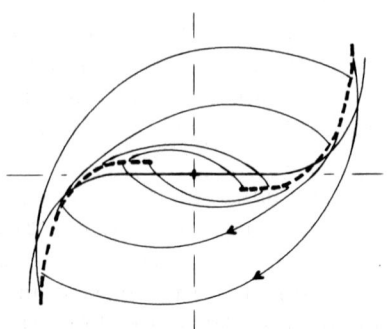 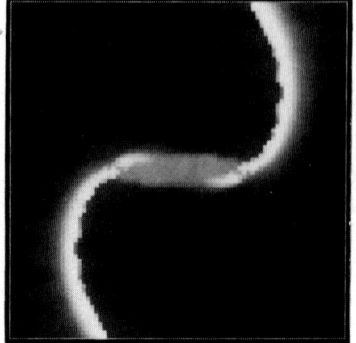

Figure 2. Gas streamlines and gas density distribution for model B.

that simulates "gaseous friction" in the inner parts of the disk. The forward shift of the shock onto the leading edge of the bar in model B produces an offset similar in some respects to the offset exhibited by the dark dust lanes observed on the leading edges of bars.

ALLEN: Did I understand correctly that your model has a gas density contrast between the arms and the interarm regions of about 20?

W.W. ROBERTS: Yes, that is correct.

RADIO OBSERVATIONS OF MOLECULES IN NEARBY GALAXIES

J.B. Whiteoak
Division of Radiophysics, CSIRO, Sydney, Australia

The observation in 1963 of absorption due to the 18 cm ground-state transition of OH in the direction of Cas A (Weinreb et al., 1963) marked the first occasion on which a molecular cloud was detected at radio wavelengths. However, it was not until the later discovery of high-intensity OH emission (Weaver et al., 1965) that attention turned to nearby galaxies. Unfortunately, searches for OH emission in the Magellanic Clouds (McGee et al., 1965; Radhakrishnan, 1967) and in more distant galaxies (Roberts, 1967) were unsuccessful. The first detection of an OH transition, in absorption against the radio continuum of NGC 253 and 3034 (Weliachew, 1971), went almost unnoticed because the results were unconvincing. However, Whiteoak and Gardner (1973) and Nguyen-Q-Rieu et al. (1976) confirmed the existence of the absorption.

Transitions of other molecules abundant in our galaxy were later detected in nearby galaxies: the 6 cm 1_{11}-1_{10} transition of H_2CO (Gardner and Whiteoak, 1974), the 2.6 mm J = 1-0 transition of CO (Rickard et al., 1975), the 3.2 mm J = 1-0 transition of HCN (Rickard et al., 1977b), and the 1.3 cm 6_{16}-5_{23} transition of H_2O (Churchwell et al., 1977). Table I shows the molecules in chronological order of detection, for each of the 12 galaxies. Typical spectra and, for comparison, HI, Hα, H92α, and H166α profiles, are shown in Figures 1 to 3. In addition to these results there has been a possible detection of the 9 cm F = 1-0 transition of CH in the Large Magellanic Cloud (Whiteoak and Gardner, 1978), while Rickard et al. (1977b) have reported a negative search for the 3.1 mm J = 2-1 transition of CS in NGC 253 and 3034.

OH. The main-line transitions (at 1665 and 1667 MHz) of the $^2\pi_{3/2}$, J = 3/2 ground state of OH are best seen in absorption against the small-diameter (about 0'.5 arc) nuclear radio sources of dusty edge-on galaxies (Figs. 1a,b,c). Thus for NGC 253 and 4945 the observed peak line-to-continuum ratios at 1667 MHz are about 0.03 and 0.1, while for NGC 5236, an almost face-on spiral, the ratio is below 0.005. The optical depths are uncertain because the absorption distribution over the continuum is not known. In NGC 4945, for an assumed optical depth of 0.3 (Whiteoak and Gardner, 1975) and a kinetic temperature of 10 K, the total column density of OH is about $10^{17} cm^{-2}$.

Table I
Molecules observed in external galaxies

Galaxy name	Observed molecules	References
NGC 224 (M31)	CO	2, 12
NGC 253	OH, H_2CO, CO, HCN	3, 4, 9, 10, 11, 12, 13, 14
NGC 598 (M33)	H_2O	1
NGC 1068 (M77)	CO	10
LMC	CO, H_2CO, OH	7, 16, 17
NGC 3031 (M81)	CO	2
NGC 3034 (M82)	CO, OH, (HCN)	8, 9, 10, 11, 12, 13
NGC 4945	OH, H_2CO	3, 14, 15
NGC 5055 (M63)	(CO)	12
NGC 5128 (Cen A)	H_2CO, OH	5, 6
NGC 5194 (M51)	CO	2, 10, 12
NGC 5236 (M83)	CO	10

1: Churchwell et al. (1977); 2: Combes et al. (1977); 3,4,5,6: Gardner and Whiteoak (1974, 1975, 1976a,b); 7: Huggins et al. (1975); 8: Nguyen-Q-Rieu et al. (1976); 9,10,11: Rickard et al. (1975, 1977a,b); 12: Solomon and de Zafra (1975); 13: Weliachew (1971); 14,15,16,17: Whiteoak and Gardner (1973, 1975, 1976a,b).

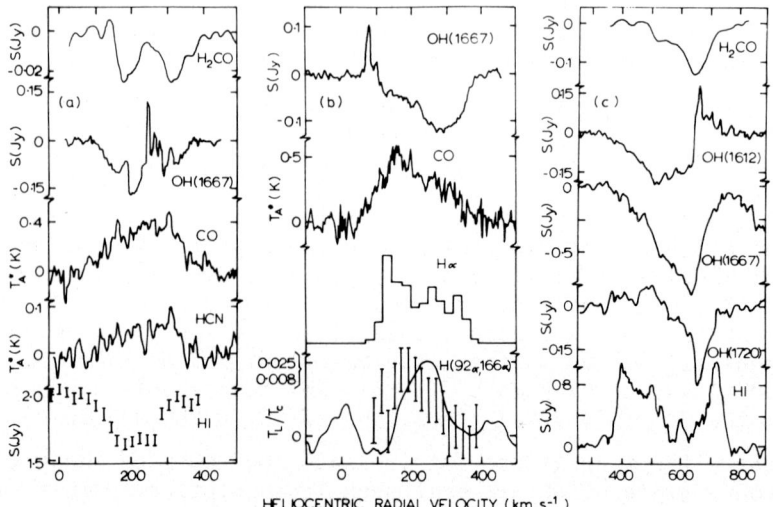

Fig. 1 - Line profiles towards the centres of the edge-on galaxies (a) NGC 253, (b) NGC 3034 (M82) and (c) NGC 4945. The vertical ordinates are flux density (S), corrected antenna temperature (T_A^*), or line-to-continuum ratio (T_L/T_c). In (b) the profile shown by error bars is for H166α; the appropriate scale is given by the lower value (0.008). The 1665 MHz profile has been excluded from (c) - it is similar in shape to the 1667 MHz profile but reduced in amplitude by a factor of 1.8.

The absorption profiles for edge-on galaxies have velocity spreads centred approximately on the systemic velocities, and widths (of about 300 km s^{-1}) not much less than the total variation in radial velocity associated with the overall rotation of each galaxy. For NGC 4945 these points are illustrated by comparison with the HI profile (Whiteoak and Gardner, 1976c), the width of which is determined by the overall rotation. However, the large widths of the absorption profiles must be the result of non-circular motions in the galaxies, because any widening due to differential galactic rotation across the extent of the radio continuum would be small (as observed in the HI absorption observations of NGC 253 by Gottesman et al. (1976)). Moreover, because the radio continuum is believed to originate close to the centre of the galaxies, the non-circular motions must consist of motions both towards and away from the centre. Optically the latter have been observed in NGC 253 (Demoulin and Burbidge, 1970) and in NGC 3034 (Heckathorn, 1972).

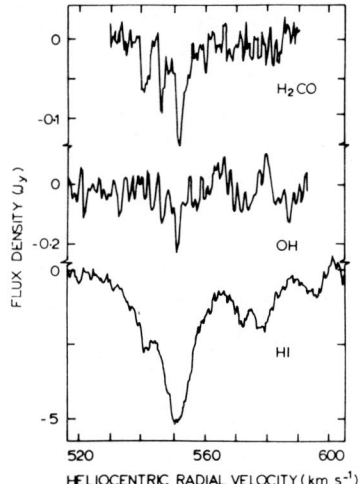

Fig. 2 - Line profiles for NGC 5128 (Cen A).

The satellite OH ground-state transitions at 1612 and 1720 MHz have been observed for NGC 253 (Gardner and Whiteoak, 1975) and NGC 4945 (Whiteoak and Gardner, 1975); in contrast to the main-line transitions they indicate systematic departures from local thermodynamic equilibrium. For both galaxies, instead of the profiles expected for LTE (fainter than the main-line profiles but similar in shape) there is a progressive change in the departures across the profiles, with enhanced 1612 MHz absorption and 1720 MHz emission at the lower velocities, and the reverse situation at the higher velocities. Similar anomalies occur in the central regions of our galaxy (Whiteoak and Gardner, 1976d), and imply transfers in population between the F = 2 and F = 1 hyperfine levels of the OH ground state. Whiteoak and Gardner (1975) used the theory of Litvak (1969) to propose a model in which the transfers are caused by intense infrared radiation located in the nucleus of the galaxy. In this model the 'infalling' material is located near the nucleus while the outward motions take place further out.

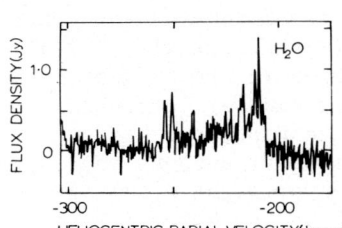

Fig. 3 - H$_2$O spectrum for IC 133 in NGC 598.

In the 1667 MHz profile for both NGC 253 and 3034 is a narrow (5 to 10 km s^{-1}) feature which is probably maser emission (Whiteoak and Gardner, 1973). Although this emission is 1 to 2 orders of magnitude stronger than the most intense high-gain emission in our galaxy (the Class I OH emission sources associated with HII regions), it lacks the features of Class I emission. It is considerably weaker at

1665 MHz and for NGC 253 at least is unpolarized, whereas Class I emission tends to be highly polarized and stronger at 1665 MHz. However, the features could result from low-gain amplification of the high brightness radiation in the galactic nuclei. The location may be settled when angular sizes of the emission become available.

In the LMC OH absorption is observed in the HII region N159 (Whiteoak and Gardner, 1976b). Despite intensive observations along the bar of this galaxy, no narrow-band emission above 0.15 Jy in flux density has been detected. At this level, brighter Class I sources of our galaxy, if located at the distance of the LMC, would have been detectable.

H_2CO. The $1_{10}-1_{11}$ transition is in absorption, and towards the nuclei of NGC 253 and 4945 the profiles are wide and generally similar in shape to the OH absorption, suggesting a common origin for the two molecules. However, the H_2CO absorption is much fainter than the OH absorption, the maximum line-to-continuum ratios being only 0.02 and 0.08 for the two galaxies. The ratio of H_2CO to OH absorption tends to increase with velocity as for the centre of our galaxy.

H_2CO clouds well away from centres of galaxies have been detected in the LMC and NGC 5128. For the former, absorption was detected only towards N159 (Whiteoak and Gardner, 1976a); the profile is simple and comparatively narrow (9 km s^{-1}), with a maximum line-to-continuum ratio (0.014) small compared with the average (0.074) for our galaxy but comparable to values for some prominent regions (e.g. the Carina nebula). For NGC 5128, the profile towards the nucleus (Gardner and Whiteoak, 1976b) shows narrow (1.8 to 3.5 km s^{-1}) H_2CO features, all with velocities exceeding the systemic velocity of 535 km s^{-1}. The H_2CO, OH and HI absorption probably arises in several 'infalling' cold clouds located in the outer dust lane which overlies the nucleus. An additional cloud was detected in this lane and 4' arc to the south-east of the nucleus; the absorption is probably against the extended continuum emission associated with the galaxy.

CO. Unlike OH and H_2CO, the CO J = 1-0 transition appears in emission, and the chances of detection should be independent of galaxy inclination and background radio continuum. However, out of 32 bright galaxies observed, Rickard et al. (1977a) detected CO emission in the nuclear regions of only 8, where the galactic nuclei showed optical emission lines or contained radio continuum or strong infrared emission.

Figures 1(a),(b) contain the CO profiles observed towards the centre of NGC 253 (Rickard et al., 1977b) and NGC 3034 (Rickard et al., 1977a). Compared with the OH and H_2CO absorption the CO spectra are broader, and the profile shapes, although possibly showing some similar large-scale structure, differ in detail. A difference is to be expected, because the OH absorption occurs in front of the small-diameter nucleus whereas the CO clouds are located anywhere along the lines of sight within a 65" arc beam. For NGC 3034 the CO profile resembles the Hα profile constructed by Rickard et al. (1977a) from the observations of Heckathorn (1972), suggesting a location for the CO in the dust of the optical filaments. The CO profile is also roughly similar to the H92α and

H166α profiles (Chaisson and Rodriguez, 1977; Shaver et al., 1977). The profile widths of NGC 5194 and 5236 are only about half those shown in the figures - probably because of the high inclination of the planes of these galaxies to the line of sight.

Maps of the CO distribution in NGC 253 and 3034 (Rickard et al., 1977a) show that the emission is highly concentrated towards the nuclei - the half-intensity widths of the distributions were only 1½ to 2 times the 65" arc beamwidth. The distribution maximizes at a position on the major axis close to but displaced from the nucleus (by about 0'.5 arc) and has a velocity centroid that varies in position with a sense consistent with the general galactic rotation. These features, and the large profile widths (Solomon and de Zafra, 1975), are similar to those for the CO emission in our galaxy. Rickard et al. (1977a) have claimed a further similarity for the abundance ratios of CO to OH and H_2CO, although the calculations were based on some questionable assumptions.

CO emission has been detected in N159 in the LMC (Huggins et al., 1975), and along the dusty sides of spiral arms in NGC 224, 3031 and 5194 (Combes et al., 1977). The emission bandwidths are narrow (typically 20 km s^{-1}) and the velocities are similar to HI, OH and H_2CO velocities in the same directions, although in NGC 224 there may be a slight systematic difference between the CO and HI velocities (F. Combes, private communication). Combes et al. (1977) have suggested that the CO cloud density is similar to that within the region 4 to 8 kpc from the centre of our galaxy, but the claim was based on a sample of only seven CO detections.

HCN. The J = 1-0 transition was detected in emission towards the centre of NGC 253 and possibly NGC 3034 (Rickard et al., 1977b). The profile for NGC 253 (Fig. 1a) is much fainter than the CO profile but similar in shape. Calculations based on an assumption (probably incorrect) of low optical depths yielded an abundance ratio of HCN to CO close to the value for our galaxy.

H_2O. Maser emission at 22 GHz (Fig. 3) was detected in one HII region (IC 133) out of a total of 16 observed in NGC 598 by Churchwell et al. (1977). The brightness temperature of the strongest line would be about 10% to 15% of that for W49 during its more quiescent periods, if W49 were at the distance of IC 133. However, IC 133 does not resemble the bright compact HII regions with which many of the H_2O sources in our galaxy are associated.

This review has shown that a total of five molecules have been detected at radio wavelengths in 12 nearby galaxies. Most transitions have been observed towards the centres of dusty galaxies - the absorption profiles for edge-on galaxies provide unquestionable evidence for the existence of non-circular motions. There is little doubt that other molecules will be detected as better receiving equipment becomes available.

Chaisson, E.J. and Rodriguez, L.F.: 1977, *Astrophys. J. Lett.* 214, L11.
Churchwell, E., Witzel, A., Huchtmeier, W., Pauliny-Toth, I., Roland, J. and Sieber, W.: 1977, *Astron. Astrophys.* 54, 969.
Combes, F., Encrenaz, P.J., Lucas, R. and Weliachew, L.: 1977, *Astron. Astrophys.* 55, 311.
Demoulin, M.H. and Burbidge, E.M.: 1970, *Astrophys. J.* 159, 799.
Gardner, F.F. and Whiteoak, J.B.: 1974, *Nature* 247, 526.
Gardner, F.F. and Whiteoak, J.B.: 1975, *Mon. Not. R. Astron. Soc.* 173, 77P.
Gardner, F.F. and Whiteoak, J.B.: 1976a, *Mon. Not. R. Astron. Soc.* 175, 9P.
Gardner, F.F. and Whiteoak, J.B.: 1976b, *Proc. Astron. Soc. Aust.* 3, 63.
Gottesmann, S.T., Lucas, R., Weliachew, L. and Wright, M.C.: 1976, *Astrophys. J.* 204, 699.
Heckathorn, H.M.: 1972, *Astrophys. J.* 173, 501.
Huggins, P.J., Gillespie, A.R., Phillips, T.G., Gardner, F.F. and Knowles, S.H.: 1975, *Mon. Not. R. Astron. Soc.* 173, 69P.
Litvak, M.M.: 1969, *Astrophys. J.* 156, 471.
McGee, R.X., Robinson, B.J., Gardner, F.F. and Bolton, J.G.: 1965, *Nature* 208, 1193.
Nguyen-Q-Rieu, Mebold, U., Winnberg, A., Guibert, J. and Booth, R.: 1976, *Astron. Astrophys.* 52, 467.
Radhakrishnan, V.: 1967, *Aust. J. Phys.* 20, 203.
Rickard, L.J., Palmer, P., Morris, M., Zuckerman, B. and Turner, B.E.: 1975, *Astrophys. J. Lett.* 199, L75.
Rickard, L.J., Palmer, P., Morris, M., Turner, B.E. and Zuckerman, B.: 1977a, *Astrophys. J.* 213, 673.
Rickard, L.J., Palmer, P., Turner, B.E., Morris, M. and Zuckerman, B.: 1977b, *Astrophys. J.* 214, 390.
Roberts, M.S.: 1967, *Astrophys. J.* 148, 931.
Shaver, P.A., Churchwell, E. and Rots, A.H.: 1977, *Astron. Astrophys.* 55, 435.
Solomon, P.M. and de Zafra, R.: 1975, *Astrophys. J. Lett.* 199, L79.
Weaver, H., Williams, D.R.W., Dieter, N.H. and Lum, W.T.: 1965, *Nature* 208, 29.
Weinreb, S., Barrett, A.H., Meeks, M.L. and Henry, J.C.: 1963, *Nature* 200, 829.
Weliachew, L.: 1971, *Astrophys. J. Lett.* 167, L47.
Whiteoak, J.B. and Gardner, F.F.: 1973, *Astrophys. Lett.* 15, 211.
Whiteoak, J.B. and Gardner, F.F.: 1975, *Astrophys. J. Lett.* 195, L81.
Whiteoak, J.B. and Gardner, F.F.: 1976a, *Mon. Not. R. Astron. Soc.* 174, 51P.
Whiteoak, J.B. and Gardner, F.F.: 1976b, *Mon. Not. R. Astron. Soc.* 176, 25P.
Whiteoak, J.B. and Gardner, F.F.: 1976c, *Proc. Astron. Soc. Aust.* 3, 71.
Whiteoak, J.B. and Gardner, F.F.: 1976d, *Mon. Not. R. Astron. Soc.* 174, 627.
Whiteoak, J.B. and Gardner, F.F.: 1978, submitted to *Mon. Not. R. Astron. Soc.*

DISCUSSION FOLLOWING PAPER II.5 GIVEN BY J.B. WHITEOAK

WELIACHEW: A FIRST STEP TOWARD THE RADIAL DISTRIBUTION OF CO IN M31
Emission from carbon monoxide has been systematically searched along the major axis of M31, in the region of spiral arms where the first detection was done (Combes et al. 1977, A.A. 55, 331). A partial radial distribution of CO emission is derived showing that carbon monoxide is unambiguously located on the inner side of the HI spiral arms. This can be explained in the frame of the density wave theory of spiral structure. As we are inside the co-rotation radius molecular clouds are formed by the compression of the interstellar medium on the inner side of a spiral arm, assuming trailing arms. All CO and HI (Emerson 1976, M.N.R.A.S. 176, 321) velocities agree within the uncertainties in their determination.

BALDWIN: Your peak temperature seen in CO of 0.3 K occurs at a point where the projected density of HI is also at maximum. Comparison of the CO and HI at the individual points where the measurements were made might modify the conclusion about the displacement between CO and HI.

EMERSON: With respect to the relative intensities of CO and HI in the N and S of M31, and the relative positions of HI and CO emission, I do not think that your comparison with Guibert's major axis HI distribution is valid. Your CO points in the S of M31 are not typical points in HI, being exceptionally bright "hot-spots", and are not represented adequately in a major axis HI cut. It is not meaningful to make a comparison of the intensities and positions of your individual CO points observed with a 1' beam, with the global radial distribution of HI deduced from Guibert's Nançay observations made with a beamwidth of 4' x 24'.

[see Chapter III for other papers on M31]

"Gas always goes down."

F.H. Shu in Discussion II.2

HOT GAS IN THE GALAXY: HOW EXTENSIVE IS IT?

Frank H. Shu
University of California, Berkeley, CA 94720, USA

1. INTRODUCTION

Recent satellite detections of O VI absorption and soft X-ray emission leave little doubt that, <u>locally</u>, a substantial fraction of the volume of space between interstellar clouds must be filled with rarified and highly ionized gas at temperatures ranging from 2×10^5 to $> 10^6$ K. (See the reviews of Spitzer and Jenkins 1975, and Kraushaar 1977.) The physical state of this gas contrasts sharply with the theoretical picture of a largely neutral, warm, intercloud medium at $\sim 10^4$ K developed by Pikel'ner (1967), Field, Goldsmith, and Habing (1969), and Spitzer and Scott (1969). My purpose here is to review the evidence, observational and theoretical, concerning how extensive hot gas at $\sim 10^6$ K might be.

2. IMPLICATIONS FOR GALACTIC SHOCKS

Hot gas which fills a large fraction of interstellar space would have important consequences for galactic shocks. The response of any component of a galaxy to forcing by a spiral gravitational field supported by the disk stars depends very sensitively on the effective dispersive velocity of that component. Shown in Figure 1 are the theoretical variations in surface density of matter with velocity dispersions of 8, 32, and 128 km/s when this material is exposed to the spiral field believed to be present at the solar circle. (See Roberts 1969; Shu, Milione, and Roberts 1973.) To fix ideas, we may think of the curves labelled 8 and 128 as the responses, respectively, of hypothetical intercloud media at 10^4 and 10^6 K to the forcing of the spiral density wave supported by the disk stars — curve 32. The response of the warm neutral medium is very strong and involves the production of galactic shocks. The response of the hot ionized medium is not detectable on the scale of this graph. If a distribution of interstellar clouds were embedded in the former, the sharp increase of pressure inside spiral arms can be expected to induce the more massive clouds in the distribution to collapse to form clusters of stars (Shu et al. 1972).

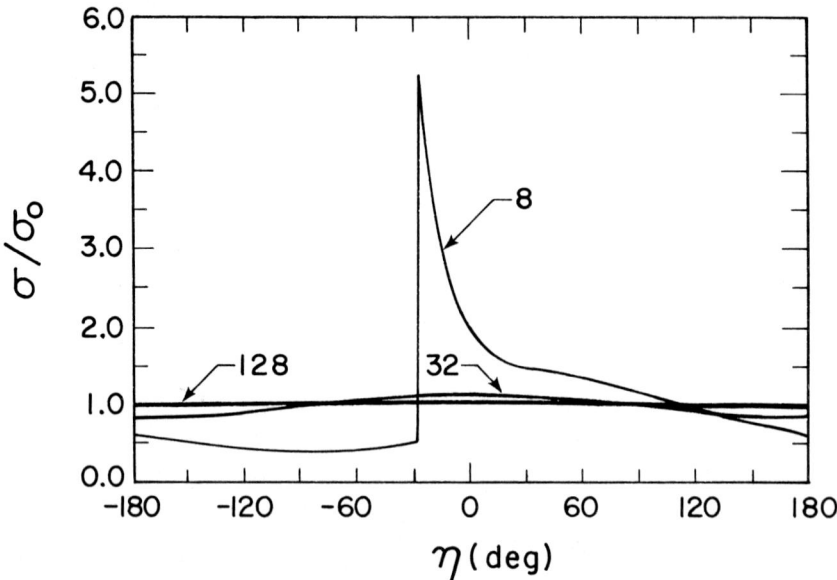

In contrast, if interstellar clouds were embedded in an intercloud medium at 10^6 K, no significant increase of their surface pressure can be expected. Does the observed predominance of massive stars in spiral arms then prove that 10^6 K gas cannot be pervasive in most spiral galaxies?

Unfortunately, the answer to this question is more complex than appears at first sight. The observed velocity dispersion of interstellar clouds is also about 8 km/s; hence, if their collision mean-free-path is short enough to treat the ensemble of clouds as a dissipative continuum, the curve labelled 8 in Figure 1 should also approximate the variation of the number of clouds per unit area. The large increase of the number of clouds per unit area inside spiral arms would then give rise not only to the dust lanes, but also to an increase in the frequency of cloud-cloud collisions. The latter process may enhance the fraction of massive clouds susceptible to gravitational collapse at a <u>fixed</u> external pressure. Observationally, only $\sim 10^7$ yr elapse between the time of the formation of the dust lane ("shock") and the appearance of massive OB stars (Mathewson, van der Kruit, and Brouw 1972). Such a timescale may seem too short to trigger star formation by the process of building up big clouds from little ones, but our knowledge of such processes is too insecure to claim that the timescale is prohibitive.

Another serious obstacle facing proponents of an extensive hot intercloud medium is the observational evidence from radio continuum and gamma-ray studies that the <u>entire</u> interstellar medium -- and not just the collection of cloud centers -- suffers an overall compression in spiral arms (van der Kruit and Allen 1976; Kniffen, Fichtel, and Thompson

1977). The compression of magnetic field and cosmic rays is most easily accomplished by directly compressing the intercloud gas, but it may also be possible to accomplish this task indirectly by only bringing cloud centers closer together if most of the interstellar magnetic flux threads clouds. We conclude that a 10^6 K intercloud medium is difficult but perhaps not impossible to reconcile with the observed compression of the interstellar medium by spiral density waves. Thus, to address definitively the question of how pervasive such hot gas might really be, we must turn to the observations.

3. REVIEW OF THE OBSERVATIONS AND INTERPRETATIONS

3.1 H I observations

The strongest evidence for the intercloud medium being warm and neutral comes from the H I radio observations of 21-cm line emission and absorption (Clark 1965; Hughes, Thompson, and Colvin 1971; Radhakrishnan et al. 1972). Emission line profiles show broad wings which are absent in the absorption lines produced in the directions toward discrete radio continuum sources. This fact has been widely interpreted to indicate the presence of two components of H I: a cold cloud component with temperatures $\sim$ 70 K which is seen both in emission and absorption, and a warm intercloud component with temperatures in excess of 1000 K which is seen only in emission.

3.2 O VI observations

Ultraviolet line observations have revealed substantial column densities of O VI in the absorption spectra of bright OB stars (Jenkins and Meloy 1974). Interpreted as thermal broadening, the line widths suggest interstellar gas at temperatures in excess of 2×10^5 K. Ionization equilibria considerations suggest that $10^{5.5}$ K probably represents a good estimate for the state of most of the observed O VI gas. This temperature is lower than the best estimates of the temperature of the soft X-ray emitting regions; hence, it is likely that the O VI gas and soft X-ray observations refer to different gas (cf. Shapiro and Field 1976).

Gas at a few hundred thousand degrees is near the peak of the theoretical cooling function of a collisionally ionized gas (Cox and Tucker 1969). Thermal stability and energy requirements then suggest that the filling factor of O VI gas cannot be very large. The two main theories which attempt to explain the O VI observations place the O VI gas in special locations -- namely, at the conductive interfaces between cool or warm gas and hot gas. In the model of interstellar bubbles driven by stellar winds (Castor, McCray, and Weaver 1975; Weaver et al. 1977), this conductive interface occurs between shocked stellar wind (T $\sim 10^6$ K) and the swept-up outer shell of interstellar H II gas (T $\sim 10^4$ K). In McKee and Ostriker's (1977) elaboration of a model proposed by Cox and Smith (1974), the conductive interface occurs between H I clouds and a 10^6 K intercloud medium generated by overlapping supernova remnants (SNR).

3.3 Soft X-ray observations

The diffuse X-ray background above 2 keV in energy is almost certainly extragalactic (see Silk 1973); however, the excess emission below 1 keV is predominantly Galactic. Indirect evidence for this contention exists in the correlation of the soft X-ray emission with known Galactic objects -- e.g., with the radio continuum emission from Loop I or the North Polar Spur (Bunner et al. 1972). Direct evidence exists in the failure to detect measurable soft X-ray absorption by M 31 (Margon et al. 1974) and by the Large and Small Magellanic Clouds (Rappaport et al. 1975; Long, Agrawal, and Garmire 1976; McCammon et al. 1976).

The accepted explanation for the diffuse soft X-ray flux is that it arises by thermal emission, mostly in unresolved X-ray lines, from a hot rarified gas ($T \sim 10^6$ K, $n \sim 0.003$ cm^{-3}). The strong differential absorption by interstellar gas implies that most of the observed soft X-ray flux originates from a local region whose column density is $\lesssim 10^{20}$ cm^{-2} (Kraushaar 1977). Thus, the observations suggest that we are locally inside a region containing million degree gas, but they do not tell how common such regions are in a typical spiral galaxy.

3.4 Theoretical estimates

Theoretical estimates of the filling fraction f of 10^6 K gas have been made on the basis of the model of overlapping SNR (Cox and Smith 1974, McKee and Ostriker 1977). The fundamental idea is straightforward. Let f(t) represent the filling fraction of SNR at time t in a statistically homogeneous medium. Moreover, let τ = the average lifetime of a SNR, V = the average volume occupied by a SNR, and r = the rate per unit volume of supernova explosions. The probability f(t+dt) that a point is inside a SNR at time t+dt is equal to the probability f(t) that it was inside one at time t, multiplied by the probability (1-dt/τ) that this SNR has not died; plus the probability [1-f(t)] that the point was not inside a SNR at time t, multiplied by the probability rVdt that a SNR has since been created at its location:

$$f(t+dt) = f(t)(1-dt/\tau) + [1-f(t)]rVdt. \qquad (1)$$

In a steady state, f(t+dt) = f(t) ≡ f. Solving for f, we obtain

$$f = Q/(1+Q) \quad \text{where} \quad Q \equiv rV\tau. \qquad (2)$$

In our definition of the dimensionless parameter Q, the volume V and lifetime τ represent averages after taking into account the possibility that a significant fraction of interstellar space might already be filled with SNR (cf. Cox and Smith, McKee and Ostriker 1977). Complications enter in making reliable estimates for f because Q in equation (2) is itself a function of f. In particular, Q(f) clearly increases for increasing f because larger and longer lasting SNR can be blasted if a bigger fraction of the interstellar medium is filled with rarified gas. Cox and Smith rely on basically this "bootstrapping" mechanism to form a network of supernova "tunnels" with f ≃ 0.5 even though the unperturbed "porosity" q = Q(f = 0) may be fairly modest. Smith (1977) has revised this estimate to f > 0.3 inside spiral arms. McKee and Ostriker, using a different set of assumptions, obtain a filling factor f ≃ 0.8; thus, their overlapping SNR constitute the entire intercloud medium.

Simple but generous estimates for the important parameters (e.g., $r = 10^{-13}$ pc^{-3} yr^{-1}, $V = 10^6$ pc^3, $\tau = 10^7$ yr) yield marginal values for Q -- i.e., values on the order of unity. Hence, it is clear that more sophisticated analyses can give almost any value of f, large or small, that one wants. As an example, if one's prejudices lie in a small general value of f, one might follow Scott, Jensen, and Roberts (1977) and suppose that supernova explosions are highly correlated in space and time since massive stars tend to be found in OB associations, which in turn concentrate in gas-rich spiral arms. Repeated explosions at the same isolated places would shred neighboring big clouds into many small ones (see, e.g., Woodward 1975). The subsequent conduction of heat from the hot gas of the blast wave into the small fragments may evaporate the latter, and may lead to large but isolated "holes" of HI. Such holes have in fact already been seen in the high resolution studies of M101 (Allen 1974). These holes would be important and interesting for theories of the interstellar medium -- especially if we happen to sit in a hole (see Weaver 1977) -- but they may be too few in number to influence appreciably the large-scale galactic dynamics.

4. CONCLUSIONS

Definitive determinations of the filling factor of 10^6 K gas in spiral galaxies await future observations, particularly in the radio and soft X-ray portions of the electromagnetic spectrum. High resolution 21-cm line investigations should reanalyze the question of whether the properties of individual clouds change systematically inside and outside spiral arms (see Weaver 1970). Especially important would be indications of whether the surface pressure of interstellar clouds increase or not inside spiral arms. A detailed investigation of the large holes of H I seen occasionally in the spiral arms of external galaxies would also be informative.

If hot gas fills almost the entire volume of interstellar space, high resolution studies should detect enhanced soft X-ray emission in the directions toward external spiral galaxies. Relevant observations already exist for the Large Magellanic Cloud (LMC). Rappaport et al. (1975) and Long et al. (1976) differ on the interpretation of the data at energies between 0.4 and 1.5 keV, but neither group finds excess emission from the LMC at energies ∿ 1/4 keV -- energies which characterize the observed radiation of the hot gas in the solar neighborhood. It would be important to see if enhanced X-ray flux at 1/4 keV is also absent from normal spiral galaxies. Experiments with sufficient angular resolution probably await the launch of the HEAO-B satellite.

ACKNOWLEDGEMENT

I thank Stu Bowyer, Dick McCray, Chris McKee, Jerry Ostriker, and Harold Weaver for informative discussions. This research was supported in part by NSF grant AST 75-02181.

REFERENCES

Allen, R.J.: 1974, in La Dynamique des Galaxies Spirales (ed. L. Weliachew), Paris, CNRS, p. 157.
Bunner, A.N., Coleman, P.L., Kraushaar, W.L., and McCammon, D.: 1972, Astrophys. J. Lett. 172, L67.
Castor, J., McCray, R., and Weaver, R.: 1975, Astrophys. J. Lett. 200, L107.
Clark, B.G.: 1975, Astrophys. J. 142, 1398.
Cox, D.P. and Smith, B.W.: 1974, Astrophys. J. Lett. 189, L105.
Cox, D.P. and Tucker, W.H.: 1969, Astrophys. J. 157, 1157.
Field, G.B., Goldsmith, D.W., and Habing, H.J.: 1969, Astrophys. J. Lett. 155, L149.
Hughes, M.P., Thompson, A.R., and Colvin, R.S.: 1971, Astrophys. J. Suppl. 23, 323.
Jenkins, E.B. and Meloy, D.A.: 1974, Astrophys. J. Lett. 193, L121.
Kniffen, D.A., Fichtel, C.E., and Thompson, D.J.: 1977, Astrophys. J. 215, 765.
Kraushaar, W.L.: 1977, Invited Lecture at 149th Meeting of Am. Astron. Soc., Honolulu.
Kruit, P.C. van der, and Allen, R.J.: 1976, Ann. Rev. Astron. Astrophys. 14, 417.
Long, K.S., Agrawal, P.C., and Garmire, G.P.: 1976, Astrophys. J. 206, 411.
Margon, B., Bowyer, S., Cruddace, R., Heiles, C., Lampton, M., and Troland, T.: 1974, Astrophys. J. Lett. 191, L117.
Mathewson, D.S., Kruit, P.C. van der, and Brouw, W.N.: 1972, Astron. Astrophys. 17, 468.
McCammon, D., Meyer, S.S., Sanders, W.J., and Williamson, F.O.: 1976, Astrophys. J. 209, 46.
McKee, C.F. and Ostriker, J.P.: 1977, Astrophys. J., in press.
Pikel'ner, S.B.: 1967, Astron. Zh. 44, 1915.
Radhakrishnan, V., Murray, J.D., Lockhart, P., and Whittle, R.P.J.: 1972, Astrophys. J. Suppl. 203, 15.
Rappaport, S., Levine, A., Doxsey, R., and Bradt, H.V.: 1975, Astrophys. J. Lett. 196, L15.
Roberts, W.W.: 1969, Astrophys. J. 158, 123.
Scott, J., Jensen, E., and Roberts, W.W.: 1977, Nature 265, 123.
Shapiro, P.R. and Field, G.B.: 1976, Astrophys. J. 205, 762.
Shu, F.H., Milione, V., Gebel, W., Yuan, C., Goldsmith, D.W., and Roberts, W.W.: 1972, Astrophys. J. 173, 557.
Silk, J.: 1973, Ann. Rev. Astron. Astrophys. 11, 269.
Smith, B.W.: 1977, Astrophys. J. 211, 404.
Spitzer, L. and Jenkins, E.B.: 1975, Ann. Rev. Astron. Astrophys. 13, 133.

Spitzer, L. and Scott, E.H.: 1969, Astrophys. J. 157, 161.
Weaver, H.F.: 1970, in IAU Symp. No. 39, Interstellar Gas Dynamics (ed. H.J. Habing), Reidel, Dordrecht, p. 22.
Weaver, H.F.: 1977, in preparation.
Weaver, R., McCray, R., Castor, J., Shapiro, P., and Moore, R.: 1977, preprint.
Woodward, P.R.: 1976, Astrophys. J. 207, 484.

DISCUSSION FOLLOWING PAPER II.6 GIVEN BY F.H. SHU

GALLAGHER: Do you consider the North Galactic Spur local, and is it not evidence for a hot tunnel type of model?

SHU: Yes, I think that it is local, and that the evidence of hot tunnels is an indication for this.

VAN DER LAAN: In your simple derivation of the hot gas filling factor the product $rV\tau$ occurs, where τ is the SNR duration. It seems to me that τ is ill defined in the case of overlapping SNRs. After all, τ does not then have the old meaning of age till the SNR loses its identity among interstellar clouds, but rather the duration of the hot (Sedov) gas' persistence. This may be much longer than 10^7 years, in which case f goes to 1 and the hot gas will escape as a galactic wind in the z-directions. The HEAO-B imaging X-ray telescope will hopefully map such situations in nearby spirals in a year or so, or it will dismiss the whole scenario.

SANCISI: Are you aware of the Perseus-Taurus "hole" and that we may be inside it?

SHU: Weaver favours the idea that we are inside a big hole, and I think there is no question that we must be quite close to gas of a million degrees. Whether the rest of the interstellar medium looks like that I don't know. Neither do I know of observations bearing on that question.

BURBIDGE: If we live in a large hole you are putting us in a special position. Apart from the philosophical problems this might mean that hot gas is rare in our Galaxy and other galaxies.

GIOVANELLI: I would like to point out that high velocity clouds may cause the production of soft X-rays, especially in the northern hemisphere. If HVCs are infalling material interacting with the galactic disk, post-shock temperatures on the order of 10^6 K would be achieved. According to Chow and Savedoff (1972, Nuovo Cimento 88) the emission measure could be high enough to contribute a significant fraction of the soft X-ray background.

III   NEARBY GALAXIES OF LARGE ANGULAR SIZE

"*A difficulty in studying M31 is its rather closely edge-on view. We may take comfort from the fact that astronomers in M31 would have the same problem in studying our Galaxy*"

P.C. van der Kruit in Discussion III.6

RADIO CONTINUUM OBSERVATIONS OF M31 AND M33

Elly M. Berkhuijsen
Max-Planck-Institut für Radioastronomie, Bonn, F.R.G.

1. INTRODUCTION

The two nearest spiral galaxies, M31 and M33, have been extensively studied both at optical as well as at radio wavelengths. Reviews of radio continuum observations are given in van der Kruit and Allen (1976) and in von Kap-herr et al. (1978), respectively. In Table 1 new radio observations in various stages of progress are listed, all of which show — or are expected to show — spiral structure. In this paper the Effelsberg maps at 11 cm of M31 and at 6 cm of M33 are discussed and compared with optical data. Some attention is given to the separation of thermal and nonthermal emission which is essential to any discussion of the origin of the radiation.

2. M31

In Figure 1 the 11-cm map of Berkhuijsen and Wielebinski (1974) is shown superimposed onto an optical picture (Lick Observatory). Intense radio emission is seen from the nucleus and the optically bright arms, especially from arms 4 and 5 (Baade 1958). Several point sources in the radio map are coincident with known HII regions (Baade and Arp 1974), but most of the point sources are probably unrelated to M31. Berkhuijsen (1977; hereafter referred to as Paper II) subtracted the point sources from the map and made a detailed comparison of the distribution of the extended radio emission at 11 cm with the distributions of HI, HII regions, OB associations and blue light. Some conclusions of that paper are:

(1) Apart from the nucleus the general distributions of the 11-cm continuum emission, HII regions, young stars and HI are very similar, all of them being characterised by the intense ring at about 9 kpc from the centre. This is evident from Figures 4, 5 and 6 in Paper II and can also be seen from the radial distributions shown in Figure 4 below.
(2) The lines of maximum surface brightness in the distributions of the radio continuum, HI and HII are generally coincident with each other and often follow the dust lanes. Along the arms peaks in the

Table 1. RADIOCONTINUUM OBSERVATIONS SHOWING SPIRAL STRUCTURE

Galaxy				M31		M33	
Telescope	λ (cm)		HPBW (')	Area (R in ')	Noise in T_b	Area (R in ')	Noise in T_b
Cambridge	200	TP	3 × 5	200	< 20 K	–	–
	75	TP	4.0	80[a]	1 K	–	–
Westerbork	49	TP	1.0×1.5	120* } 3 fields	1 K	60*	1 K
		POL	3.6×5.4	120[b]	60 mK	–	–
	21	TP	0.4×0.6	12,20[c]	1 K	–	–
	21	TP	0.4×0.6	160 } 15 fields	1 K	20[e]	1 K
		POL				–	–
	6	TP	0.1×0.2	–	–	N604	
Effelsberg	21	TP	9.2	130*	20 mK	70*	22 mK
		POL		130*	11 mK	70*	12 mK
	11	TP	4.8/4.5	125,105[d]	4.5 mK	60*	4 mK
		POL		90	2.5 mK		
	6	TP	2.6	20*	2.3 mK	45[f]	2.3 mK
		TP		90			

a) Pooley 1969; b) Segalovitz et al. 1977; c) van der Kruit 1972;
d) Berkhuijsen, Wielebinski 1974; e) Israel, van der Kruit 1974;
f) von Kap-herr et al. 1978; * observed and partly reduced;
rest: accepted proposals, partly observed
Area: determined by R = distance from center along major axis.

continuum emission are typically 2 kpc apart; they are usually found
near to but not coincident with peaks in the distributions of HI or HII.
These properties are in agreement with the predictions of Mouschovias
et al. (1974).
(3) The large-scale correlations between the various constituents can
be described by power laws. For example it was found that the number
of HII regions in the radio beam $N_{HII} \propto T_b^{2.9}$ (T_b = observed brightness
temperature due to thermal + nonthermal emission along the line of
sight in M31). Recent independent measurements at 11 cm (Beck, Berkhuijsen, Wielebinski), however, suggest a different zerolevel for the
11-cm map than has been assumed in Paper II; this may change the power
law to $N_{HII} \propto T_b^{\sim 2.0}$. If N_{HII} is proportional to the total thermal
emission (i.e. including diffuse emission and undetected HII regions)
then a power law with an exponent > 1 indicates that the nonthermal
radiation is more smoothly distributed than the thermal radiation. A
steepening of the spectrum with increasing distance from the center is
then to be expected.

A first attempt to find the spectral index distribution across M31
is being made by Beck et al. (1978). Comparison of the 21-cm map
recently observed in Effelsberg with the 11-cm map smoothed to the

III.1 RADIO CONTINUUM OBSERVATIONS OF M31 AND M33

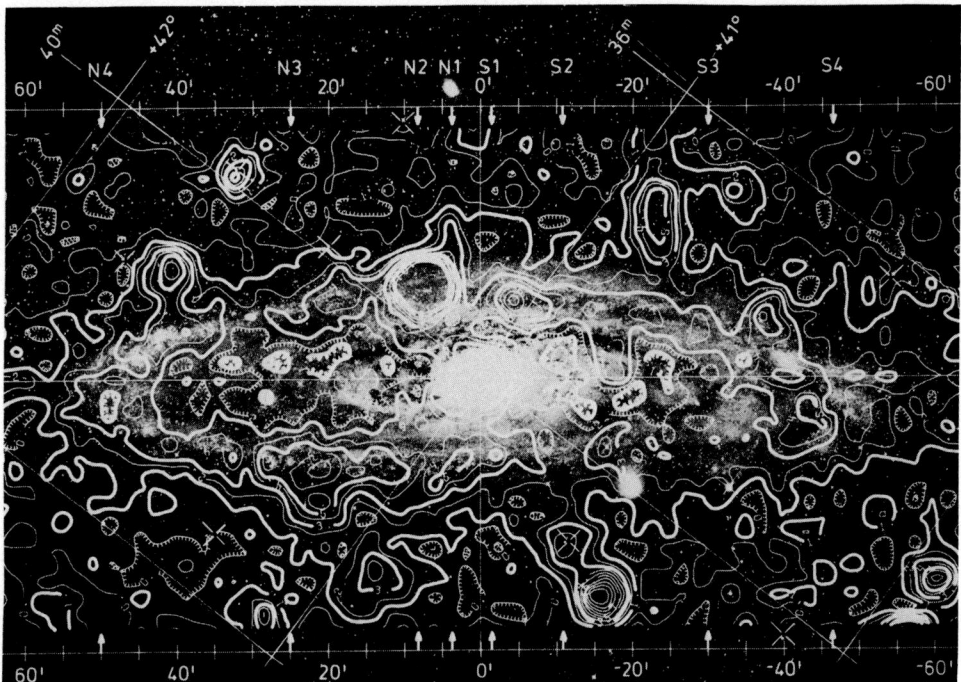

Fig. 1. Contour map of M31 at 11 cm including unrelated point sources (contour interval = 15 mK in T_b) superposed onto a Lick photograph. The scale along the minor axis is the same as that along the major axis.

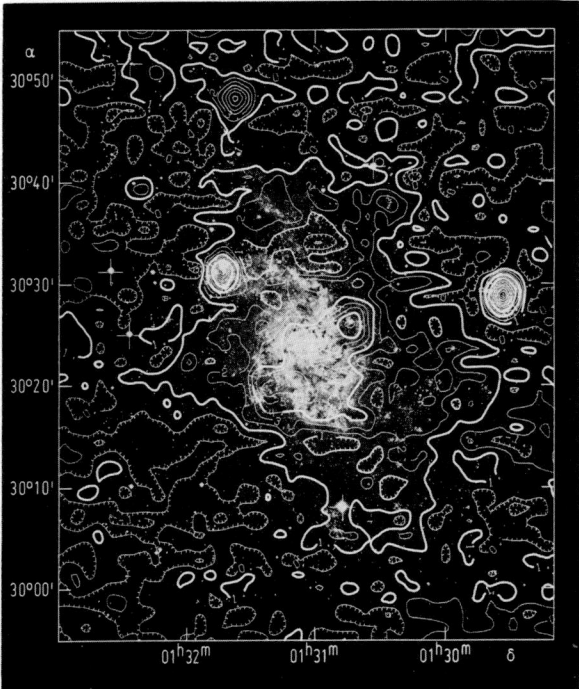

Fig. 2. Contour map of M33 at 6 cm superposed onto a Lick picture (contour interval = 8 mK in T_b). A plus indicates the position of the optical center (von Kap-herr et al. 1978).

beamwidth at 21 cm does indeed show a steepening of the spectrum towards the outer regions of M31. The preliminary results indicate a striking difference between the northern and the southern half of M31: in the north the spectrum is nearly constant, while in the south it steepens strongly towards the outer regions. We are presently investigating whether this difference might be related to asymmetries between north and south in other constituents that are known to exist in M31 (Paper II).

3. M33

The 6-cm map of M33 of von Kap-herr et al. (1978) superimposed onto a Lick photograph is shown in Figure 2. Contour 1 (= 8 mK in T_b) encloses an area of extended radio emission roughly centered on the galaxy and partly coinciding with optical features. The strongest emission inside contour 1 comes from the HII region complexes NGC 604 and NGC 595, and from some of the optically brightest parts near the center.

The catalogue of HII regions in M33 of Boulesteix et al. (1974) gives emission measure and size for each region, from which the expected flux at 6 cm can be derived. The total flux of these optically detected regions appears to contribute 26% of the integrated flux at 6 cm within 40' from the center in the plane of M33. The observed diffuse emission of emission measure $E \simeq 50$ cm^{-6} pc within 20' from the center (Boulesteix, Laval 1975) adds 32% of thermal radiation. Hence, within 40' from the center the thermal fraction of the integrated flux at 6 cm is $\gtrsim 0.58$; this is a lower limit since absorption effects, missed HII regions and diffuse emission weaker than observed are not taken into account.

The thermal fraction as a function of distance to the center $f_{th}(R)$ was studied by comparing the expected thermal emission in the Effelsberg beam at 6 cm with the observed radio emission, both averaged in rings of 2' width around the center. It was assumed that the thermal fraction at the center ≤ 1.0, and that the total thermal emission has the same radial distribution as the detected HII regions. Figure 3a shows that $f_{th}(R)$ generally decreases with increasing R. In this case $f_{th}(R < 40') = 0.68$ for the integrated flux, only slightly larger than the lower limit of 0.58, suggesting that absorption and other effects are not very large. With $f_{th}(R < 40') = 0.68$ and a spectral index of the integrated emission $\alpha = 0.52$ (von Kap-herr et al. 1978) the spectral index of the nonthermal emission $\alpha_n = 0.85$ is found. If α_n is constant for $R < 40'$ the expected variation of the total spectral index $\alpha(R)$ can be derived from $f_{th}(R)$ (Figure 3a); we eventually hope to check $\alpha(R)$ by comparing with observations at other frequencies.

Knowing $f_{th}(R)$ the thermal emission T_{th} and nonthermal emission T_n can be separated (Figure 3b). The radial distribution of T_n is significantly flatter than that of T_{th}; this suggests that either the objects X that produce the relativistic electrons have a distribution quite different from that of the HII regions or, if the objects X are

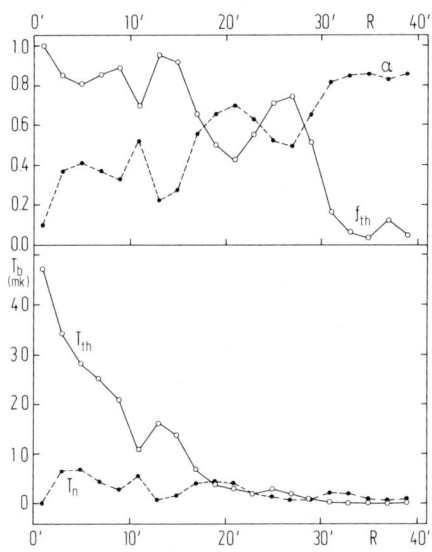

Fig. 3. Radial distribution in M33 of a (top): thermal fraction f_{th} and expected total spectral index α; b (bottom): thermal emission T_{th} and non-thermal emission T_n.

distributed like the HII regions, strong diffusion of relativistic electrons in the disk outwards is taking place. A similar conclusion was reached by van der Kruit (this volume) for M51.

4. RADIAL DISTRIBUTIONS

The radial distributions of various constituents in M31 and M33 are shown in Figures 4 and 5, respectively. M31 is an Sb-galaxy with both a significant nuclear bulge as well as a disk component. Both components are visible in the curves of the radio continuum, blue light and total mass. The disk component has its maximum at the bright ring at $R \simeq 9$ kpc as is evident from the curves of HII regions, OB associations and HI. The surface brightness (or surface density, depending on which constituent is considered) of the disk component decreases exponentially from $R \simeq 9$ kpc outwards (see Paper II for details).

M33 is of Hubble type Sbc and consists mainly of a disk component which has its maximum near the center. For $R = 0$ to $R \simeq 7$ kpc the surface brightness or surface density of total mass, blue light, radio continuum and HII regions decreases exponentially, while that of HI remains constant. Beyond $R \simeq 7$ kpc the surface density of HI and of HII regions drops suddenly. For comparison a distribution of supernovae is shown. Note, however, that this curve does not represent the specific distribution of SN in M31 or M33, but a sum distribution of some 150 galaxies; the 2 values closest to the center are unreliable.

If the surface brightness or surface density decreases as $\sigma(R) = \sigma(0)\, e^{-R/L}$ the scale length L normalised with the Holmberg radius, L/R_{Ho}, can be derived for the disk component. In both galaxies the exponential part of the disk component (i.e. $8 < R \lesssim 20$ kpc in M31, and $0 < R < 7$ kpc in M33) in blue light yields $L/R_{Ho} = 0.25 \pm 0.01$. To within the errors the other constituents have this same scale length apart from HI in M33, and the HII regions (0.13 ± 0.01) and OB associations (0.17 ± 0.01) in M31. These observations may be a challenge to galaxy-model makers.

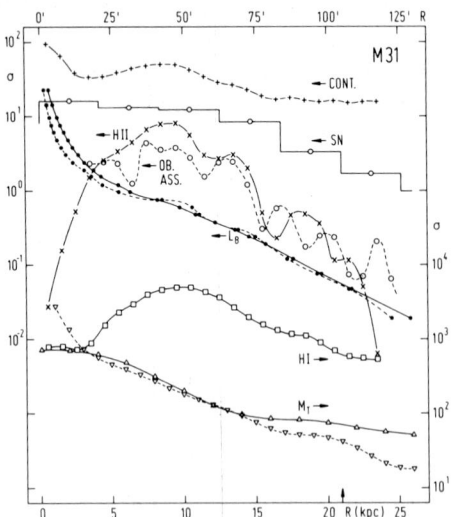

Fig. 4. Radial distribution of various constituents in M31:
Cont - 11-cm radio continuum (Berkhuijsen 1977), SN - supernovae (Iye, Kodaira 1975), HII - HII regions (Baade, Arp 1964), OB.ASS. - OB associations (van den Bergh 1964; Richter 1971), L_B - blue light (de Vaucouleurs 1958), HI - neutral hydrogen (Emerson 1974), M_T - total mass (Δ Roberts, Whitehurst 1975; ∇ Emerson 1976).

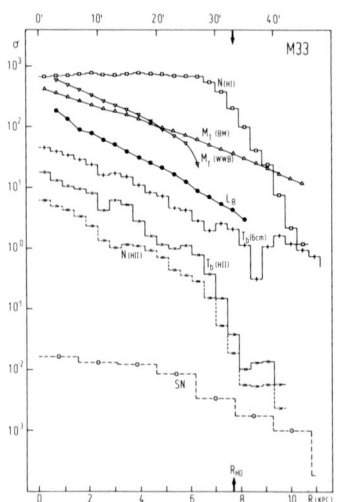

Fig. 5. Radial distribution of various constituents in M33:
N(HI) - neutral hydrogen (Rogstad et al. 1976), M_T - total mass (Δ Boulesteix, Monnet 1970; ∇ Warner et al. 1973), L_B - blue light (de Vaucouleurs 1959), T_b (6 cm) - 6-cm radio continuum (von Kap-herr et al. 1978), T_b (HII) and N(HII) - brightness temperature and surface density of individual HII regions only (Boulesteix et al. 1974), SN - supernovae (Iye, Kodaira 1975).

5. CONCLUSIONS

Both in M31 as well as in M33 a large-scale similarity in the distribution of the various constituents appears to exist. In both galaxies the surface brightness of the disk component decreases exponentially from its maximum outwards; the normalised scale length of blue light and of several other constituents is 0.25.

The origin of the radio continuum radiation is still unclear. In M33 at least 60% of the integrated flux for R < 40' is thermal. The radial distribution of the nonthermal emission is much flatter than that of the thermal emission and of supernovae suggesting that supernovae are not the only sources of relativistic electrons. In M31 the situation is similar. The minimum in the radial distribution of the continuum radiation at R ≃ 4 kpc in this galaxy indicates that diffusion of relativistic electrons from the nucleus into the spiral arms cannot significantly contribute to the emission from the arms.

Finally: much remains to be observed! For instance, in the radio continuum the distribution of polarised radiation at various wavelengths could tell us more about the nonthermal radiation and magnetic fields; the distribution of the thermal emission could be derived from observations at $\lambda \simeq 2$ cm. Combination of the latter data with the distributions of the total flux (i.e. integrated along the line of sight) in $H\alpha$ and/or $H\beta$ would give information on the distribution of diffuse HII emission and of optical absorption; data on the distribution of young stars and of the surface brightness in blue light (and other colours) could be improved; the distribution of optical polarisation is still unknown. The proximity to M31 and M33 makes these observations — although time consuming — extremely interesting.

REFERENCES

Baade, W.: 1958, Ric. Astron. Specula. Astron. Vatic. 5, 1.
Baade, W., Arp, H.C.: 1964, Astrophys. J. 139, 1027.
Beck, R., Berkhuijsen, E.M., Baker, J.R., Wielebinski, R.: 1978, in preparation.
Bergh, S. van den: 1964, Astrophys. J. Suppl. Series 86, 65.
Berkhuijsen, E.M.: 1977, Astron. Astrophys. 57, 9.
Berkhuijsen, E.M., Wielebinski, R.: 1974, Astron. Astrophys. 34, 173.
Boulesteix, J., Courtès, G., Laval, A., Monnet, G., Petit, H.: 1974, Astron. Astrophys. 37, 33.
Boulesteix, J., Laval, A.: 1975, private communication.
Boulesteix, J., Monnet, G.: 1970, Astron. Astrophys. 9, 350.
Emerson, D.T.: 1974, Monthly Notices Roy. Astron. Soc. 169, 607.
Emerson, D.T.: 1976, Monthly Notices Roy. Astron. Soc. 176, 321.
Israel, F., Kruit, P.C. van der: 1974, Astron. Astrophys. 32, 363.
Iye, M., Kodaira, K.: 1975, Publ. Astron. Soc. Japan 27, 411.
Kap-herr, A. von, Berkhuijsen, E.M., Wielebinski, R.: 1978, Astron. Astrophys. 62, 51.
Kruit, P.C, van der, Allen, R.J.: 1976, Ann. Rev. Astron. Astrophys., p. 417.
Mouschovias, T.Ch., Shu, F.H., Woodward, P.R.: 1974, Astron. Astrophys. 33, 73.
Richter, G.A.: 1971, Astron. Nachr. 292, 275.
Roberts, M.S., Whitehurst, R.N.: 1975, Astrophys. J. 201, 327.
Rogstad, D.H., Wright, M.C.H., Lockhart, I.A.: 1976, Astrophys. J. 204, 703.
Vaucouleurs, G. de: 1958, Astrophys. J. 128, 465.
Vaucouleurs, G. de: 1959, Astrophys. J. 130, 728.
Warner, P.J., Wright, M.C.H., Baldwin, J.E.: 1973, Monthly Notices Roy. Astron. Soc. 163, 163.

DISCUSSION FOLLOWING PAPER III.1 GIVEN BY E.M. BERKHUIJSEN

VAN DER KRUIT: How sensitive is your estimate that in M33 about 60% of the flux density at 6.2 cm is thermal to an assumed optical absorption in the foreground, to absorption internal to the HII regions and to variations of absorption across the disk?

BERKHUIJSEN: Optical absorption did not enter my estimate. I find that 58% of the integrated flux density of M33 is thermal by just adding the flux densities of the optically detected HII regions and of the detected diffuse emission. So 58% is a lower limit. However, this lower limit does also depend on the total integrated flux density of M33 which is uncertain by a factor of 0.2. Therefore the lower limit for the thermal fraction is 0.58 ± 0.12.

VAN WOERDEN: How many hours of observation are required to map M31 at 6 cm?

BERKHUIJSEN: With the present system about 200 hours under perfect weather conditions.

BURKE: How complete is the correspondence between the radio maxima of the smooth component at 11 cm and the optically dark regions in M31? Is the correspondence significant or chance?

BERKHUIJSEN: There is not a one-to-one correspondence but I don't think it is chance. On the edges of the dust lanes most of the HII regions are found, and I have derived an empirical relation for the number of HII regions in the radio beam and the brightness temperature at 11 cm. This relation takes the form $N_{HII} \propto T_b^{2.9 \pm 0.6}$. Furthermore, the peak lines in the HI distribution derived by Emerson, with an angular resolution of about 2', closely follow the dust lanes. After smoothing of his HI distribution to the 11-cm beamwidth of 4.8, in the bright ring the HI peak lines coincide very well with the peak lines in the distribution of the emission at 11 cm [see also Bystedt's comment below].

BECK: VARIATION OF SPECTRAL INDEX ACROSS M31
    Using the 11-cm map of Berkhuijsen and Wielebinski (1974, A.A. 34, 173) and our new 21-cm map (Beck, Berkhuijsen, Baker, Wielebinski) I have calculated the distribution of the spectral index across M31 after subtraction of point sources and careful correction of the baselevels of the maps. There is a large-scale asymmetry between the northern and the southern half of M31 out to 15 kpc from the center. The average temperature spectral index β in the north is 0.36 ± 0.05 lower than in the south ($T_b \propto \nu^{-\beta}$). Baseline errors cannot explain this effect. From data averaged in rings an increase of the spectral index with radius is evident in the southern half of M31, but the spectral index is constant within the errors in the northern half.

BALDWIN: How large were the spectral variations in M31 before making serious corrections to the baselevel, i.e. we need to know how serious is serious.

BECK: Before corrections to the baselevels were made the spectral variations were even larger than I have shown.

BERKHUIJSEN: In other words: we cannot explain the steepening of the spectrum with the errors we can think of.

DAVIES: Can the difference in spectral index between the north and south of M31 be explained quantitatively by the known asymmetry of HII regions?

BERKHUIJSEN: For M31 we only know the number-density distribution of HII regions, not the distribution of emission in Hα. Therefore, we cannot yet calculate the variation of the thermal fraction across the disk which is the relevant parameter that influences the spectral index.

VAN DER KRUIT: It must be realized that M31 and M33 have weak disks in the radio emission compared to other galaxies; thermal emission is definitely going to have a very strong effect.

BYSTEDT: M31 AT 49 CM WAVELENGTH

Observations at 610 MHz with the Westerbork telescope (Israel, de Bruyn and Bystedt) of M31 clearly show the bright continuum ring associated with the strongest optical arm. There is a strong asymmetry between the NW arm which is narrow and closely follows a dust lane, and the SE arm which is much broader and not clearly related to any detailed optical features. The highest peaks in both arms are of about the same strength.

There also seems to be emission extending from the nuclear region to the ring approximately in the minor axis direction.

ALLEN: Can you be sure that the connection between the arms and the nucleus in the radio maps is not merely a projection effect?

BYSTEDT: In no other place do we see emission inside the dark ring, but a projection effect cannot be ruled out.

VELOCITY DISPERSION IN THE BULGE OF M 31 ; DYNAMICAL MODEL

G. MONNET, A. PELLET and F. SIMIEN
Observatoires de Marseille et de Lyon

I - BASIC DATA

1) Photometry from de Vaucouleurs 1958 shows a bulge obeying the $r^{1/4}$ law up to less than 5 pc from the center with an effective radius r_e = 17'5. In the region from 0.01 to 0.2 r_e, equal luminosity curves are well approximated by similar ellipses of axial ratio 0.68. The reduced spatial density ν^+ and the reduced gravitationnal potential ϕ^+ (i.e. for a mass to luminosity ratio f equal to 1) can then be easily computed (Monnet and Simien 1977).

2) To compute the contribution of the nucleus to the overall gravitationnal potential we use the mass $4.5\ 10^7\ M_\odot$ found by Light et al. 1974.

3) Pellet 1976 has made spectrographic observations of stellar absorption lines in the range 4200 - 4400 Å at 1.3 Å resolution, and derived the mean rotation velocity integrated along the line of sight $< \Theta(x) >$ on the major axis. The tilt angle of M 31 is sufficiently small to permit the use of Bertola and Capaccioli 1975 formula for the computation of the mean rotationnal velocity of the stars Θ_m in the equatorial plane. It is quite linear from 0.01 to 0.2 r_e with a slope of 350 km s^{-1} in units of r_e.

4) The same observations, both on the major and the minor axis, are used to obtain the dispersion of radial velocities σ_v. A constant value of 140 ($\pm$ 20) km s^{-1} is found in the interval 0.01 - 0.1 r_e, slightly higher than Morton et al. 1977 value (110 km s^{-1}) and using the same reduction process (fitting to an enlarged K_0 III spectrum).

II - MODEL

We suppose that the galaxy is stationnary, axisymetric, and that there is no third integral. We also assume - as a working hypothesis - that the M/L ratio f (in the B band) is constant. The Jean's equations in the adimensionnal cylindrical coordinates r, θ, z can then be written :

(1) $\frac{\partial}{\partial z}(\nu^* \sigma_r^2) = f \nu^* \partial \phi^* / \partial z$

(2) $\frac{\partial}{\partial r}(\nu^* \sigma_r^2) + \frac{\nu^*}{r}(\sigma_r^2 - \sigma_\theta^2) = \nu^* \frac{\Theta_m^2}{r} + f \nu^* \partial \phi^* / \partial r$

where the σs' are the r.m.s. dispersions of the velocity components.
The gradients of the reduced potential ϕ^* are computed from ν^* (§ I, 1), using Schmidt 1956 formulae. We add the contribution of the nucleus (§ I, 2). The contribution of the disk is negligible. Equation (1) - with the limiting condition $\sigma_r(r, \infty) \equiv 0$ - gives then $\sigma_r(r, z)/f$ over all space (figure 1). The dispersion of the radial velocity integrated along the line of sight on the minor axis is :

$\sigma_v^2(0, z) = \int \sigma_r^2(u, z) \nu^*(u) du / \int \nu^*(u) du$

Its computed values from 0.01 r_e to 0.2 r_e are shown in figure 1. The best fit to the experimental range (§ I, 4) occurs for f = 8 (± 2), which is compatible with the gas velocities on the North-East side.

Next we determine from equation (2) $\sigma_\theta(r, 0)$. Since, from 0.01 to 0.2 r_e log $\nu^* \simeq -7.32 (r^{1/6} - 1)$ within 7 %, it can be written :

$\sigma_\theta^2 = \sigma_r^2 (1 - 2.81 r^{1/6}) + r \partial \sigma_r^2 / \partial r - \Theta_m^2 - f \partial \phi^* / \partial r$

$\sigma_\theta(r, 0)$ computed for f = 8 is shown in figure 2. $\sigma_\theta(r, 0)$ and $\sigma_r(r, 0)$ then give the dispersion of the radial velocity integrated along the line of sight on the major axis : $\sigma_v(r, 0)$. σ_v is given in figure 2.

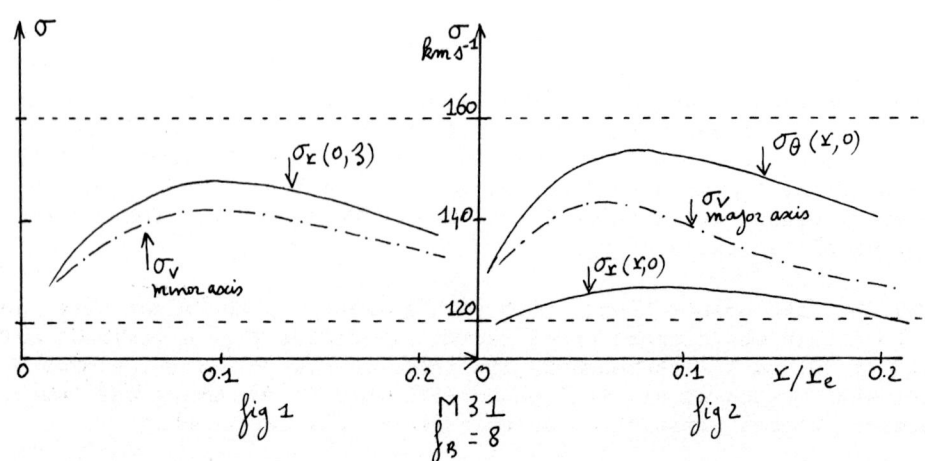

Figures 1 and 2 :
σ_θ and the σ_r s' are given for a pure $r^{1/4}$ bulge law.
The σ_v s' include the nuclear contribution.

The good agreement with the experimental range confirms the M/L ratio adopted from the minor axis data. Correcting for a B Galactic absorption of 0.44 mag (Heidmann et al. 1961), the value is :

$$(M/L)_B = 5.3 \pm 1.3 \quad \text{for} \quad r < 0.2\, r_e \;.$$

REFERENCES

Bertola, F., Capaccioli, M.: 1975, Astrophys. J. 200, 439.
Heidmann, J., Heidmann, N., Vaucouleurs, G. de: 1971, Mem. Roy. Astron. Soc. 75, 85.
Light, E.S., Danielson, R.E., Schwarzschild, M.: 1974, Astrophys. J. 194, 257.
Monnet, G., Simien, F.: 1977, Astron. Astrophys. 56, 173.
Morton, D.C., Andereck, C.D., Bernard, D.A.: 1977, Astrophys. J. 212, 13.
Pellet, A.: 1976, Astron. Astrophys. 50, 421.
Schmidt, M.: 1956, Bull. Astr. Inst. Netherlands 13, 15.
Vaucouleurs, G. de: 1958, Astrophys. J. 128, 465.

THE SPIRAL STRUCTURE OF M 31

E. Athanassoula
E.S.O., Geneva, Switzerland
and Observatoire de Lyon

RELEVANT OBSERVATIONAL MATERIAL

- The H I, H II, and OB-association densities, as a function of radius, have the same structure, reminding one of a ring with a maximum at $r \simeq 50'$.

- There is an H I excess on the NE side of the galaxy.

- From H I measurements along the major axis there is an evident asymmetry between the N and S sides. The positions of the maxima on one side correspond to the minima on the other, and vice versa.

- The rotation curve has a steep rise near the centre, then remains quite flat.

- There are systematic differences between the rotation curves of the two semi-major axes: for radii smaller than $r = r_1$, SP rotates faster than NF. At $r = r_1$ the two rotation curves cross, and for $r > r_1$ the values for the NF side are larger than the corresponding ones for SP. The values of r_1, as given by various authors, vary between $40'$ and $60'$. Similarly, an asymmetrical behaviour is found for other opposite semi-diameters of the galaxy.

- Kalnajs (1975) Fourier-analysed i) the H II regions given by Baade and Arp (1964); ii) the 106 brightest of the above regions; iii) the OB-associations given by van den Bergh (1964); iv) the H I surface brightness distribution observed by Emerson (1974). In all cases he found a one-armed leading spiral. The corresponding Fourier transform $A_1(a)$ of the function $r^2\sigma$ has a maximum at $a = -11$ (pitch angle $\simeq 5°$). In the case of the H I he also found a dominant peak at $a = 0$, expressing the H I excess on the NE side of the galaxy.

THEORETICAL APPROACH

We will try to reproduce the observational data by assuming that M 31 is forced by an orbiting retrograde companion galaxy. Our model has been described previously (Athanassoula 1977, hereafter paper I). M 31 is supposed to be a disk of zero thickness, and no z-motions are considered. The companion's orbit is assumed to be on the plane of the disk and its parameters are taken such as to mimic M 32. Our model is linear, self-consistent, non-local, and non-asymptotic (the spiral can be open). Both stars and gas are taken into account.

The orbit of M 32 gives rise to an inner Lindblad resonance. The existence of such a resonance is important since the mass of M 32 is only a few percent of that of M 31 and yet the amplitude of the gaseous spiral reaches $\gtrsim$ 30% of the axisymmetric gaseous background. This can be achieved only with the help of a resonance. For a retrograde orbit of the companion, no m = 2 resonance is possible.

We want to stress here that our model cannot describe the full complexity of the causes of the spiral structure in M 31. Non-circularity of the companion's orbit, the effect of other companions, e.g. NGC 205, and especially motions in the z-direction can modify the picture.

The spirals in the gas and in the stars are found as the solutions of two coupled integral equations. The method and the parameters and symbols used can be found in paper I. To simulate the effect of the observed variation of the gaseous axisymmetric density with the radius, we have multiplied the gaseous density by the radial distribution of H I, taken from Emerson's data. This weighting affects the spiral only slightly, attenuating low-level extensions in the inner parts and accentuating low-level noise in the outer parts.

The force field from the stellar spiral is a significant fraction of the total tidal field felt by the gas. It is thus possible to achieve a good approximation of the form and orientation of the observed spiral (e.g. Fig. 1), which is impossible if the stars and self-consistency of the gas are neglected (Kalnajs 1965). The pattern speed is Ω_p = -10.5 km sec^{-1} kpc^{-1}.

The perturbation in the stars forms a much more open spiral (a = -3) with its maximum well within resonance. The perturbation contrast in the stars is roughly 0.1 of that seen in the gas. This can explain why the spiral structure is not evident in the isophotes.

There is angular momentum exchange between the spiral wave and M 32. The wave loses positive angular momentum while M 32 gains it, and since it is in a retrograde orbit it spirals inwards. Its eccentricity grows in time.

REFERENCES

Athanassoula, E. 1977, Preprint
Baade, W. and Arp, H. 1964, Astrophys. J. 139, 1027
Bergh, S. van den 1964, Astrophys. J. Suppl. 9, 65
Emerson, D.T. 1976, Monthly Notices Roy. Astron. Soc. 176, 321
Kalnajs, A.J. 1975, in Weliachew (ed.), La Dynamique des Galaxies Spirales, 103

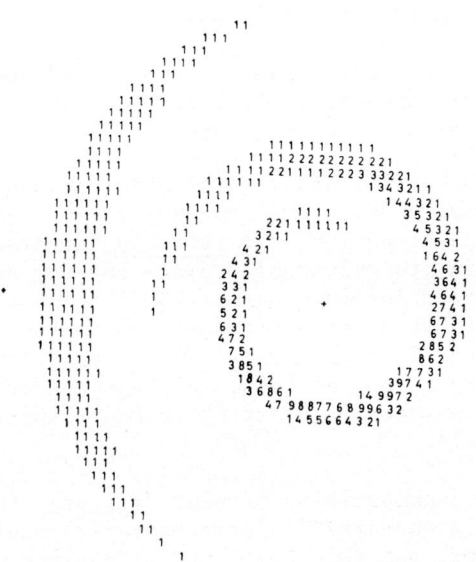

Figure 1. The gaseous spiral corresponding to the parameters: $r_R = 50'$, $u_R = -0.5$, $\Gamma = \sqrt{2}$, $\varepsilon = 0.1$, $d_0 = 0.3$, $b_s = 0.1$, $b_g = 0.005$, $C_1 = 0.01$. (see paper I for the definition of symbols).

DISCUSSION FOLLOWING PAPER III.3 GIVEN BY E. ATHANASSOULA

OORT: Baade found about six arms crossing each side of the major axis. This is a rather different picture than that of the one-arm spiral you derived.

VAN WOERDEN: Baade found six arms crossing with each half of the major axis, not "six arms". With a small pitch angle, one can obtain many axis crossings with one arm.

COURTÈS: It is obvious that a small pitch angle gives a fundamental ambiguity but it is strange to see that the one arm leading model fits better with the HII region distribution (Courtès, Mancherat, Monnet, Pellet and Simien, presented to A.A.).

ATHANASSOULA: The two-armed trailing pattern with a $7°$ pitch angle of Baade and Arp gives a reasonable fit only near the major axis. The corresponding fit obtained with a $5°$ one-armed spiral is much better.

VAN DEN BERGH: Inspection of the optical spiral arm tracers shows that they form a number of rather short arcs. Some imagination is required to fit these arcs into an "Andromeda-wide" one or two arm spiral pattern. Interpretation of the data is complicated even more by the fact that the fundamental plane of M31 is warped.

ATHANASSOULA: The optical picture is quite complicated due, mainly, to the superposition of many components. It becomes, however, much clearer after a Fourier series decomposition.

MARK: It seems that such Fourier analysis of observations is difficult to do properly and especially so for M31 because of its inclination to us. Take the hypothetical situation of a strictly two-armed spiral which is not sinusoidal in its azimuthal profile but say narrower near the peaks due to young stars forming there. Fourier analysis of this two-armed structure will give contributions to the Fourier spectrum at multiarms and with a rather flat spectrum. Extra care should be taken especially when the galaxy is so inclined. Perhaps the procedure should first be applied to a more face-on galaxy whose major spiral structure is more directly observable.

ATHANASSOULA: A Fourier decomposition of your "strictly two-armed" example (or of a "strictly one-armed" similar example) would give contributions at multiarms, but this would not affect the $m = 1$ component we are studying here. As regards the application of the Fourier decomposition procedure, it has already been applied to M51.

VAN WOERDEN: Does not the diagram for HI peaks at a pitch angle zero suggest that the gas is predominantly arranged in circular rings?

ATHANASSOULA: No. There is a peak, at $a = 0$, of the $m = 1$ Fourier component ($A_1(a)$) obtained from the HI distribution (small values of a

correspond to open spirals and large values to tightly wound ones). This peak at a = 0 comes from the HI excess on the NE part of the galaxy, which can be considered as a "one-armed bar".

COURTÈS: The interpretation in multiarms corresponds to the first HII region survey of Baade and Arp. The situation is a little clearer now after the study of associations of OB stars of van den Bergh (1964, Ap.J. Suppl. 9, 65) and our deeper HII region survey (in press) which show some evidence of continuity. I agree that anyway the structure of M31 remains very ambiguous.

ATHANASSOULA: The logarithmic spiral obtained from a fit of the deeper Hα survey you mentioned agrees completely with that presented here.

"I don't take this model seriously. I think it is a convenient way to remember the numbers."

                      W.W. Shane in Discussion III.5

THE KINEMATICS WITHIN M31

Morton S. Roberts, Robert N. Whitehurst,[*] and Tom R. Cram
National Radio Astronomy Observatory[†]
Green Bank, West Virginia

Abstract   A new hydrogen line survey of M31 is described. A rotation curve of the galaxy is derived from these data. The northern half of the rotation curve can be traced to 27 kpc, the southern to 30 kpc. They are in general agreement; both sides show an extensive region of essentially constant rotational velocity.

A high sensitivity 21-cm survey of M31 was made with the Effelsberg 100-m telescope over a region $\sim 5° \times 1.5°$. Velocity profiles were obtained at approximately 1400 positions spaced at half-beam (4!5) intervals. These data have been prepared in various formats for publication as a 21-cm catalogue of M31 (Cram, Roberts, and Whitehurst, in preparation) and include profiles, ℓ-v contour diagrams, and grey-scale displays. All of the data emphasize the inadequacy of a thin, flat disk model for M31. This is illustrated in Fig. 1, an ℓ-v contour diagram for a line parallel to the major axis but displaced 18' to the SE. The bifurcation is due to double-peaked profiles, one velocity component is due to planar HI, the other to HI extending out of the plane at a different radius and hence a different observed radial velocity. A single line of sight senses both HI features. The lower, less intense feature (at $-\lambda$ values) is consistent with the rotation curve derived from major axis data.

Detailed modeling of the spatial (x,y,z) distribution of HI in M31 is discussed in the following paper. Here we derive the rotation curve for M31. The major axis is the only region for which this can be done unambiguously. There are two reasons for this: (1) radial motions (expansion or contraction) are not sensed on the major axis, the

---

[*] On leave from the Department of Physics and Astronomy, the University of Alabama.
[†] Operated by Associated Universities, Inc., under contract with the National Science Foundation.

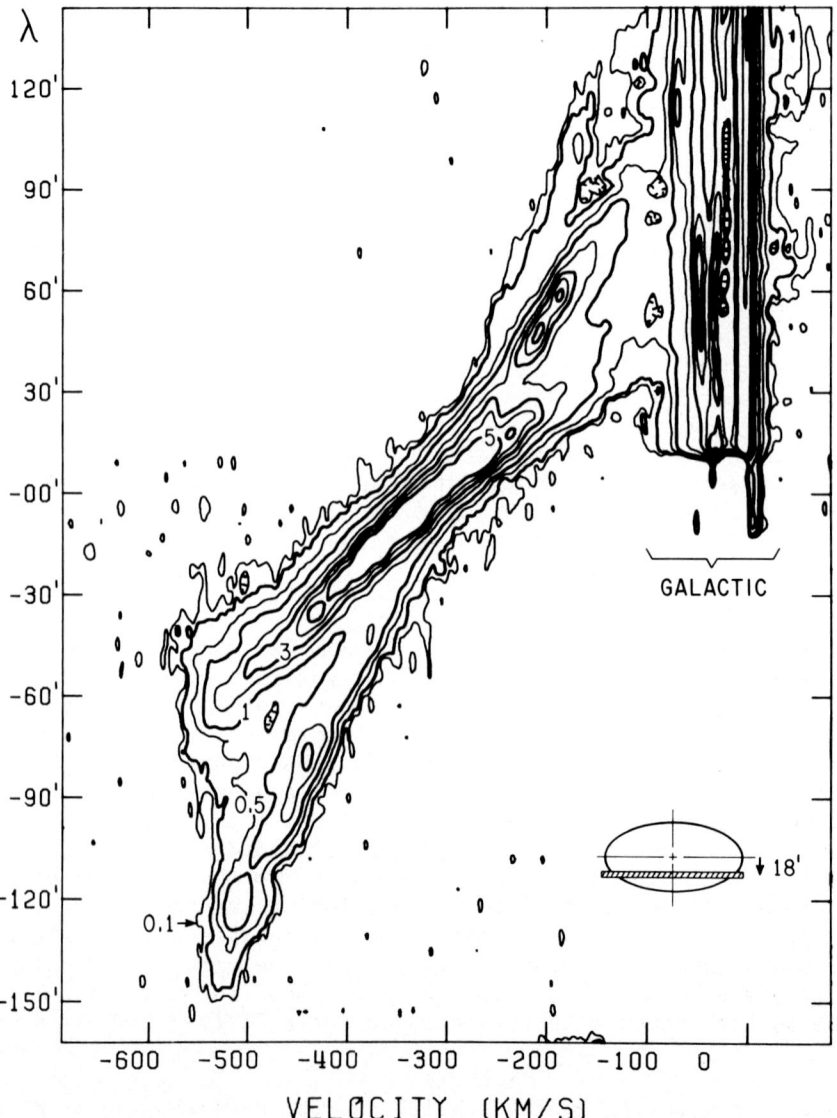

Fig. 1 - An ℓ-v diagram for a line parallel to the major axis (λ axis) displaced 18' to the SE, see the insert in the lower right. A thin disk model predicts a single diagonal feature rather than the bifurcation shown. Contour levels are from 0.1 to 5 K antenna temperature.

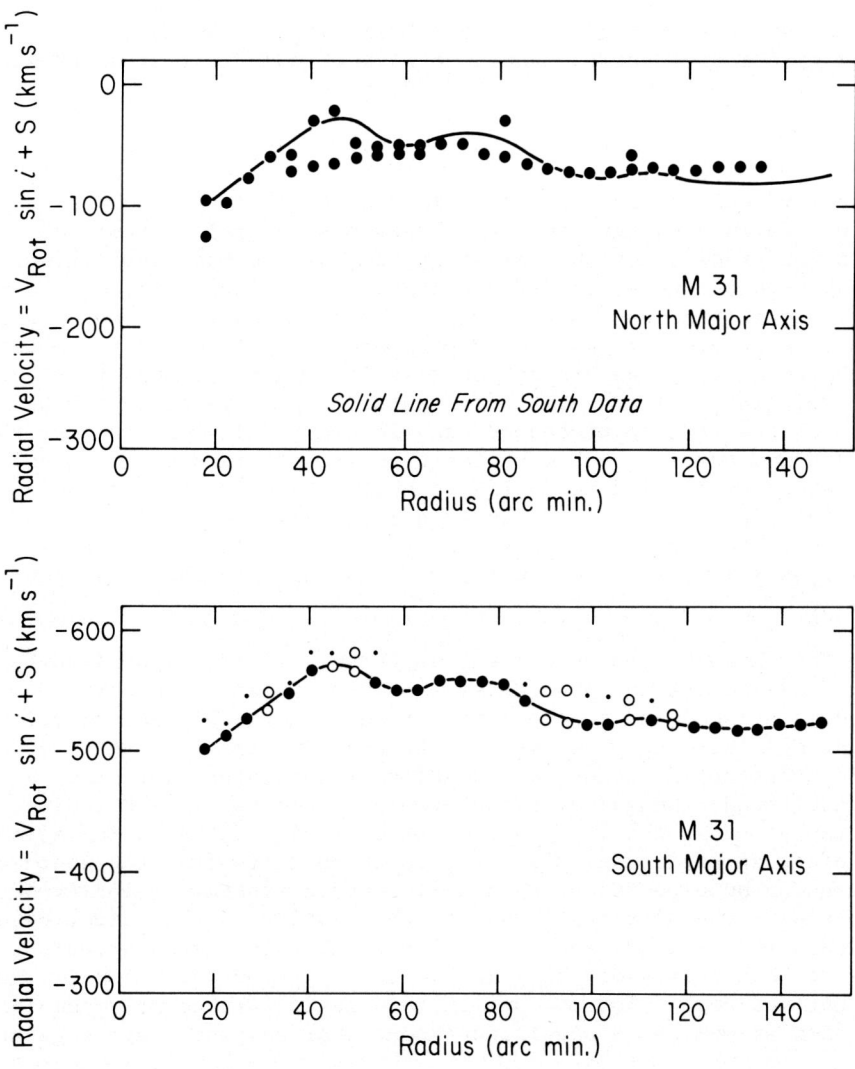

Fig. 2 - Rotation curve for M31 derived from major axis data. The upper panel is for the northern half of the Galaxy, the lower panel is for the southern half. The ordinate is observed radial velocity. At a distance of 690 kpc, 10' corresponds to 2 kpc. See the text for an explanation of the different symbols. The smoothly varying line is taken as the rotation curve of M31 and has been used for all modeling analysis.

projection yields Doppler velocities given by $E(R) \sin i \sin \theta$, where $E(R)$ is the radial term and θ is zero on the major axis. And (2), even though line of sight confusion may be present, features with radial velocities greater than the rotational velocity on the major axis, $V_{rot} \sin i$, cannot occur in a pure rotational model; or even one with moderate radial motions, except at large θ.

Thus, for complex profiles on the major axis, which is generally the case for $R \lesssim 100'$, the problem becomes one of identifying the highest velocity feature present. This is accomplished by decomposing complex profiles into enough Gaussian components to yield a residual noise level in the profile region similar to that of the baseline region. Generally 4 to 6 components are necessary. The highest velocity component is identified as ($V_{rot} \sin i$). The area under each component, i.e., the integrated brightness temperature, is not considered unless the highest velocity feature has an area less than a tenth of the next highest velocity feature. In these cases the peak velocity may be an artifact of the fitting procedures and is ignored in constructing the final rotation curve. Such weak features are retained in the display and are shown as small dots in Fig. 2 while the much stronger and next highest velocity feature is shown as a larger filled circle at the same radius. When two features are shown of similar velocity but differing by less than a factor of ten in their areas both are shown as open circles.

The northern half of the rotation curve, Fig. 2, is derived by a similar Gaussian-fitting procedure. But here Galactic foreground hydrogen is present and the approach of seeking the highest velocity component does not apply as easily. By studying the continuity of Galactic hydrogen in various ℓ-v displays a consistent set of data points are found--though, at times, two possible values are indicated. All values of ($V_{rot} \sin i$, R) are plotted in Fig. 2, including alternate choices at particular R's. The rotation curve from the southern half is shown superposed (using a heliocentric systemic velocity of -300 km s^{-1}). The agreement, even to the ripples in the curve, between south and north is good, and we conclude that the same rotation curve applies to both halves of the galaxy. The rotation curve for $R < 20'$ is not derived here. The motions in this region are clearly complex; contraction as well as a significant nonplanar component are suggested.

DISCUSSION FOLLOWING PAPER III.4 GIVEN BY M.S. ROBERTS

VAN WOERDEN: A sequence of double-peaked profiles like that shown by Dr. Roberts may, in principle, also be caused by gas in an infinitesimally thin disk. Consider two circular filaments of radii $R_1 = 90'$ and $R_2 = 110'$, both having the same rotation speed V_c. With $\cos i = 0.2$, their separation will be less than 5' over the range $0' < x < 60'$ (x = coordinate along major axis), and both will be within the 9' beam for $y = 18'$ over this range of x. Observed velocities V_1, V_2 are given by $V_i = V_c \sin i \, x/R_i$; hence one beam will see two peaks whose separation is proportional to x. The velocity separation exceeds the smearing caused by the extent of the beam in x for x > 40'.

M.S. ROBERTS: The highly idealized model van Woerden proposes will yield double-peaked profiles. However this model fails completely when one looks at the quantitative data, i.e. velocity separation of the peaks (up to 100 km/s), spatial extent, and signal strength.

SHANE: High resolution Westerbork HI data which will be discussed in a later contribution [see Discussion III.5] support the interpretation as given by Roberts.

COURTÈS: The problem of superposition on the major axis is maybe not so severe on the Andromeda nebula because of the large hole in the HI distribution. The inner HI arm for example is narrow and very far from the center of M31.

M.S. ROBERTS: Even on the major axis there is a wide spread of velocities of well over 100 km/s. You cannot explain this in terms of beam effects. Everything is consistent with a thick model.

LANDECKER: HI SURVEY OF M31

A survey of neutral hydrogen in the whole of M31 has been made with the Synthesis Telescope at Penticton, Canada. The field of view of the instrument is 2° and the synthesized beam is 2' x 3'. Low order spacings have been supplied from 25-m telescope observations. Velocity resolution is 5 km/s. The noise on line maps is 2 K r.m.s. and on the continuum map 0.1 K. M31 has been covered in 4 regions centered at ±45' and ±125' from the optical center.

"The best of all people have shown us innumerable bends in galaxies. I'm not saying how that is done; all I do is bend the galaxy."

        M.S. Roberts in Discussion III.4

THE THREE-DIMENSIONAL DISTRIBUTION OF NEUTRAL HYDROGEN IN M31

Robert N. Whitehurst,* Morton S. Roberts, and Thomas R. Cram
National Radio Astronomy Observatory[†]
Green Bank, West Virginia

Given the wealth of data, the rotation curve, and the necessity for out-of-plane hydrogen demonstrated in the preceding paper, it seems desirable to attempt to establish a systematic procedure for determining the three-dimensional distribution of hydrogen. Under the assumption of cylindrical rotation this is, in principle, possible for most of the galaxy.

The standard equation for the radial velocity of a point in cylindrical rotation may be inverted to yield

$$(V_{rot} \sin i)/R = (V_R - V_S)/x,$$

where the x axis is chosen to coincide with the major axis of the projected ellipse and x is constant for a line of sight. With a known rotation curve, $(V_{rot} \sin i)/R$ as a function of R and its inverse $R((V_{rot} \sin i)/R)$, single valued over most of its range, can be determined. Thus, from the measured quantities V_R, V_S, and x we determine R. From R and x we find y, with an ambiguity in sign. From y and the sky coordinates (λ, β) of the line of sight it is easy to calculate z, the distance from the nominal plane of the galaxy. Thus a radial velocity observed in a line of sight at position on the sky (λ, β) can be associated with a point (actually one of two points) in the rotating system.

Figure 1(a) shows a series of profiles taken on M31 at constant λ or x, here 10 kpc from the minor axis toward the southwest. Note that there are at least three prominent velocity features and that some profiles are complex, indicating that the line of sight is intercepting more than one concentration. The velocities expected for an in-plane material are indicated by tick marks on the profiles. Figure 1(b) is

---
* On leave from the Department of Physics and Astronomy, the University of Alabama.
[†] Operated by Associated Universities, Inc., under contract with the National Science Foundation.

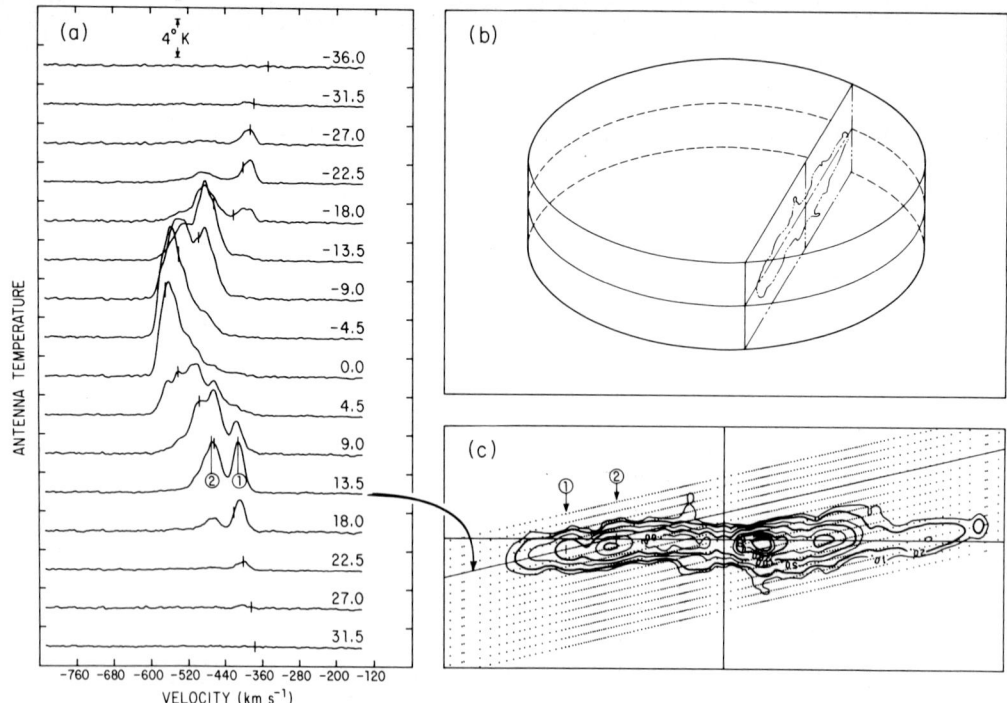

Figure 1(a). Profiles taken at $\lambda = -49°\!.5$. Tick marks on profiles indicate velocities expected of in-plane material.
    (b). Schematic representation of the plane sampled by the observations of (a).
    (c). Contour map of volume density of neutral hydrogen in the plane indicated in (b) from modeling the profiles of (a). See text for meaning of symbols.

a schematic representation of the geometry. The series of observations defines a plane perpendicular to the major axis. Each line of sight may intercept hydrogen at any point along its path through the galaxy, and each point will have its own radial velocity.

    The antenna temperature of a given velocity channel is proportional to the beam-averaged surface density of hydrogen atoms having velocities within that channel and thus, neglecting dispersion, within a fixed line-of-sight range. The surface density may, therefore, be converted to a volume density. Figure 1(c) is a contour map of the volume density in the plane indicated in (b) determined by a channel-by-channel modeling of the set of profiles of (a). The horizontal, or y, axis lies in the nominal plane of the galaxy parallel to the minor axis. The vertical, or z, axis is parallel to the axis of rotation. The negative (south preceding) major axis points through the origin out of the paper. The

III.5 THE THREE-DIMENSIONAL DISTRIBUTION OF NEUTRAL HYDROGEN IN M31

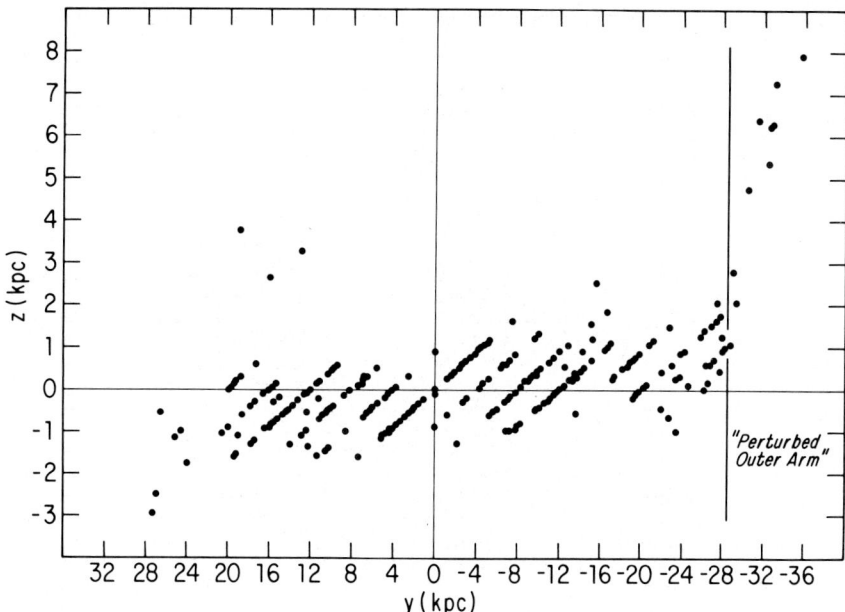

Figure 2. Points in galaxy associated with peaks in velocity profiles as viewed from direction of negative major axis.

dotted lines represent lines of sight, each dot at the point associated with a particular velocity channel.

Consider the line of sight at $\beta = 13°\!.5$ and the resulting profile. The association between hydrogen concentrations and peaks in the profiles is clear. Two such associations are identified by the circled numbers. The most obvious characteristic of this cross section is its thickness, here approximately doubled by the finite beam. Measurements made through hydrogen concentrations give a typical z extent to half density, corrected for beam size, of about $\pm$ .7 kpc. To a density of about one atom per liter, a typical z extent is about $\pm$ 2 kpc, and measurable hydrogen is sometimes found more than 5 kpc from the conventional plane.

We have applied the same techniques to the determination of points in the galaxy corresponding to peaks in the velocity profiles. Although density information is lost, these points should indicate locations of significant hydrogen densities. Figure 2 is a view of this array of points from the direction of the negative major axis. Again, the horizontal axis lies in the conventional plane. Several characteristics stand out. Again, hydrogen is found at considerable distance from the conventional plane. The "edge" is typically a kiloparsec from the plane. Secondly, the hydrogen appears to be inclined slightly to the conventional plane. It appears that an inclination of 79° is a better fit to the hydrogen distribution over much of the galaxy than is the 77° assumed here. Finally, there are a number of points which depart

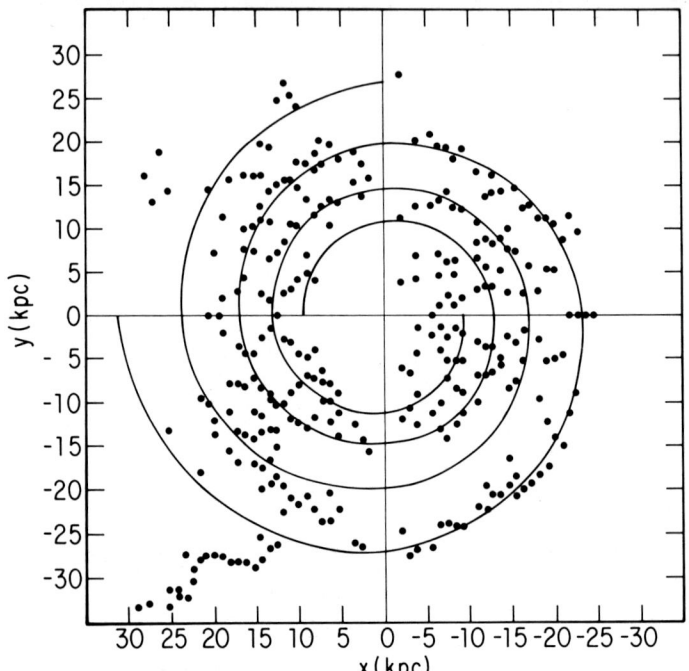

Figure 3. Face-on view of same array of points as in Figure 2. Spiral is R = 85 exp(0.0962θ).

rather violently from the plane. The three isolated points in the upper left are identified with isolated features on profiles and probably do not belong to Andromeda. At a velocity of about 115 km s^{-1} they may represent a high-velocity cloud. The series of points running up and to the right, dubbed the "perturbed outer arm", represents a feature which appears in each form of data representation as a smooth extension of normal features. For this reason, we feel that it probably is connected with M31 but probably does not follow the general rotation field.

Figure 3 presents a face-on view of the array of points. Some structure appears, particularly a feature which appears to be Baade's S7 arm and which may be traced here beyond the minor axis. The series of points which diverges from this is the "perturbed outer arm" group. We have attempted to fit a spiral to these points. The fit is at best only fair, with apparent bridges between arms. The spiral, strictly empirical, is somewhat tighter than that fitted by Arp to the HII regions, but his spiral is a poorer fit to our data.

III.5 THE THREE-DIMENSIONAL DISTRIBUTION OF NEUTRAL HYDROGEN IN M31

DISCUSSION FOLLOWING PAPER III.5 GIVEN BY R.N. WHITEHURST

GIOVANELLI: You mentioned that the "outer perturbed arm" could be anywhere along the line of sight; therefore it could be either foreground gas <u>escaping</u> from Andromeda or behind the disk and <u>falling</u> into it. Is that correct?

WHITEHURST: Yes, that is correct.

GIOVANELLI: In addition, would that change the location that you proposed for it in the (x,y) plane?

WHITEHURST: Yes, because there is an ambiguity in the sign of y.

VISSER: You said that the full thickness of the arm at half-intensity is approximately 1.4 kpc. Does the velocity dispersion of the gas support this?

WHITEHURST: I did not estimate the velocity dispersion from the observations. This model is a purely kinematic representation; it gives velocity dispersions of up to 10 km/s. A dynamicist might tell if this supports the number for the arm thickness.

TOOMRE: As Visser implied, surely it takes a vertical r.m.s. speed of 30 or 50 km/s (or its pressure equivalent) to thicken your inner gas layer to the 1.4 kpc that you estimate. And yet horizontally you seem to imply or assume a dispersion only of the order of 10 km/s. How do you reconcile those numbers?

M.S. ROBERTS: Clearly, there are two problems: how do you get the gas up there, and how do you keep it.

WHITEHURST: I think all we maintain is: it is there!

KERR: The thickness of 1400 pc is worrisome. The treatment is based on an assumption of cylindrical rotation. Have you considered how things would change if this assumption is relaxed?

M.S. ROBERTS: A lower velocity (as for the globular cluster system in our own Galaxy) will reduce the z-extent.

WESTERHOUT: With an "effective" beamwidth of 2 kpc, it seems rather impossible to be able to distinguish between a 1.4 kpc halfwidth of the layer and a much thinner layer with 1 - 2 kpc outriggers in the z-direction (like our own Galaxy). Any thin layer will simply be smoothed in with the wider distribution, and therefore any value of scale height derived from this work should be taken with a massive grain of salt.

M.S. ROBERTS: In this sort of modeling we have both radial velocity information as well as extensive positional information. Values of the z-extent have been corrected for a finite beam. The resultant value which is about half the beam size should be little effected by the finite beam size.

EMERSON: With reference to your deduced thickness (in the z-direction) of the HI distribution, the high resolution Cambridge observations imply an HI thickness increasing more or less linearly with radius, being $\sim$ 500 pc at R = 12 kpc and 2 or 3 kpc at R $\simeq$ 25 kpc.

SANCISI: The question of the scale height of the gas in normal disk galaxies has already come up several times in discussions during this symposium. The best observational conditions for determining the z-structure are, of course, offered by a galaxy which is seen almost completely edge-on. This is the case of NGC 891, where the HI disk in the inner part is essentially unresolved by the 26" (= 1.7 kpc) Westerbork beam. The full width at half-intensity of the neutral hydrogen layer must therefore be less than 1 kpc; a thickness of 250 pc, as found in our galaxy, would be consistent with the present observations. A similar thickness is also found for the narrow component of the radio continuum disk in NGC 891 at 6 cm by Allen et al. (A.A., in press). In the outer parts of the system the HI layer seems to become thicker: the full width may go up to about 2 or 3 kpc.

SHANE: ANOMALOUS MOTIONS OF SPIRAL ARMS IN M31

The central region of M31 has been observed in the 21-cm line using the Westerbork Synthesis Radio Telescope. The observations were made by Dr. E. Bajaja and a preliminary report has been published by Shane (1975, La Dynamique des Galaxies Spirales, ed. L. Weliachew, p. 257). There are 15 maps at velocities between -600 and -40 km/s at intervals of 40 km/s. The present results were derived from maps convolved to a 54" (FWHP) circular beam, on which the noise level was about 0.4 K.

The double-peaked profiles discovered by Bajaja (see Shane 1975) can be seen on these maps to arise from two HI components, easily distinguishable over most of the north-preceding side of the galaxy. One is concentrated to the spiral arms and shows rotational properties in rough agreement with published rotation curves (e.g. Emerson 1976, M.N.R.A.S. 176, 321). It shows deviations, however, in the sense that the prominent dust arm at 5 kpc from the center is moving inward with a velocity of 30 km/s whereas the bright arm at 9 kpc is moving outward at 20 km/s. These motions are seen unambiguously on the minor axis and are confirmed elsewhere. The simplest explanation invokes an explosion in the nucleus about 3×10^7 years ago, giving both arms an outward impulse. The outer arm would still be moving outward whereas the inner arm would have passed its apocenter.

The second component shows no correlation with optical features, virtually no velocity gradient perpendicular to the line of nodes, and a velocity gradient parallel to the line of nodes indicative of a distance of about 25 kpc from the center. This suggests that we are seeing a warped or thickened outer ring projected upon the central part of the disk. Analysis of single-dish measurements (M.S. Roberts, this symposium) leads to similar conclusions. Within the limited range of the WSRT survey no analogous feature is found on the south-following side of M31.

VAN DER LAAN: Have you or anyone else any suggestions as to how an explosion or anything else could transfer momentum of such large coherent magnitudes? What are the linear scales in the plane of the galaxy?

SHANE: The dark arm is about 5 kpc and the main arm 9 kpc from the nucleus. I agree that momentum transport over this distance is a problem, and this is one reason that I don't take the model seriously. I mentioned it only because it provides a simple and possibly suggestive means of representing the observations. Several more plausible (but unfortunately much more elaborate) models have been suggested.

WRIGHT: Have you seen anything which could be interpreted as high velocity clouds in M31?

SHANE: No, our sensitivity is not sufficient to really hope for that.

EMERSON: Cambridge observations do show some high velocity clouds.

WRIGHT: How much mass is involved?

EMERSON: Very little.

DAVIES: The high velocity cloud I found near M31 had dimensions of $\sim 0°.5$, a systemic velocity of -450 km/s and a central column density of a few times 10^{19} cm^{-2}.

GIOVANELLI: How far are the high velocity clouds in M31 from the disk?

EMERSON: That is a model-dependent number, so I can't say.

GIOVANELLI: Are they related to the "outer perturbed arm" mentioned by Dr. Whitehurst?

EMERSON: No, there are two isolated blobs in the extreme NE without any relation to other arms.

A WARP IN THE HI DISTRIBUTION AT THE EXTREME NE AND SW OF M31

D.T. Emerson
Max-Planck-Institut für Radioastronomie, Bonn, West Germany

K. Newton
M.R.A.O., Cavendish Laboratory, Cambridge, England

Aperture synthesis observations of the neutral hydrogen in the extreme NE and SW regions of M31 have been made using the Cambridge Half-Mile Telescope, with an angular resolution of 3.'6 x 5.'5 and a resolution in radial velocity of 16 km/s (Newton and Emerson, 1977). These observations show that a warp in the HI distribution exists in opposite directions at each end of the galaxy.

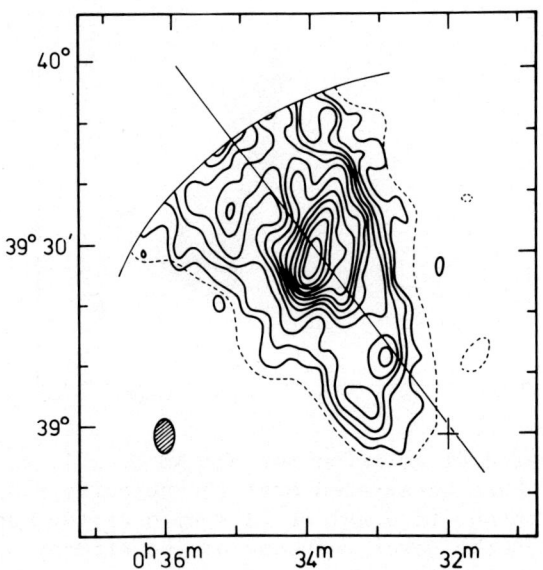

Fig. 1. The integrated hydrogen map at 3.'6 resolution of the SW of M31; the contour interval at the map centre is 42 K km/s. The outer (broken) contour is taken from 7' resolution data and shows the extent of HI detected. The major axis (P.A. 38°) is marked by a solid line.

Fig. 1 shows the HI distribution in the SW of M31. No correction has been made for the primary beam pattern of the telescope, which approximates a gaussian of 92' FWHP, centred at the point marked with a cross in Fig. 1. Note that there is a sharp edge to the observed HI emission, which is bent to the east of the major axis.

In the extreme NE (R > 24 kpc from the nucleus) of M31 previous observations have been unable to distinguish between HI in M31 and local galactic hydrogen. In the present survey the distinction is possible because of the improved angular resolution, insensitivity to structure with a scale size >2°, and the higher velocity dispersion of HI emission originating in M31. Fig. 2 shows emission which is believed to originate in M31; note that the angular extent of the emission is extremely narrow in a direction perpendicular to the major axis, certainly narrower than at corresponding positions in the SW. At large distances from the nucleus the HI deviates from the normal major axis by up to 4 kpc in the plane of the sky, an effect similar to that already found in the SW.

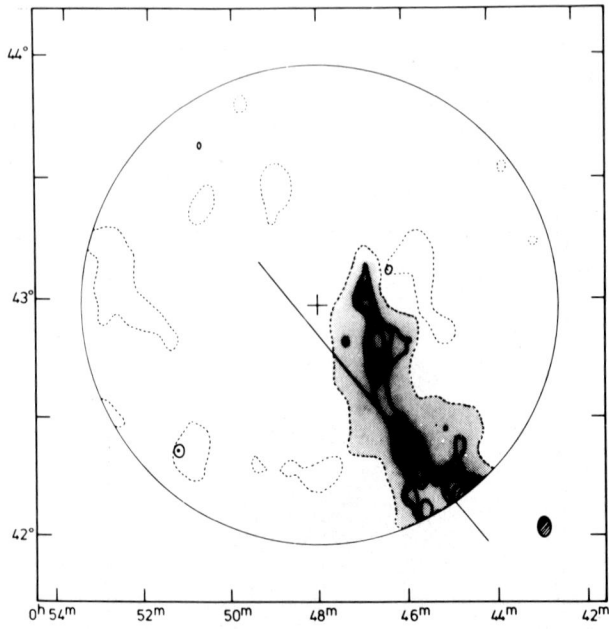

Fig. 2. The integrated HI emission for the NE of M31, as in Fig. 1. The HI emission has been integrated over the velocity range of -121 to -55 km/s, and the contour interval is 22 K km/s at the map centre. The radius (60') at which the power response of the primary beam has fallen to 0.33 is shown as a circle.

OUTER ROTATION CURVE

Fig. 3 shows the rotational velocities derived from the North and South halves of M31 along the dynamical major axis, assuming a heliocentric systemic velocity of -300 km/s. A constant inclination of 78° has been assumed, but the derived rotational velocities are insensitive to this.

III.6 A WARP IN THE HI DISTRIBUTION AT THE EXTREME NE AND SW OF M31

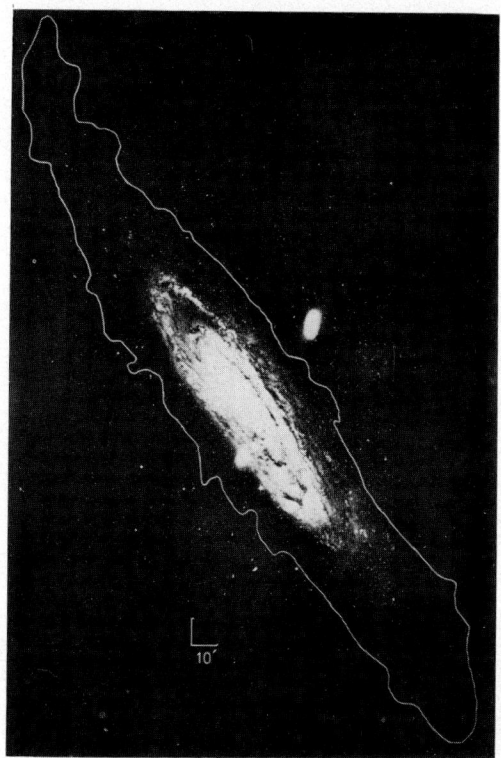

Plate I

A 48 inch Schmidt photograph of M31 (Hale Observatories) with a contour showing the extent of HI in M31 observed by the Half-Mile Telescope. The parts of the contour near the centre are taken from Emerson (1974).

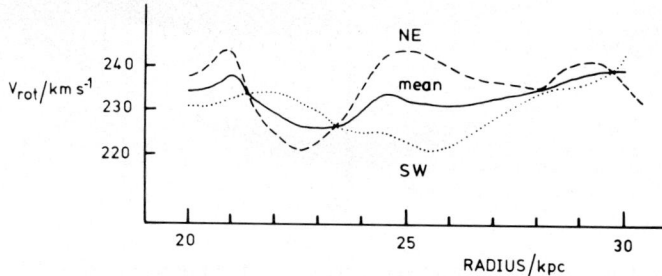

Fig. 3. Rotation velocities for R > 20 kpc, showing the NE, SW and mean curves.

The difference between the NE and SW curves at any radius is less than $\sim$ 20 km/s, which is typical of local velocity deviations seen at R < 80' by Emerson (1976). The agreement between the curves for the NE and SW suggests strongly that, to within ± 10 km/s, the observed velocities result from the true rotation law of M31, rather than reflecting local non-circular motions. At R > 20 kpc the peak deviation of the mean of the NE and SW curves from a constant VROT of 233 km/s is only ± 7 km/s; from lower resolution observations of the SW alone, Roberts and Whitehurst (1975) suggested a value of 228 km/s.

MODEL HI DISTRIBUTION

The following simple model of the distribution of HI was found to reproduce the observed velocity features well: out to 25 kpc, a flat thin disc of P.A. $38°$ and inclination $78°$, but between 25 and 30 kpc a system of low-density rings, each in circular motion around the nucleus, with P.A. and inclination varying linearly at $1°$ per kpc, from $38°$ and $78°$ at 25 kpc to $33°$ and $83°$ at 30 kpc. The simple model was found to match the observed velocities better after including the effects of parallax (the nearer edge of M31, assumed to be the NW, is $≈$ 9 per cent closer to us than the SE edge).

Fig. 4 gives a 3-dimensional impression of the sense and degree of bending of the plane of M31 suggested by the present observations. Shown in Plate I, is the boundary of HI in M31 detected by the Half-Mile Telescope. Although the model was derived from the observed *velocity* field of M31, the observed HI *densities* and the limit of HI emission indicated in Plate I are also well reproduced by the model.

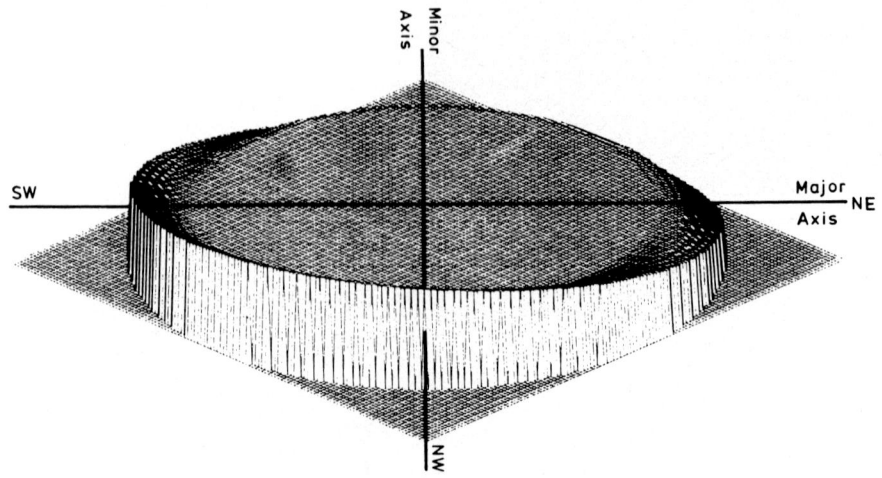

Fig. 4. A 3-dimensional representation of the warp of the plane of M31, based on the model described in the text. The figure shows the distance of the HI distribution from an arbitrary flat plane parallel to and 6 kpc below the main disc of M31, viewed at an angle of 25 degrees to the plane.

REFERENCES

Emerson, D.T.: 1974, *Monthly Notices Roy. Astron. Soc.* 169, 607
Emerson, D.T.: 1976, *Monthly Notices Roy. Astron. Soc.* 176, 321
Newton, K., Emerson, D.T.: 1977, *Monthly Notices Roy. Astron. Soc.* 181, 573.
Roberts, M.S., Whitehurst, R.N.: 1975, *Astrophys. J.* 201, 327

III.6 A WARP IN THE HI DISTRIBUTION AT THE EXTREME NE AND SW OF M31

DISCUSSION FOLLOWING PAPER III.6 GIVEN BY D.T. EMERSON

WELIACHEW: As you have shown, the outer parts of M31 are strongly warped. Then, in order to derive a "rotation curve" in those outer parts you need to make simplifying geometrical assumptions regarding space and velocity coordinates. What assumptions have you made?

EMERSON: It is assumed that the gas is always in circular motion around the nucleus of M31, but in orbits defined by a position angle and inclination which are a function of radius. The derived rotation curve is insensitive to the inclination (an error of ±5 km/s is implied for true inclination varying between 73 and 85 degrees). The position angles of the circular orbits are defined by the locus of maximum deviation of observed velocities from the systemic velocity.

GIOVANELLI: Is your estimate of the warp consistent with the one reported by Dr. Whitehurst? From your last figure I had the impression that yours was farther out.

EMERSON: A substantially flat disk is a good fit to the data as far out as $\sim$ 20 kpc in the plane of the galaxy. The warp becomes visible beyond 25 kpc.

SHOSTAK: If you look along the major axis in the southern half of M31, at a radius at which the rotation curve has become flat, what is your observed line width?

EMERSON: The observed line width is about 25 km/s, corresponding to a true width of 20 km/s; after allowing for velocity smearing in some parts of the galaxy there is no evidence for any significant deviation from this value.

VAN DER KRUIT: Could you comment on the effects of the warp on the determination of the rotation curve and mass models? Also, are your data on the rotation curve inconsistent with a constant mass-to-light ratio with radius?

EMERSON: The uncertainties in the true rotation curve due to non-circular motions - e.g. the difference between the NE and SW rotation curves at a given radius - are probably more important than the effects of the warp itself. The main uncertainty in derived M/L ratios results from the lack of knowledge of the rotation law beyond 30 kpc. The derived M/L ratio increases by a factor of order 3 from R = 5 to R = 25 kpc. Beyond 25 kpc the M/L ratio is essentially unknown, since the derived mass density is so sensitive to assumptions about the rotation law beyond 30 kpc.

M.S. ROBERTS: Am I correct in concluding that you are only able to obtain a constant M/L ratio with radius when you assume that the mass in M31 extends only to your last measured point?

EMERSON: The data imply a M/L ratio (uncorrected for absorption) of ~ 16 at R = 5 and 15 kpc, increasing to ~ 44 at 25 kpc. Beyond 25 kpc it is only possible to derive a lower limit; the possible values of the M/L ratio range from ~ 20 (assuming no mass beyond 30 kpc) to ~ 200 (assuming a flat rotation curve out to 50 kpc).

DAVIES: NEW OUTER HI ARMS IN M31

A high sensitivity neutral hydrogen survey of the southern region of M31 has been made (together with G.P. Davidson) with the MK IA (beamwidth = 13') and the MK II (beamwidth = 33') radio telescope at Jodrell Bank. The area further south than 90' from the center was mapped with a sensitivity better than 0.02 K. The following are the main conclusions of the study:
(a) There is no large bend in the HI distribution beyond 120' south of the center.
(b) Weak emission with a line integral of a few times 10^{18} atoms cm^{-2} is seen out to 162' from the center where it lies approximately 30' either side of the major axis.
(c) The shape and velocity of these weak outer arms suggest that they may have an inclination (i) of $\sim 70°$ whereas the main arms have i = 76° to 78°.
(d) The emission profiles at more than 120' from the center are narrow. They have full half-power widths of 20 - 25 km/s. Similar narrow emission profiles are seen along the minor axis of M31.
(e) The rotation curve does not fall beyond 80' south of the center. The total mass out to 200' (the distance corresponding to the outermost HI emission) is 5.5×10^{11} $M_\odot$.

ALLEN: I am trying to understand the interpretation of the velocity measurements furthest from the center. I think I understand your geometrical model; if we assume it is correct, what is your <u>physical</u> interpretation of the rising rotation curve at these very great distances?

DAVIES: If the rotation velocity goes up, then we must have more mass in those outer parts.

WRIGHT: What is the velocity dispersion for the outer HI distribution?

DAVIES: About 9 - 11 km/s.

VISSER: What about the possibility that you are looking at high velocity clouds of M31?

DAVIES: You then would expect velocity differences of 50 to 100 km/s, which we don't see.

BURKE: It is not clear that having clouds moving toward the plane on one side and away on the other is more <u>ad hoc</u> than symmetric infall or outflow. An intergalactic wind, condensing into flat sheets of low velocity dispersion in the vicinity of the plane might be just as physically reasonable, and you could still have a flat average rotation curve.

DAVIES: Let me give a partial answer: On the minor axis the velocity differs about 20 km/s from the systemic velocity on one side and less than 5 km/s on the other side.

BURKE: An intergalactic wind need not be uniform over a scale of tens of kiloparsecs.

M.S. ROBERTS: Three points: (1) The high values of rotational velocity you derive can be avoided by adopting relatively minor changes in the position angle of this outermost hydrogen. (2) Rotation curves can go down or up. The requirement that they always decrease after the peak value only reflects a preconceived model of the mass distribution. (3) The shape of a rotation curve tells us nothing of the three-dimensional mass distribution; there is not enough information in $V_{ROT}(R)$ for this. Thus a flat or even an increasing rotation curve does not require a halo for its explanation.

VAN DER KRUIT: A difficulty in studying M31 is its rather closely edge-on view. Astronomers in M31 would have the same problem when studying our Galaxy. This inclination makes it difficult to locate your features and to deproject the velocities. I wonder how big the error bars would be on your rotation curve for the outermost points.

DAVIES: The error on the actually measured velocities is about 5 km/s. The error on the rotation velocities is about 10 km/s.

VAN DER LAAN: In the discussion concerned with the origin of quasar absorption lines, the effective size of galaxies in the gas is a key parameter. We may conclude from your results that M31 has a diameter of at least 100 kpc at column densities $N_H \simeq 10^{18}$ cm^{-2}, as smoothed with a 10' arc beam. For the discussion alluded to, it is very important to determine how smooth or clumpy the HI surface density is in fact. So even this far out, observations with a beamwidth of $\sim$ 1' should be attempted.

THE LARGE-SCALE DISTRIBUTION OF HI IN M33 AND IC342

J.E. Baldwin
Cavendish Laboratory, Cambridge, England

Distortions in the distribution of the HI in the outer parts of M33 are thought to be due to a warp in the HI plane. Measurements by Reakes and Newton with the Half Mile Telescope at Cambridge provide new evidence on the extent and kinematics of this gas. The integrated HI map with an angular resolution of 7 x 14 arcmin in shown in Fig 1a. The wings extend roughly symmetrically in outline to 70 arcmin (14 kpc) from the nucleus in the plane of the sky both to the NW and the SE. The radial velocity field in Fig 1b shows large deviations from normal rotation in a plane in the outer parts. A model in which the gas rotates in circular orbits whose inclination and position angle of the major axis vary with radius, fits the HI distribution and kinematics quite well. Much larger extensions in R and z are required than in the model of Rogstad et al. (1976) and with a slower trend of i and P.A. with radius.

All the spirals in the Local Group are now known to have warps in the outer parts of comparable amplitude, but much weaker in projected HI density, than those seen in more distant galaxies by Sancisi (1976). They would be very hard to detect in galaxies at a distance of 5 Mpc. So all spirals may have warps. These nearby ones are good systems for testing theories of warping such as gravitational tides. M31 and M33 are at the same distance, are quite close neighbours and have radial velocities corrected for galactic rotation differing by only 64 km/s. If their warps are due to mutual tides during a close approach in the past, then their amplitudes must be internally consistent. A crude calculation of the ratio of radii in the two galaxies at which the warps occur is consistent with a tidal origin. Alternatively one can say that all warps, including those found by Sancisi, occur at the edge of the optical image on well exposed plates. The main problem is the well-known one of maintaining the warps until the present time. The distance of closest approach of M33 to M31 in any bound orbit is about 40 kpc, near enough to raise adequate tides in both galaxies but at a time 1.3×10^9 years ago, since when the warps would be largely washed out by differential precession. Putting more mass into spherical galactic halos is one way of preserving them

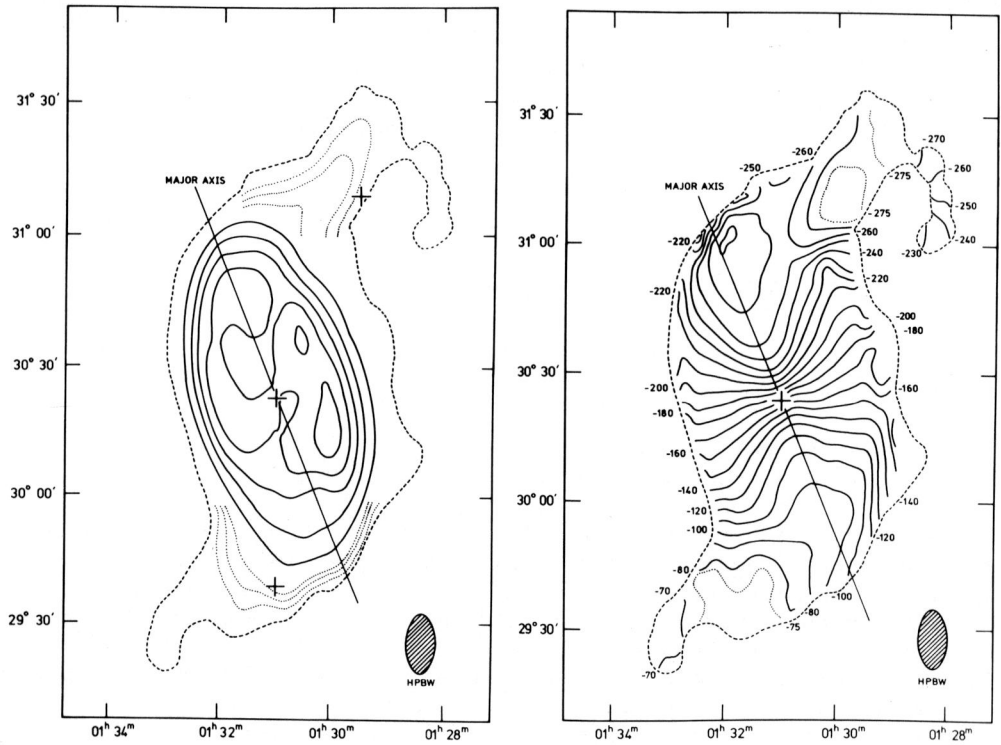

Fig 1a. HI in M33. Contours at 50, 75, 100, 125, 200, 350, 500 K km/s.

Fig 1b. Radial velocity field of M33.

for longer.

Although HI warps are detectable most clearly in edge-on systems their effects may also show up in the kinematics of face-on spirals. IC342 may be such a case. Newton has mapped the HI with an angular resolution of 1.75 arcmin and radial velocity resolution of 16 km/s. The large scale structure is best indicated by the low resolution map (7 x 7.5 arcmin) shown in Fig 2 superimposed on an optical photograph. The HI extent is roughly elliptical, with the exception of the NW quadrant which reaches 43 arcmin from the nucleus in the plane of the sky (52 kpc assuming a distance of 4.5 Mpc to IC342). Fig 3 shows the observed radial velocity field. The central region exhibits normal differential rotation, but again the outer parts show large scale deviations from normal rotation. The dynamical major axis deviates from a straight line, in opposite directions at the NW and SE ends, in a similar way to that observed in other nearby galaxies. The whole of the NW extension is apparently rotating faster than expected for normal rotation, the perturbation beginning at R=12 arcmin and increasing to R=30 arcmin but decreasing thereafter.

Fig 4 is a photographic representation of the 1.75 arcmin resolution integrated HI map. Spiral structure, with a high contrast ratio,

III.7 THE LARGE-SCALE DISTRIBUTION OF HI IN M33 AND IC 342

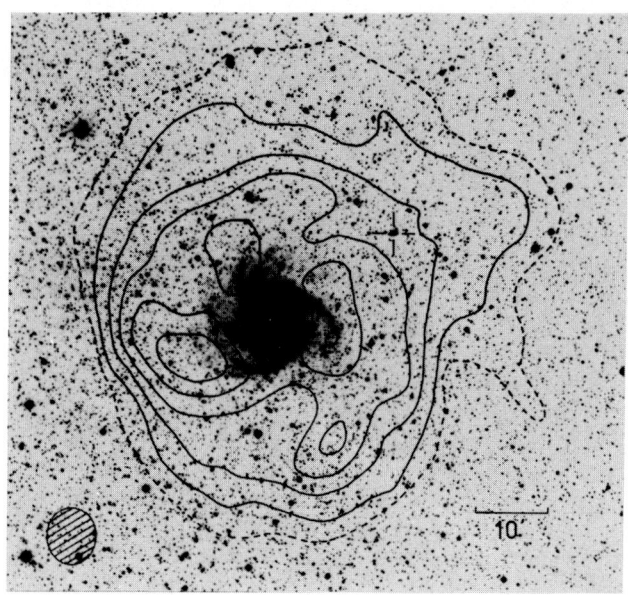

Fig 2. Integrated HI in IC342. Solid contours at 75, 175, 275, 375 K km/s.

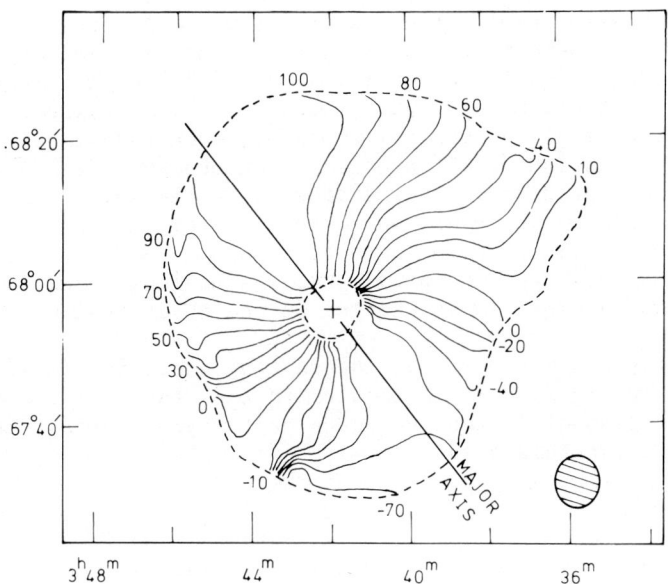

Fig 3. The radial velocity field of HI in IC342. Contour interval 10 km/s.

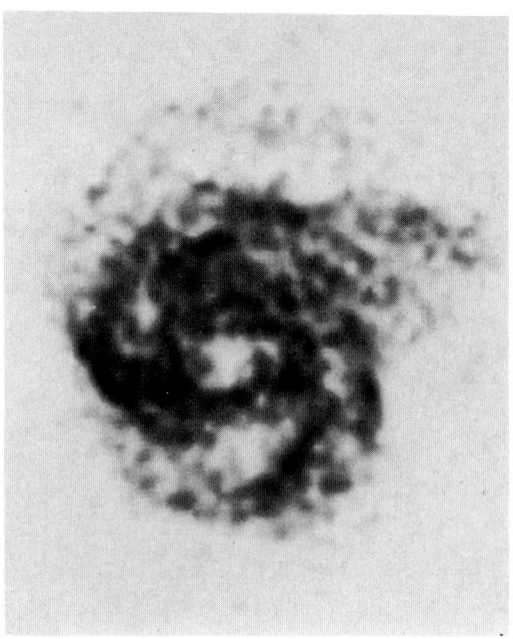

Fig 4. Photographic representation of the HI in IC342. Resolution 1.75 x 2.0 arcmin.

is clearly extending into the outermost regions of detected emission. The NW extension is also resolved into spiral arms. A likely interpretation of the velocity perturbations is that the outer spiral arms are warped away from the plane of the central disc. The low inclination of IC342 (25 degs to the line of sight) means that small changes of inclination give rise to large variations of observed radial velocity. A difference in inclination of $10°$ produces a change in radial velocity of 30 km/s. The measured radial velocities along the spiral features in the NW may be explained by hydrogen moving in orbits with inclination increasing with radius (ie becoming more edge-on). However, the magnitude of the warp varies for each spiral arm, the maximum effect being not in the outermost arm but in the centre of the NW extension.

The most likely candidate for an interacting companion to IC342 is marked by a cross in Fig 2. It is located in the centre of the NW extension in the region of largest velocity perturbation but at present we have no redshift for this galaxy.

References

Rogstad, D.H., Wright, M.C.H. & Lockhart, I.A., 1976. Astrophys. J., 204, 703.
Sancisi, R., 1976. Astron. Astrophys., 53, 159.

DISCUSSION FOLLOWING PAPER III.7 GIVEN BY J.E. BALDWIN

HUNTER: Baldwin has suggested that all spiral galaxies may be warped, and have integral sign shapes. There must be projection effects to be considered. Not all edge-on galaxies with integral sign shapes can be expected to be viewed from the angle at which the warp is most noticeable. Hence Sancisi's observations (1976, A.A. 53, 159) may be giving conservative indications of the true warps, and the galaxy in which he observed no warp may really be warped, but with the warp not visible because we are viewing along the integral sign.

SANCISI: I would like to emphasise that in all cases of edge-on galaxies I have studied, and especially in NGC 5907 and 4565, the bending of the HI layer is not entirely restricted to the two extreme ends of the galaxy, perpendicular to the line of sight (i.e. in the line of nodes of galactic plane on sky plane), as the shown pictures seem to suggest. In fact the bending continues through a large range of azimuth angles away from the line of nodes. In the case of NGC 891, which has been reported as the exception (no warp), there is indeed HI emission at large z-distances on both sides of the plane, which might be interpreted as due to a bending in the line of sight direction (at $90°$ to the line of nodes). But there are reasons for preferring alternative explanations such as a thickening of the HI layer in the outer parts of the system.

As regards the statistics of HI warps in galaxies it should be noted that the present sensitivity limits for most HI observations of galaxies are so high ($N_{HI} \simeq 5 \times 10^{19}$ cm^{-2} with resolutions of several kiloparsecs) that in the outermost parts of the disks of galaxies the HI density may become too low for detecting the bending. There may be more cases of warps like M33 and M31 in more distant galaxies, which could not be detected at present. Also for NGC 5907 and 4565 the present observations probably show only the beginning of the warp in the high density region; the warped layer may extend at lower brightness temperatures much farther from the optical plane and from the center.

SHOSTAK: I might remark that HI Westerbork syntheses of two nearly edge-on spirals, NGC 3556 and NGC 7640, show no warps. These galaxies have no obvious companions for interaction.

A SENSITIVE SINGLE-DISH HI-SURVEY OF THE GALAXY M33

W.K. Huchtmeier
Hamburger Sternwarte

Extended HI-distributions around late-type galaxies have been found at a surface density of roughly 10^{20} atoms cm^{-2} (e.g. Roberts 1972, Davies 1973). For the northern part of M33 a low density HI-component ($\sim 10^{19}$ cm^{-2}) was observed to extend to about 1.5 Holmberg radii (Huchtmeier 1973). As part of a program to study the HI-distribution and kinematics of HI-shells (i.e. the HI outside the Holmberg diameter, d_H) and to search for high-velocity cloud phenomena in a dozen nearby late-type galaxies the 100-m radio telescope in Effelsberg has been used to map the neutral hydrogen in and around M33 with high sensitivity to the limit where sidelobe contributions become important. Unpublished studies of the antenna pattern and that of Reich et al. (1976) place this limit to about 10^{18} cm^{-2} in the case of this galaxy. This limit can be reached in one hour of observing time. In 1973 an area of approximately $2°5$ by $2°5$ has been observed, which has been undersampled with the chosen grid separations of 9' in δ and $12!9$ in α. Integration times inside d_H were considerably shorter than outside.

The integrated HI-distribution (over the velocity range -350 to -50 km/s heliocentric) is given in Fig. 1. The peak of the distribution reaches 856 K km/s (1.6×10^{21} atoms cm^{-2}). The integral over this map yields under the usual assumptions of low optical depth in the HI-line a total HI mass of 1.35×10^9 M$_\odot$ ($\pm$ 7%) of which 10% are outside the Holmberg limit. The most striking features of this HI-distribution are its large extent ($\sim$ 2.2 Holmberg radii) and the different orientation of the lower contours compared to the optical image of the galaxy. This is reflected in the orientation of the velocity field, too. There are some obvious indications of structure in the HI-shell of M33, for example a number of local density maxima. In the South a separate condensation is located on the major axis but clearly separated from the galaxy by its radial velocity. In the Northwest the wing splits into a northern and a western branch. In the Northeast an extended region (6' x 9') shows multiple line structure. The lowest contour in Fig. 1 is dotted in three places where significant HI signal is present at the border of the observed grid.

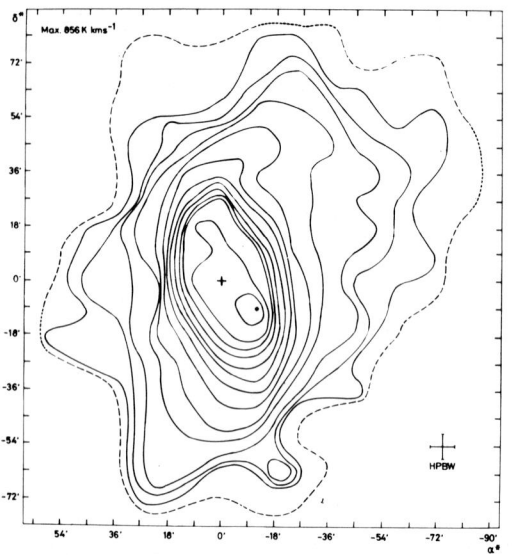

Fig. 1. HI distribution around M33. Contours are in 10% of the maximum; lower contours correspond to 5, 2, 1, and 0.5%. The broken line is the limit to which HI has been found.

The total width of the double-peaked global profile is considerably larger than twice the maximum rotational velocity (the rotation curve is given by Huchtmeier 1975). Most of the HI at the excess velocities is located inside the 20% contour of Fig. 1 and not in the wings. Within observational errors the absolute value of the excess is the same at low and high velocities. There is definitely a low intensity wing at velocities lower than -286 km/s (i.e. at the 20% level). At -328 km/s a significant HI signal is still observed. The high velocity wing is disturbed by local hydrogen. A cautionary remark may be due for all those procedures taking the total line widths of profiles as a criterium (for total masses or for distance measures) as the noted discrepancy seems to be different between galaxies.

The model of tilted rings (i.e. different inclinations and position angles of concentric outer rings of the galaxy) by Rogstad et al. (1976) does not predict these excess velocities. In that model velocities of the highly inclined outer rings are always closer to the systemic velocity than the bulk of the HI. Another difficulty of that symmetric model is the asymmetry of the HI-shell.

A consequence of our finite spatial resolution should be broad profiles in the central part of our map because the steep velocity gradient there is the most important cause for line broadening. Once the maximum rotational velocity has been reached local effects like streaming motions and turbulence determine the line width. Half-power widths of profiles near the border of the shell are of the order of 25 km/s (similar values have been observed for M31, i.e. Roberts and Whitehurst 1975). In Fig. 2 the distribution of line widths at the 25% level is presented. This distribution is rather unusual showing unexpected maxima in the HI-shell in addition to the expected maximum in the centre. Great values of line width can originate in a great velocity gradient like in the centre of the galaxy and in those regions NE and SW (e.g. Huchtmeier 1973) as seen in high resolution observations (Warner et al. 1973). Large profiles in the wing areas probably do not correspond to steep velocity gradients but rather to different HI complexes along the line of sight. There are several local maxima in line width in the HI-shell. Multiple profiles and the fact of several

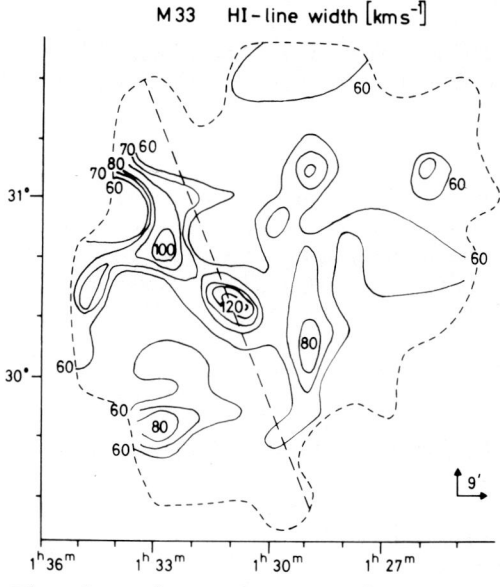

Fig. 2. Lines of equal width at a level of 25% of the maximum of the profiles are presented. Major axis and HI-extent (0.25% of the maximum) are given by broken lines.

peaks per channel map are in favour of different HI complexes in the shell.

An exception seems to be the HI complex at the lower end of the major axis. Its velocity is clearly separated from the bulk of the galaxy by 20 to 30 km/s. Its limited diameter of $\sim 7'$ corresponds to 1 kpc linear extent at the distance of M33 and its HI mass to about 10^6 $M_\odot$. The line width (corrected for instrumental broadening: 12 km/s) is typical for single HI complexes and HVC's. It should be noted that this relatively weak signal would escape detection when projected onto the bulk of the HI in this galaxy. We do not know whether this cloud is local or proper to M33. The fact that there are a number of associations between HI-complexes and nearby galaxies (Huchtmeier 1976) is taken in favour of an extragalactic solution.

A measure of the HI-shell of a galaxy is the ratio of the HI-radius (at a constant surface density) to d_H. For a number of galaxies we can plot this ratio as a function of the velocity difference Δv = total profile width (at 20% of maximum) minus twice the maximum rotational velocity. This excess velocity has not been corrected for inclination as a certain amount of it is due to an "expected" line width (10 to 25 km/s) and as it is not clear that the corresponding HI is located in the plane of the galaxy. From Fig. 3 it is evident that for small values of the shell extent (i.e. $\gamma = r_{HI}/r_{Ho} < 2$) there are no excess velocities greater than 25 km/s (or about the "expected" widths for low resolution observations in spiral galaxies). For extended HI-shells ($\gamma > 2$) we always observe velocity excesses greater than 30 km/s. The relation does not seem to be type-dependent. Irregular galaxies cover the whole range of observed

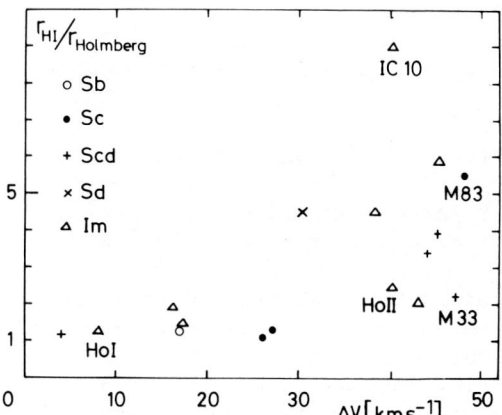

Fig. 3. The ratio r_{HI}/r_{Ho} is given as a function of the excess velocity Δv.

values. For example: Holmberg I does not seem to have any HI-shell whereas IC10 has one of the greatest shells and a large velocity excess. This phenomenon is not limited to irregular galaxies. The giant Sc I-II galaxy M83 has the greatest known HI-shell so far and the greatest excess velocity.

Unfortunately the sample of galaxies with reliable rotation curves and with sensitive observations is only small and suffers severely from selection effects. However, it seems clear that the shell phenomenon can reach important dimensions and can be a phenomenon related to either high velocity clouds or intergalactic hydrogen. To a given limiting sensitivity (surface density of 3×10^{18} to 6×10^{18} atoms cm^{-2}) the shell phenomenon is not found in all galaxies.

REFERENCES

Davies, R.D.: 1973, in IAU Symp. No. 58 "The Formation and Dynamics of Galaxies", J.R. Shakeshaft (ed.), p. 117.
Huchtmeier, W.K.: 1973, Astron. Astrophys. 22, 91.
Huchtmeier, W.K.: 1975, Astron. Astrophys. 45, 259.
Huchtmeier, W.K.: 1976, Circular No. 16 "Working Group IAU Comm. 28", p. 4, Instituto de Astronomia, Mexico.
Reich, W., Kalberla, P., Neidhöfer, J.: 1976, Astron. Astrophys. 52, 151.
Roberts, M.S.: 1972, in IAU Symp. No. 44 "External Galaxies and Quasi-Stellar-Objects", D.S. Evans (ed.), p. 12.
Roberts, M.S., Whitehurst, R.N.: 1975, Astrophys. J. 201, 327.
Rogstad, D.H., Wright, M.C.H., Lockhardt, I.A.: 1976, Astrophys. J. 204, 703.
Warner, P.J., Wright, M.C.H., Baldwin, J.E.: 1973, Monthly Notices Roy. Astron. Soc. 163, 163.

DISCUSSION FOLLOWING PAPER III.8 GIVEN BY W.K. HUCHTMEIER

TOOMRE: Do you gentlemen corroborate the claims by Rogstad et al. (1976, Ap.J. 204, 703) about the more slowly rotating "weak component" in M33?

BALDWIN: We would not be able to detect such a weak component.

HUCHTMEIER: Because of our relatively large beam width this component would be just in the wing of the profile, and therefore invisible.

WRIGHT: Recent, as yet incomplete, observations at Arecibo show that the narrow feature extending south from the north-preceding wing of M33 continues as a narrow clumpy HI feature. This feature is separate from the main HI distribution but its velocity follows that of the M33 rotation isovelocities. The peak surface density is $\sim 10^{20}$ atoms cm^{-2} and a typical clump has an HI mass of 5×10^5 M$_\odot$.

D'ODORICO: SUPERNOVA REMNANTS IN M33

The emission line intensity ratios in HII regions and supernova remnants (SNR) in the Galaxy and in the LMC have been compared to predict the [SII]/Hα intensity ratio in SNR in M33. The different value of this ratio in normal HII regions and in SNR allows the separation of the two types of nebulae. On this criterion we have identified 3 SNR candidates in an area 8' in diameter centered on the main southern arm of M33 by comparing Hα and [SII] narrow filter photographs taken with the Asiago 1.82-m telescope (D'Odorico et al. 1977, A.A., submitted). The most striking example is a half circle 4 arc sec in diameter (10 pc) facing an HII region. This object, which resembles IC 443 in the Galaxy, suggests that the shock wave from a SN explosion has given origin to an optical emission when interacting with the denser medium associated with the HII region. The three SNR candidates have been detected in the high resolution, 21-cm radio survey of M33 by Israel and van der Kruit (1974, A.A. 32, 363). The fluxes are consistent with the values expected from galactic SNR of that size.

IV NEARBY ACTIVE GALAXIES AND THEIR NUCLEI

"Several years ago, before it was fashionable to feed the gas-eating monsters which are supposed to lurk in the nuclei of some galaxies ...."

R.J. Allen in Discussion IV.2

RADIO PROPERTIES OF ACTIVE NEARBY SPIRAL GALAXIES

A. G. de Bruyn
Hale Observatories
Pasadena, California, USA

1. INTRODUCTION

Before reviewing the radio properties of active spiral galaxies it is appropriate, as well as instructive, to think about a working definition of an active galaxy. This may seem a trivial point when we consider classes of objects like Seyfert galaxies or radio galaxies but this is much less so when we inspect the nearby systems. In a sample of nearby galaxies one is likely to encounter much more milder forms of activity than that known from studies of more distant galaxies. This is, of course, due to the fact that the latter have been <u>selectively</u> taken from a much larger reservoir of objects. (Although this may seem a disadvantage from an observational point of view, there are some clear advantages as well. Firstly, we can obtain a much better linear resolution once we have been able to isolate the active nearby galaxies and, secondly, we eventually may hope to learn more about the physical situation that existed prior to the onset of the active period and how the surrounding medium reacts to the activity.)

Before proceeding then, first some comments about the qualitative observational definition of an active galaxy. <u>Optically</u> a galaxy is called active when its nucleus, which is a still poorly defined small volume in the central region, (1) emits significant amounts of non-stellar or non-thermal radiation, (2) contains excessive amounts of ionized gas, or (3) exhibits gas motions at velocities much larger than those reasonably ascribable to circular rotation. In the radio domain non-thermal emission is the rule rather than the exception and the classification of a spiral galaxy as active on the basis of its radio emission alone, requires the knowledge of what constitutes a "normal" amount of radio emission. The answer to that question is intimately related to the origin of relativistic particles in spiral galaxies and its variation with galactocentric distance. Since our knowledge of that subject is, at present, in a quite unsatisfactory state we have to employ a mostly empirical approach to the question of the radio continuum activity in spiral galaxies.

High-resolution one- and two-dimensional studies of spiral galaxies have indicated that the radio emission is generally dominated by a fairly axisymmetric disk component whose surface brightness increases towards the centre. (For a review see van der Kruit and Allen, 1976.) The disk-averaged brightness temperature at a wavelength of 21 cm lies in the range from 0.1 to 10 K (van der Kruit, 1973; Ekers, 1974b).

In Section 4 I will discuss several galaxies with large-scale disk structures with brightness temperatures considerably in excess of 10K. These structures are probably causally related to past nuclear activity although the nuclei do not necessarily exhibit any activity at present. The radio properties of presently optically active spiral galaxies will be discussed in Section 3. First, however, I will consider the radio data on the central regions of nearby galaxies to assess whether they contain indications of ongoing nuclear activity.

2. THE RADIO PROPERTIES OF THE CENTRAL REGIONS OF SPIRALS

Recently Crane (1977) has completed an extensive survey of spiral galaxies with the NRAO interferometer. For an optically defined sample of 181 galaxies he has collected complete data about the total radio emission and that of central sources less than 1 kpc in diameter and stronger than 20 mJy at a wavelength of 11 cm. In Table 1 I have collected the properties of all those galaxies in his sample that satisfy the criterium that at least 25% of their radio emission originates in a region of less than 200 pc diameter. With this somewhat arbitrary criterium I do not claim to have separated the galaxies with active from those with quiescent nuclei but I think that this group certainly gives a fair representation of the radio and optical properties of spiral galaxies with possibly active radio nuclei. Note that the above-mentioned selection criterium implies that the brightness temperature of these central sources is at least one thousand times higher than that of the surrounding disks. Clearly some extra-ordinary processes must be going on. For comparison I also have listed the properties of the ultra-compact radio nuclei of M 82 and the Galaxy.

Although the number of objects in Table 1 is rather small I wish to draw attention to a few interesting correlations:

1) Active radio nuclei are found among all optical morphological types but there may be a tendency to avoid late-type spirals.
2) Only one galaxy in Table 1 has Seyfert characteristics (NGC 1068) and only one other (NGC 3504) emits intense optical emission lines from its

IV.1 RADIO PROPERTIES OF ACTIVE NEARBY SPIRAL GALAXIES

Table 1: Galaxies with strong radio nuclei

Galaxy		Dist.	Radio Size		α [a]	P_{11cm} [b]	% [c]	Ref.
			Angular	Linear				
NGC#	Type	(Mpc)	(")	(pc)				
1068	Sb	23	<0.5	< 56	−0.5	472	30	1
2655	SOp	32	<1	<155	−0.8	64	>65	1
3031	Sb	3.25	<0.0004	< 0.006	+0.2	0.8	25	2,3
3079	Sc	24	<1.3	<151	−0.4	56	30	4,5
3504	S(B)b	30	<0.2	< 29	−0.7	86	30	6,7
4579	S(B)b	19.5	<1	< 95	−0.3	13	50	1
4594	Sa	19.5	≤0.0002	≤ 0.02	+0.3	36	100	2,8
3034	Irr	3.25	∼0.0014	∼ 0.022	−1.0	3.3	6	9,10
Sgr A	S	0.01	∼0.017	∼ 0.0008	0	0.0001	0.0005	11,12

Notes: [a] Spectral index defined as $S \propto \nu^\alpha$
[b] In units of 10^{19} W Hz^{-1}Ster^{-1}
[c] Fraction of galaxy emission coming from central region

References: 1. Crane (1977); 2. de Bruyn et al. (1976); 3. Kellermann et al. (1976); 4. de Bruyn (1977b); 5. Seaquist, private communication; 6. de Bruyn and Wilson (1976); 7. Crane, private communication; 8. Kellermann, private communication; 9. Kronberg and Wilkinson (1975); 10. Geldzahler et al. (1977); 11. Lo et al. (1975); 12. Kellermann et al. (1977).

central region (Burbidge and Burbidge, 1962). Optical and radio activity therefore do not necessarily go together.

3) Although several galaxies in Table 1 have bright radio structures in the inner few kpc they do not have exceptional disk emission. This suggests that galactic nuclei do not contribute significantly to the cosmic ray reservoir in spiral galaxy disks.

Table 1 is nearly complete for galaxies with central sources having a luminosity greater than 10^{20} W Hz^{-1} Ster^{-1} (since 90% of the galaxies in Crane's sample have a distance less than 25 Mpc). Of course, many more galaxies may exist that radiate more than 25% of their radio emission from a region less than 200 pc in diameter, but these then must have both intrinsically faint nuclei and disks.

The radio nuclei of NGC 3031 and NGC 4594 stand out in Table 1 because of their inverted spectra. These are shown in Figure 1. Their exceptional

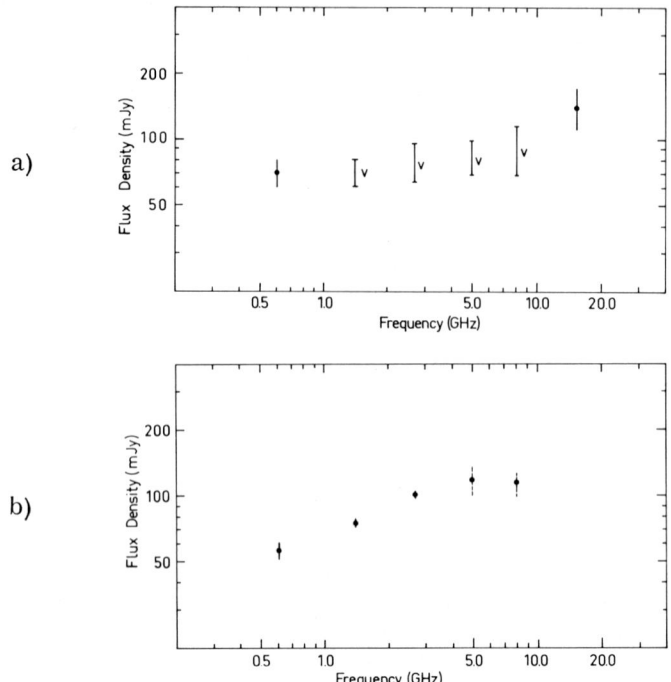

Fig. 1. a) Radio spectrum of the nucleus of NGC 3031. The vertical bars at frequencies from 1.4 to 8.1 GHz indicate the range over which the flux density varies. b) Radio spectrum of the nucleus of NGC 4594 (from de Bruyn et al., 1976).

character has been pointed out already several years ago by Ekers (1974a). Since the radio nuclei of these galaxies are also among the best studied we will now discuss their properties in more detail. The inverted spectra are most likely interpreted as due to self-absorption in non-uniform synchrotron sources with linear dimensions of a few times 10^{-3} pc (de Bruyn et al. 1976). An important property of these sources is their truly isolated nature with no significant emission coming from scales ranging upwards to several kpc. The source in NGC 3031 has been observed for nearly ten years now (the earliest data are from Wade (1968)), and this combined with the VLBI size limit indicates that the source cannot undergo an overall expansion at velocities exceeding about 500 km/s. This would appear to rule out an interpretation as a single supernova remnant and suggests a fixed source structure. The same conclusion probably applies to the radio nuclei of NGC 4594 and M 82 (Kronberg and Wilkinson, 1975; Geldzahler et al., 1977).

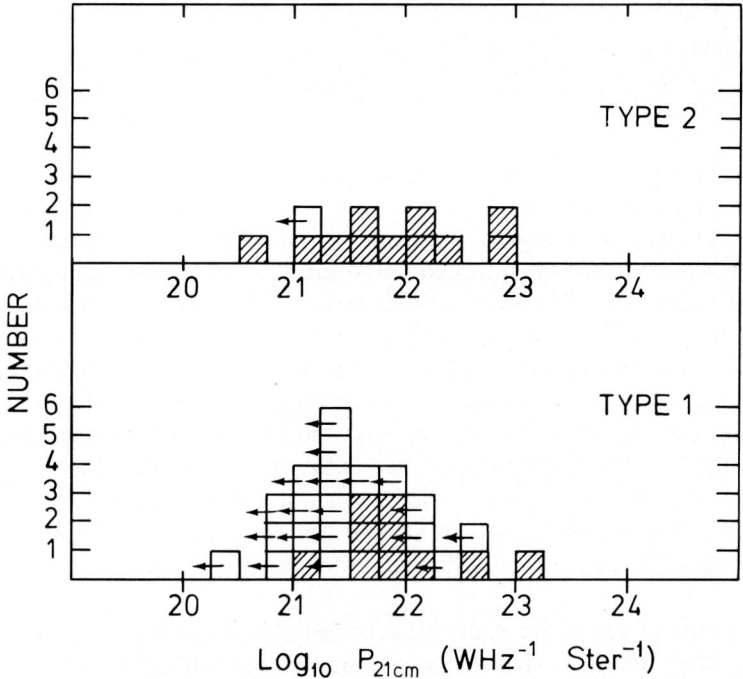

Fig. 2. Histograms of the luminosity distribution at 21 cm continuum of 12 type 2 and 29 type 1 Seyfert galaxies. Open boxes with arrows indicate upper limits for undetected galaxies (from de Bruyn and Wilson, 1978).

3. RADIO EMISSION FROM OPTICALLY ACTIVE SPIRAL GALAXIES

A surprising property of the nuclei of the galaxies described in the preceding section is that they exhibit little or no associated optical activity. In this section we will look at this problem from another way and investigate the radio properties of galaxies that are known or thought to be active on the basis of their optical properties.

3.1 Sersic galaxies

A group of galaxies that has been known for some time to possess peculiar optical nuclei are the so-called hot-spot nucleus galaxies investigated by Sersic and Pastoriza (1965) and Sersic (1973). Lequeux (1971) found a positive correlation between the presence of this type of nucleus and a radio nucleus. The study by Crane (1977) however, which involved a much larger group of Sersic galaxies, indicates no statistically significant correlation between the Sersic galaxies and radio emission from either disk or central region. The radio data therefore do not support a picture in which the hot spots would be the result of nuclear activity.

3.2 Seyfert galaxies

From the work of Wade (1968), Lequeux (1971) and van der Kruit (1971) it is known that the classical Seyferts have, on average, stronger nuclear radio sources than normal spiral galaxies. This result has recently been substantiated and extended to include also the tens of Seyfert galaxies discovered in Markarian's surveys (Sramek and Tovmassian, 1975; Sulentic, 1976; de Bruyn and Wilson, 1976). Figure 2 shows the distribution of radio luminosity of all the Seyfert galaxies in the first four lists of Markarian. Type 2 Seyferts are, on average, stronger radio emitters than type 1 Seyferts, a surprising result in view of the fact that the latter have stronger non-thermal optical continua. The Seyfert galaxies, both types taken together, occupy the upper end of the distribution of radio powers of normal spirals and frequently have luminosities exceeding by a factor of ten those of the most luminous normal spirals (Crane, 1977; de Bruyn and Wilson, 1978).

The sizes of the radio sources associated with Seyfert galaxies span the range from less than 1 parsec to more than 100 kpc although the majority have a size of a few tenths of a kpc to several kpc (de Bruyn and Wilson, 1978). The apparent scarcity of ultra-compact opaque radio sources -- of the type seen in NGC 3031 and NGC 4594 -- need not be intrinsic since it may be due to the effects of free-free absorption by the large amounts of ionized gas known to exist in Seyfert nuclei.

An interesting but puzzling result is that none of the optically discovered Seyfert galaxies shows the double structure characteristic of so many other energetic extra-galactic radio sources (see also Section 4).

4. LARGE-SCALE RADIO DISTURBANCES IN SPIRAL DISKS

Ever since the discovery of double radio sources associated with elliptical galaxies radio astronomers have searched for similar phenomena around spiral galaxies. The original studies by de Jong (1965, 1966) and Tovmassian (1968) apparently indicated an excess of so-called satellite radio sources around spiral galaxies. More recent studies (summarized by Willis, 1976), however, failed to confirm these early results. I think it is true to say that no single case is known as yet where from the structure and/or alignment of such satellite sources a link with the central galaxy has been convincingly demonstrated.

This absence of double radio sources is now also found to extend to the

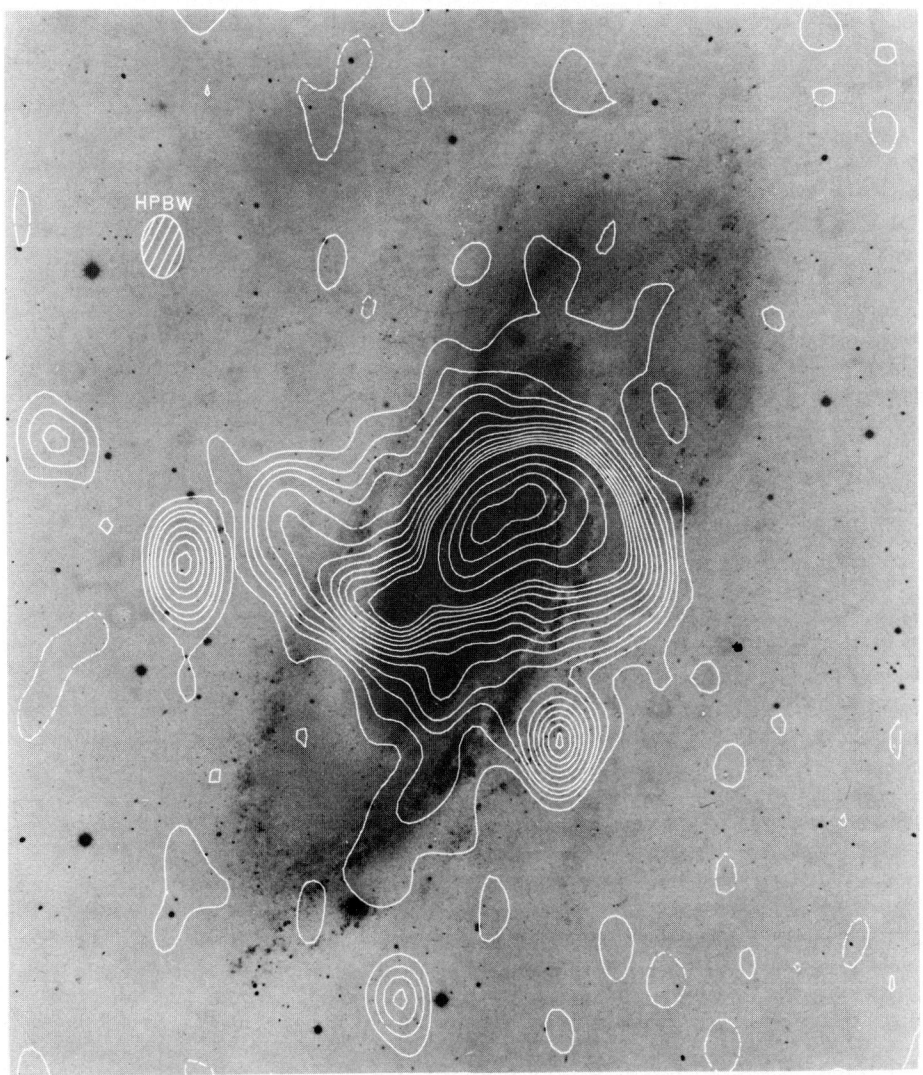

Fig. 3. Contour map of the 49 cm radio emission of NGC 4258 superimposed on an optical photograph (de Bruyn, 1977b). Contours are in units of 2.25 K brightness temperature up till 9 K, then 4.5 K till 45 K and 22.5 K till 135 K.

Seyfert galaxies, many of which may be classified as spiral galaxies. Most Seyferts are too distant to investigate with enough linear resolution whether this double structure in the radio remains absent when we get closer in to the seat of activity. We know that Seyfert galaxies eject large amounts of gas at velocities large enough to allow it to escape from the central region and one wonders what the fate of this gas is and how it interacts with the surrounding

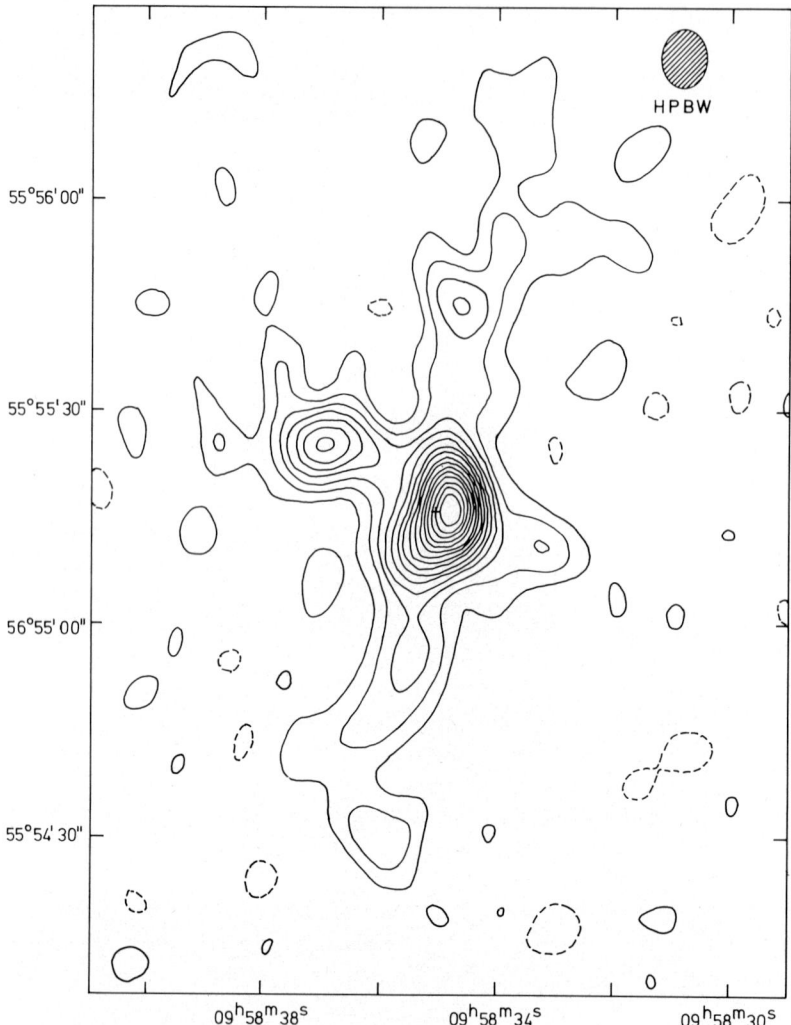

Fig. 4. Contours of the 6 cm radio emission of NGC 3079 (de Bruyn, 1977b). The contour interval is 1.5 K brightness temperature.

disks. With these questions in mind I will now present some high-resolution radio continuum maps of spiral galaxies whose radio morphology does not fit into the "standard" picture. Most of these results are from recent observations by the author with the Westerbork radio telescope (de Bruyn, 1977b).

4.1 The case of NGC 4258

This galaxy has been the subject of extensive investigations ever since the paper by van der Kruit et al. (1972) was published (van der Kruit, 1974;

van Albada and Shane, 1975; de Bruyn, 1977b). Its properties have been reviewed at various occasions by Oort (e.g. Oort, 1974) and I will not repeat them here. A radio contour map is shown in Figure 3. Note the very large contrast in radio surface brightness between the normal spiral arms and the anomalous arms which are thought to be due to high-speed gas ejection from the nucleus. Although the latter explanation is not generally accepted no alternative theories have been put forward. The main difficulty with the expulsion hypothesis is perhaps the fact that no <u>direct</u> evidence of the ejected gas has been found although it is very unlikely that all ejected gas has been slowed down to velocities less than 50 - 100 km/s. Inspection of the map of total neutral hydrogen given by van Albada and Shane (1975) indicates that of the order of 10^8 $M_\odot$ of gas is "missing" in the outer H I disk near the anomalous arms. This gas may have been swept out of the galaxy or shock-ionized to very high temperatures. Its detection by means of optical or X-ray emission would constitute a powerful argument in favor of the ejection hypothesis.

4.2 The radio structure of NGC 2146 and NGC 3079

In an attempt to find more cases like NGC 4258 I have recently made extensive radio observations of about ten galaxies with strong radio emission. The galaxies that come closest to NGC 4258 in terms of their radio properties are NGC 2146 and NGC 3079 which I will discuss in turn below.

The radio emission of NGC 2146 is dominated by a narrow ridge of intense emission centered on the nucleus and extending several kpc to either side of it (de Bruyn, 1977b). The brightness temperature in this ridge is typically of the order of 1000 K at 21 cm wavelength, which is more than three orders of magnitude higher than that of the surrounding disk or that of most other spiral galaxy disks for that matter. It can be argued that the relativistic electrons in the ridge are younger than about 10^6 years. The large extent of the radio ridge and its symmetry with respect to the nucleus then suggest that these electrons are coming from the nucleus. Alternatively, the relativistic electrons may have been produced <u>in-situ</u> by a violent excitation of the interstellar medium due to gas ejection from the nucleus. It is probably relevant that the galaxy also looks very disturbed optically. Recently neutral hydrogen has been detected at distances up to 100 kpc from the centre (Fisher and Tully, 1976), but the relation between this gas and the optical and radio continuum disturbances is unclear.

NGC 3079 has probably the most peculiar radio structure of any spiral galaxy studied thusfar. A radio contour map is shown in Figure 4. In addition to a bright radio nucleus (compare Table 1) it shows radio emission structured in the form of a cross (de Bruyn, 1977b; Seaquist et al., 1978). The arms of this cross in position angles 80° - 260° are directed along the minor axis and reach to the projected edge of this nearly edge-on galaxy. The interpretation

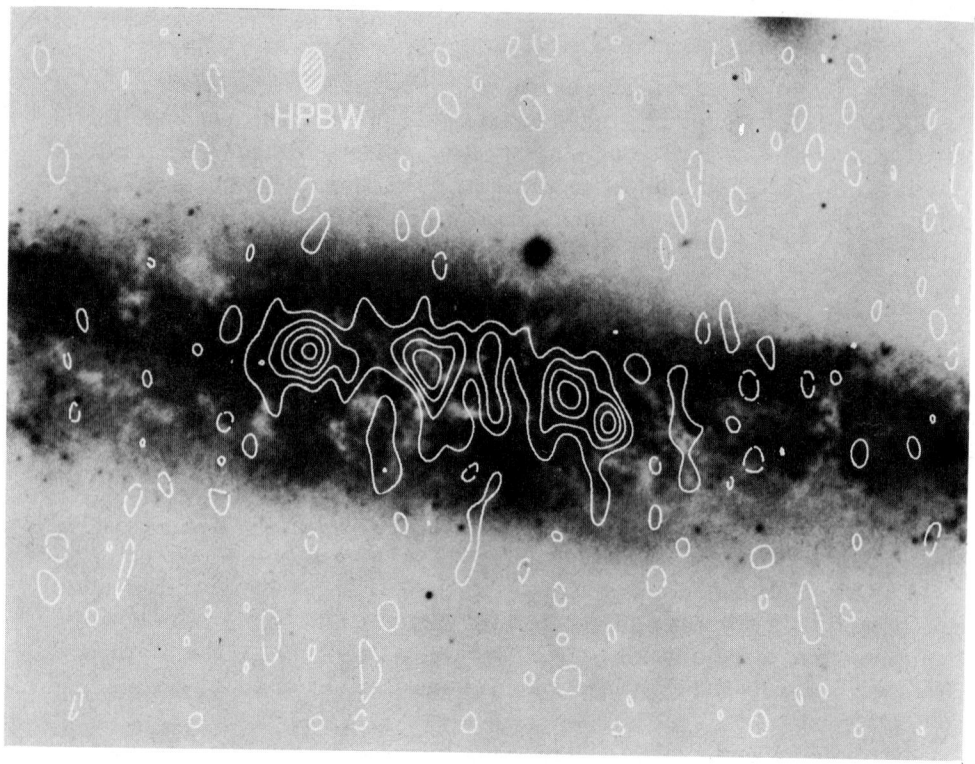

Fig. 5. Contour map of the 6 cm radio emission of NGC 4631 superimposed on an optical photograph (de Bruyn, 1977b). The contour interval is 1.0 K.

of this phenomenon depends on whether the radio structure lies in the disk or is oriented perpendicular to the disk but in either case nuclear involvement seems likely. Also in this galaxy there are several optical features suggestive of nuclear activity.

4.3 Other galaxies with peculiar disk radio features

Among the tens of galaxies that have now been mapped with high spatial resolution NGC 4631 and NGC 4736 are noteworthy because of the presence of intense radio features in the central region that are fairly symmetrically located with respect to the nucleus (van der Kruit, 1971; de Bruyn, 1977a, b). A radio map of NGC 4631 is shown in Figure 5. NGC 5005 and NGC 5033 may also belong to this category although in these galaxies the double structures are more complex and their brightness enhancement over the base disk less dramatic (Seaquist, private communication). It is estimated that several percent of all spiral galaxies have radio components similar to those seen in these galaxies.

Although the radio structures of these galaxies may be related to nuclear activity this interpretation is by no means unique and much more observational and theoretical work is needed. In particular it would be interesting to find out how tidal interactions between galaxies effect the relativistic component and magnetic fields in the disks of spiral galaxies.

Acknowledgements: I am indebted to Drs. P. C. Crane, K. I. Kellermann, E. R. Seaquist and G. D. van Albada for allowing me to use their data in advance of publication. This review was prepared while I held a Carnegie Fellowship and I am grateful to Dr. H. W. Babcock for his hospitality at the Hale Observatories.

References:

van Albada, G. D., Shane, W. W. 1975, Astron. Astrophys. 42, 433.
de Bruyn, A. G. 1977a, Astron. Astrophys. 54, 491.
de Bruyn, A. G. 1977b, Astron. Astrophys. 58, 221.
de Bruyn, A. G., Crane, P. C., Price, R. M., Carlson, J. B. 1976, Astron. Astrophys. 46, 243.
de Bruyn, A. G., Wilson, A. S. 1976, Astron. Astrophys. 53, 93.
de Bruyn, A. G., Wilson, A. S. 1978, Astron. Astrophys. (in press).
Burbidge, E. M., Burbidge, G. R. 1962, Astrophys. J. 135, 694.
Crane, P. C. 1977, The Radio Emission from Normal Spiral and Irregular Galaxies, Ph. D. Thesis, M.I.T.
Ekers, R. D. 1974a, in IAU Symposium No. 58 "Formation and Dynamics of Galaxies", Ed. J. Shakeshaft.
Ekers, R. D. 1974b, in "Structure and Evolution of Galaxies", Ed. G. Setti, D. Reidel Publ. Co., Dordrecht.
Fisher, J. R., Tully, R. B. 1976, Astron. Astrophys. 53, 397.
Geldzahler, B. J., Kellermann, K. I., Shaffer, D. B., Clark, B. G. 1977, Astrophys. J (Letters) 215, L5.
de Jong, M. L. 1965, Astrophys. J. 142, 1336.
de Jong, M. L. 1966, Astrophys. J. 144, 555.
Kellermann, K. I., Shaffer, D. B., Pauliny-Toth, I. I. K., Preuss, E. Witzel, A. 1976, Astrophys. J. (Letters) 210, L121.
Kellermann, K. I. Shaffer, D. B., Clark, B. G., Geldzahler, B. J. 1977, Astrophys. J. (Letters) 214, L61.
Kronberg, P. P., Wilkinson, P. N. 1975, Astrophys. J. 200, 430.
van der Kruit, P. C. 1971, Astron. Astrophys. 15, 110.
van der Kruit, P. C. 1973, Astron. Astrophys. 29, 263.
van der Kruit, P. C. 1974, Astrophys. J. 192, 1.
van der Kruit, P. C., Oort, J. H., Mathewson, D. S. 1972, Astron. Astrophys. 21, 169.
van der Kruit, P. C., Allen, R. J. 1976, Annual Review Astron. Astrophys. Vol. 14, 417.

Lequeux, J. 1971, Astron. Astrophys. 15, 30.
Oort, J. H. 1974, in IAU Symposium No. 58, "The Formation and Dynamics of Galaxies", Ed. J. Shakeshaft.
Seaquist, E. R., Davis, L., Bignell, R. C. 1977, preprint.
Sersic, J. L. 1973, Publ. Astron. Soc. Pacific 85, 103.
Sersic, J. L., Pastoriza, M. 1965, Publ. Astron. Soc. Pacific 77, 287.
Sramek, R. A., Tovmassian, H. M. 1975, Astrophys. J. 196, 339.
Tovmassian, H. M. 1968, Astrophysics 4, 32.
Wade, C. M. 1968, Astron. J. 73, 876.
Willis, A. G. 1976, Astron. Astrophys. 52, 219.

DISCUSSION FOLLOWING REVIEW IV.1 GIVEN BY A.G. DE BRUYN

ALLEN: The total radio flux density spectra of galaxies which are commonly called "active" very often have quite horizontal or even upturned and lumpy shapes at short centimeter wavelengths. I do not see how you find evidence for activity from the very straight spectra like $S \propto \nu^{-0.7}$ which you have shown us for the last few galaxies.

DE BRUYN: The very fact that the spectra don't curve at high frequencies means that the electrons do not loose radiation to synchrotron and inverse Compton processes. The absence of a break in the spectrum indicates that the sources of the electrons must be very young, $\sim 10^6$ years; combined with the large linear extent and the high surface brightness of these galaxies this is an argument for activity.

BURBIDGE: How dependent is this result on the equipartition assumption?

DE BRUYN: Not very dependent, because the inverse Compton losses are of the same order as the energy in the stellar photons. I have taken a magnetic field strength of 10^{-5} Gauss which is equal to the equivalent field strength provided by the stellar photons. If the field strength is larger, the age limits become smaller than 10^6 years.

EKERS: Could one not apply the same argument, i.e. the lack of a high frequency break in the radio spectrum, to the disk emission in which case we would call the whole disk "active"?

DE BRUYN: The spectra of the disks of quiet spirals, as I would call them, are not known at high frequencies. Also, the magnetic field strength and the inverse Compton losses are less in these galaxies, so the age limit for NGC 891 for example is at least a factor of 5 higher. This, combined with the fact that quiet spirals are brighter, would make them very exceptional.

VELUSAMY: The straight spectra shown for the nuclear emission of active galaxies are similar to those of the disk emission of spirals. Models for the disk suggest continuous injection and a leakage time of $< 10^7$ years. Similarly, the nuclear sources with continuous activity could be much older than you estimated.

DE BRUYN: If the electrons would stay in the active galaxies 10^7 years and would not be reaccelerated, you would see a break in the spectrum at a frequency far below 15 GHz. This we don't see. Hence, the age limit for the electrons in the active galaxies is probably a factor of 10 less than that in the disks of spirals.

PREUSS: THE NUCLEUS OF NGC 1275 AT 2.8 CM WAVELENGTH

The nucleus of the Seyfert galaxy NGC 1275 (3C 84) is one of the strongest extragalactic radio sources at cm wavelengths. Between 1960 and 1975 its flux density increased by more than a factor of 5, and has since remained relatively constant. High-resolution interferometric observations made at 1.3, 2, 2.8, and 6 cm show that there are three main centers of emission with an overall extent of 0.006 arc sec (3 pc), aligned roughly along position angle -9 degrees. Figure 1 shows the structure of 3C 84 based on measurements made at 2.8 cm (Schilizzi et al. 1975, Ap.J. 201, 263; Pauliny-Toth et al. 1976, Nature 259, 17; and unpublished) at 5 epochs between March 1972 and July 1976.

During this period any relative motion of the components has been with an angular velocity < 50 micro arc sec per year. If the source has expanded from a common point the velocity of the components would have had to be greater by a factor $\gtrsim$ 10 during at least part of this time. The present low velocity of separation implies either deceleration of the components, or independent activity in the components. Although there is no evidence for component motion, the size of the individual components appears to have increased at a rate $\sim$ 30000 km/s with the peak brightness temperature (of the central component) decreasing from 4×10^{11} K to 1×10^{11} K.

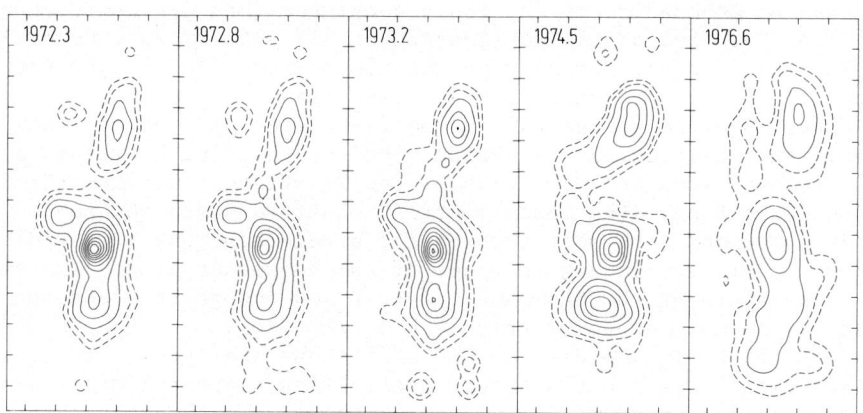

Figure 1. The structure of the radio nucleus of NGC 1275 as derived from VLBI data for five epochs. The spacing between tick marks is 0″.001. The solid contours represent steps of 3×10^{10} K brightness temperature, the dashed contours 0.5 and 0.25 of this value.

EKERS: Cannot many of the triple and double sources in the central region of active galaxies be just as well described as somewhat irregular disks?

DE BRUYN: I agree that in some cases, like NGC 4736 and 5033, the structure might be due to spiral arms. I haven't looked at optical photographs to check this. But I think NGC 1505 and 4631 have bright sources in the central regions.

SANCISI: No! The complex structure in the central region of the edge-on galaxy NGC 4631, which is found at 6 cm, could be adequately explained by a 3 kpc diameter edge-on disk. The two sources on opposite sides of the center, shown at 21-cm resolution, could simply be due to spiral arms or ring structure of that small central disk.

DE BRUYN: Well, I disagree.

BURBIDGE: If you were given a sample of weak radio sources to map in this way, some associated with QSOs, some with radio galaxies, some with spirals, and some unidentified, could you from the maps identify the class of optical objects?

DE BRUYN: No, I could not. The only difference sometimes seen is a larger smooth background in the case of spirals due to the disk emission. But if you would take that smooth component away, some of these may look like extragalactic radio sources.

EKERS: Just for the record: I disagree.

VAN ALBADA: HI IN NGC 4258
 NGC 4258 is an S(B)bc spiral galaxy with anomalous gaseous arms that have been observed in the radio continuum (van der Kruit et al. 1972, A.A. 21, 169) and in Hα (Courtès et al. 1961, Compt. Rend. Acad. Sci. Paris 253, 218). These arms cut almost radially through the entire disk in diametrically opposed directions.
 I have observed the galaxy in the 21-cm line of neutral atomic hydrogen. The observations were made with the Westerbork Synthesis Radio Telescope and with the 100-m dish of the Max-Planck-Institut für Radioastronomie in Bonn. The most important observed features are:
(1) In agreement with most theoretical predictions there is a distinct lack of HI emission in and in front of the anomalous arms. The emission temperature of the anomalous arms is low enough that absorption cannot be the main cause of this lack of emission.
(2) Outside 10 kpc the disk of NGC 4258 shows nearly circular rotation, together with a small but distinct contraction along the minor axis. On theoretical grounds it is impossible to explain this contraction as the aftereffect of an explosion in the nucleus.
(3) Inside 10 kpc very large non-circular motions begin to occur. The contraction along the minor axis reaches velocities up to at least 100 km/s. A velocity jump of at least 200 km/s occurs at a distance strongly reminiscent of the velocity field seen in theoretical computa-

tions of gas response to a bar. The presence of a bar at the correct position angle is suggested by the total HI map and by some optical photographs. The morphological evidence is rather weak however. The velocity jump lies near the front (leading) edge of the anomalous arms, and could also be explained, at least in part, by the same explosive mechanism that is used to explain the anomalous arms.

BALDWIN: What criteria were used to define the major and minor axes in NGC 4258?

VAN ALBADA: The axes were determined from a best fit to the velocity field beyond 10 kpc from the center. The deviation from the axes is about 10 km/s for both the major and the minor axis, but the deviation from the minor axis is very consistent over a large distance.

BURKE: The form of the Hα inner arms of NGC 4258 and the early model calculations gave strong support for nuclear ejection as a mechanism. It is surprising, though, that the overall velocity field of the HI shows so little perturbation. Rather than appealing to nuclear ejection, therefore, can the theorists find a catastrophic mode of spiral arm formation that produces the arms directly through the infalling material forming a shock or hydraulic jump with enough energy to ionize the medium? The general inflow along the minor axis might be evidence for material falling into the gravitational perturbation of the density wave.

SHU: Ten km/s is much too small to ionize hydrogen completely in a shock. Velocities of order 50 – 100 km/s are needed.

VAN DER KRUIT: Velocity differences in NGC 4258 are much larger than 10 km/s and spectra show that the line strengths are consistent with collisional excitation.

DE BRUYN: The fact that we hardly ever see noncircular motions in HI exceeding a value of about 100 km/s in the disks of spiral galaxies may be partly due to a "selection" effect. Namely, if gas would move at much higher velocities it will be collisionally ionized due to interaction with the interstellar medium, reach a very high temperature, and may expand from the disk before it could recombine to produce observable optical line emission or 21-cm line radiation.

OORT: One should point out that in the inner parts of the anomalous synchrotron arms in NGC 4258 ionized hydrogen arms are observed which fit so closely to the synchrotron arms that they are undoubtedly connected with these. This shows that at least in this part of the synchrotron arms the gas has been ionized. It is plausible that further out the gas would have still higher temperatures and would therefore not be visible in HI. It is likely that what we do observe in HI is connected with subsidiary stages.

"Mr. Chairman, in the outskirts of this room it is difficult to follow the nuclear activity up front."

H. van Woerden in Discussion IV.2

RADIO PROPERTIES OF THE NUCLEI IN ELLIPTICAL, SO AND SPIRAL GALAXIES

R.D. Ekers
Kapteyn Astronomical Institute
University of Groningen

When Heeschen (1968) found that the nuclear sources in the two elliptical galaxies NGC 4278 and NGC 1052 contained flat spectrum unresolved sources he conjectured that these might be weaker versions of the compact optically thick and strongly variable sources found in the quasar nuclei. This conjecture appears to be correct since these are both now known to have diameters <0.1 pc (Cohen et al. 1971) and to be variable on time scales of order one year. This variability has recently been confirmed by observations with the VLA (Heeschen, private communication).

From further surveys of elliptical and SO galaxies we now have about 20 more galaxies with similar flat spectrum compact nuclear sources. Furthermore, similar nuclear sources have also been found in many radio galaxies. Most of the elliptical galaxies that contain such nuclear sources also show optical emission lines in their central regions.

The radio properties of the nuclei of the elliptical galaxies are quite different from those just discussed by de Bruyn for the spiral and Seyfert galaxies. For example, almost all the radio sources in the nuclei of elliptical and SO galaxies are less than a parsec in size whereas most of the radio emission from the spiral and Seyfert galaxies usually comes from a region many hundred parsecs in size (Ekers 1977). There is also a clear difference between the spectral index distribution for the nuclei of elliptical and spiral galaxies (Ekers 1974). With the larger sample now available (de Bruyn 1976, Crane 1977, Ekers et al. 1978, Hummel private communication) we can make a finer division into Hubble types, as is shown in the Figure. We see a fairly smooth trend from the elliptical galaxies which have mostly flat spectrum nuclear sources to the late-type spirals with only steep spectrum nuclear sources.

A new intriguing correlation is suggested by the observation of HI gas in the two prototype elliptical galaxies with radio core sources: NGC 4278 and NGC 1052 (Gallagher et al. 1977, Knapp et al. 1977, Fosbury et al. 1977). This point has been checked further by comparing the radio continuum data (Hummel private communication) for six elliptical galaxies

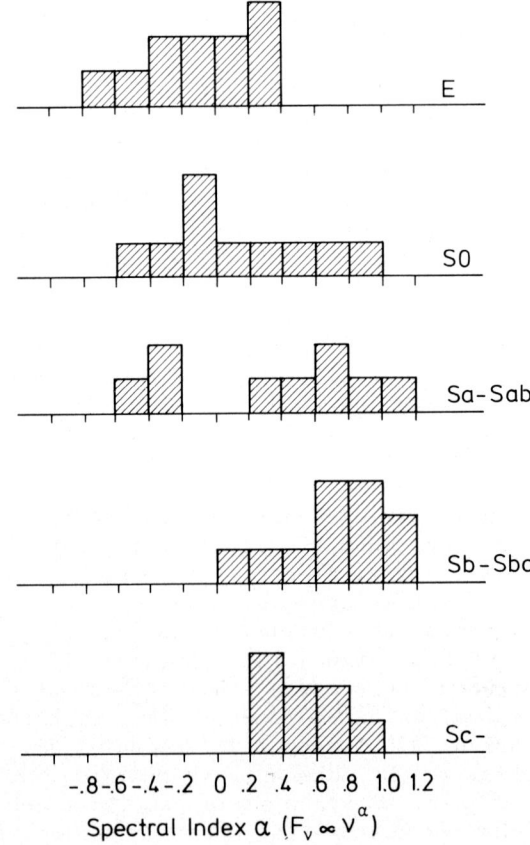

Figure: Distribution of radio spectral index, α ($F_\nu \propto \nu^\alpha$), for galactic nuclei subdivided by Hubble type.

with detected HI with that for a comparison sample consisting of the six elliptical galaxies with lowest HI limits in the literature. While four of the six elliptical galaxies with HI detected also have flat spectrum nuclei, none of the comparison sample do.

In conclusion, we see a striking change in the properties of the nuclei of galaxies of different Hubble type indicating that the overall morphology of a galaxy is very relevant to the state of the nucleus. There are various ways to interpret these results: perhaps the gas in a spiral galaxy absorbs the radio emission from the compact nuclei or perhaps it extinguishes the radio source by overfeeding the nuclear engine. Alternatively, it may be that greater sphericity of the ellipticals lets material sink deeper into the nucleus making a more compact source. We might also contemplate the opposite situation in which some property of a seed nucleus determines whether a spiral or an elliptical galaxy is formed in the beginning.

REFERENCES

De Bruyn, A.G.: 1976, Ph.D. dissertation, University of Leiden
Cohen, M.H., Cannon, W., Purcell, G.H., Shaffer, D.B. Broderick, J.J., Kellermann, K.I., and Jauncey, D.L.: 1971, Astrophys. J. 170, 207
Crane, P.C.: 1977, Ph.D. dissertation, M.I.T.
Ekers, R.D.: 1977, Physica Scripta, 17, 188
Ekers, R.D., Ekers, J.A., Rogstad, D.H. and Smeding, A.G.: 1978) in preparation
Fosbury, R.A.E., Mebold, U., Goss, W.M., and Dopita, M.A.: 1977, Mon. Not. Roy. Astr. Soc., in press
Gallagher, J.A., Knapp, G.R., Faber, S.M., and Balick, B.: 1977, Astrophys. J. 215, 463
Heeschen, D.S.: 1968, Astrophys. J. 151, L135
Knapp, G.R., Gallagher, J.S., and Faber, S.M.: 1977, preprint

DISCUSSION FOLLOWING PAPER IV.2 GIVEN BY R.D. EKERS

VAN DER LAAN: Two remarks connected with your hints relating Hubble types to the presence and spectrum of nuclear radio sources. (1) I think it is instructive when discussing these phenomena to distinguish "engine power" and fuelling rate. A large range of engine powers seems possible among elliptical galaxies, but in spirals big engines seem not to be found. Possibly the galaxy formation process is responsible for this difference. (2) If there is a lot of gas, a nuclear engine may not remain visible through the radio window. Not only synchrotron self-absorption but especially free-free absorption may hide activity in late spirals' nuclei. Possibly these can be uncovered by hard-X ray experiments planned for Spacelab II.

DE BRUYN: I would like to point out that the distributions of radio power and sizes of the (Markarian) Seyferts are considerably broader than the one Dr. Ekers has shown us. Therefore I am not sure whether the gap between Seyferts and radio galaxies in the diagram is real.

WELIACHEW: Is there a correlation between the presence of compact radio sources and the presence of the λ3727 lines of [OII] in elliptical galaxies?

EKERS: Yes, there is a very good correlation: about 80% of the ellipticals with compact sources have emission lines, usually λ3727. However, the optical sample of spectrographs should be made more complete.

BARTEL: What is the range of the spectral index variations in NGC 1052 and what is known about the physical processes that trigger the spectral variations?

HEESCHEN: I don't know the cause of the spectral index variations in NGC 1052. They are similar to those seen in some quasars and radio galaxies, and may have a similar cause, perhaps that suggested some time ago by Kellermann and van der Laan. The range of spectral index variation is at least ± 0.2 at a given wavelength, say around 1 cm - 10 cm.
[see Discussion I.5 for HI in NGC 1052 and 4278]

KERR: COMPACT COMPONENTS IN ACTIVE SPIRAL AND SEYFERT GALAXY NUCLEI AND GALAXY INTERACTIONS

In an interferometric study made by J.B. Carlson, University of Maryland, of 61 active spiral and Seyfert galaxy nuclei, compact nuclear components as small as 0.1 arcsec were found in 26 cases. Some of the highly compact sources are of the order of parsecs in size or less. Two-point spectral indices calculated between 2695 and 8085 MHz show that compact sources tend to have flat, positive or complex spectral indices, whereas more extended sources (> 100 pc diameter) tend to have indices clustering around -0.7.

No discernible differences were found between Seyferts and active spirals on the basis of nuclear radio properties alone, but Seyferts tend to have greater nuclear radio luminosities and greater total optical luminosities. Active Irr II galaxies also have high centimeter-wavelength nuclear radio luminosities, for a given total optical luminosity. No correlation was found to exist between the nuclear radio properties of spirals or Seyferts and the Hubble type or stage.

All but two of the 26 active galaxies in the present sample are probable members of small groups, are members of interacting pairs, or show evidence of tidal disturbance according to the DDO classification system (i.e. they have "n", "*", or "t" DDO designations). Attention is called to interacting systems such as NGC 3623/3627/3628, NGC 4631/4627/4656 and NGC 3031/3034/3077, for which high-resolution neutral hydrogen studies are available as well as data on the nuclear radio properties. These systems have been found to contain intergalactic bridges, streamers and clumps of HI and the constituent galaxies contain active compact (radio) nuclei.

It is suggested that the activation of "excited" galactic nuclei is externally initiated by interaction with another galaxy, leading to tidal disruption or similar effects. In this hypothesis, the interaction process feeds fresh gas into the nuclear region which could yield the observed nuclear activity by initiating a burst of star formation, or perhaps accretion on a black hole.

VAN ALBADA: What is the most extreme mass ratio for galaxies in a pair in which a nuclear source has been excited in at least one galaxy of that pair?

KERR: The members of a pair tend to have about equal mass.

ALLEN: Several years ago, before it was fashionable to feed the gas-eating monsters which are supposed to lurk in the nuclei of some galaxies, Ekers, Burke, Miley and I published a short paper in Nature in which we examined the correlation between total radio emission and interaction for a small sample of objects from Arp's atlas. We concluded then that, when the sample was compared with a sample of normal non-interacting galaxies there was no clear evidence either for an increased frequency of occurrence or for an increased intensity of the radio emission from the individual members of an interacting system. This conclusion was corroborated with a larger sample of interacting and non-interacting systems surveyed later by Wright with the Parkes telescope. Could you comment on the difference of your results with these?

KERR: Perhaps your statement referred to the total flux of the galaxies, while ours is very definitely for the small diameter component at the center.

EKERS: Do you know if these are really nuclear sources, or just compact emission from somewhere in the system?

KERR: They are nuclear sources because we have got positions that quite accurately coincide with the optical ones.

VAN DER KRUIT: Since the Westerbork survey of interacting galaxies by yourself and others the 5 GHz survey of Sramek and of Sulentic have appeared. We can see now that you happen to have selected some of the radio faintest Arp interacting pairs detected by them.

INFRARED, OPTICAL, AND X-RAY PROPERTIES OF THE NUCLEI OF NEARBY GALAXIES

G. Burbidge
Department of Physics
University of California, San Diego
La Jolla, CA 92093

GENERAL CONSIDERATIONS

The term "nearby galaxies" is not very precise. If we restrict ourselves to galaxies within the local group, we are really only talking about our Galaxy, M31 and M33. Since the Galactic Center has been reviewed extensively by Oort in Annual Reviews (1977), and I feel that there is nothing exceptional to say about the nuclei of M31 and M33 as far as phenomena other than their stellar content and central dynamics are concerned, to discuss interesting properties we must consider more distant objects. If we go out to the distance of the Virgo cluster, we already include objects such as NGC 5128, M82 and M87. Each of these galaxies shows or was claimed to show evidence of different kinds of violent nuclear activity. Indeed, it is obvious that within the volume occupied by the supercluster (whether or not it is really a physical entity) there must be many galaxies in which nuclear activity can be detected.

While it may be stretching things a little, the organizers have implied that NGC 1275 might be considered a "nearby" galaxy for the purposes of this discussion. Out to the distance of NGC 1275 (about 100 Mpc) there must be $\sim 10^5$ galaxies. Clearly within this sample all types of nuclear activity should be detectable.

How does one define nuclear activity of a special kind?

The distinction that I would like to make is between energy release processes which take place through the evolution of normal stars in low density systems, and processes which are more exotic than this and can only take place either through the evolution of high density stellar systems or from new physics.

This leads to two obvious distinctions which can be made observationally. For normal stars the largest energy release which can occur rapidly is in a supernova explosion. Thus, any phenomenon which can be shown to have released more than 10^{53}-10^{54} erg cannot be due to a single normal star. A second distinction is associated with the position in the galaxy where this outburst occurs. If only a single star is involved, the outburst can take place anywhere in the galaxy. However, activity involving more exotic processes is only likely to occur in the mass center -- the true nucleus.

A fundamental problem which remains in this exciting field is the very tenuous link between theory and observation. A simple view based on the observations originally put forward by Burbidge, Burbidge and Sandage (1963) is that there are such a wide variety of manifestations of activity in the nuclei of galaxies that it could be supposed that all galaxies have active nuclei throughout their lives, and that it is only the level of activity and the problem of detection that are the variables from system to system. We were thinking in those days that the violent nuclei were the results of single violent events (explosions) whose effects could last for $\sim 10^6$-10^8 years, but that such events could repeat.

The models for violent activity which have been developed are of several kinds:

(a) Multiple supernovae and their remnants (pulsars), with a rate of outbursts chosen to explain the observed level of activity.
(b) Energy released by a massive rotating superstar (spinar) similar in some ways to a pulsar.
(c) Energy released by matter falling into a massive black hole and/or surrounding accretion disk.
(d) Energy pouring out of a singularity (white hole) in the form in which it is observed.

Each of these schemes has its advocates and its periods of popularity. Currently (c) is the most popular and (d), because it involves modifications of physics and has not been worked out in any detail, is the least popular.

But when we begin to ask how far these theories go in explaining what we see, and even more, in making predictions which would discriminate between models, the answer is that very little progress has been made. We observe some, or all, of the following:

(i) Nonthermal optical and radio continua which are generally thought to be incoherent synchrotron radiation.
(ii) A hot gas emission-line spectrum, the lines being exceedingly broad ($\lesssim$ 10 000 km sec^{-1}) almost certainly due to mass motions.
(iii) Very large infrared fluxes extending in some cases to 100-300 μ. In many cases the IR flux is thought to be thermal.
(iv) The ejection of large masses of gas, and in some cases what appear to be coherent objects.
(v) Extended radio sources.

One of the reasons why it is so difficult to relate the "machine" to its observational consequences is that the scales are so different. Since gravitational energy must be the ultimate energy source in all conventional models, we believe, without understanding very clearly, that the energy must be released fairly close to the Schwarzschild radius which, even for a 10^{10} $M_\odot$ object, is only $\sim 10^{15}$ cm and is proportional to the mass. Now the observational phenomena which we have to explain take place on scales which range from sizes that may be as small as this, up to dimensions of kiloparsecs (for optical phenomena) and megaparsecs (for radio phenomena). But only from measures of variability and light travel time can we measure small sizes optically at present (down to $\sim 10^{15}$ cm) and very little is known about light variations in the nuclei of galaxies. VLBI techniques allow us to measure 10^{-3} to 10^{-4} arc sec corresponding to scales ≤ 0.5 pc for distances ≤ 100 Mpc, and the size of the very small resolved radio source in the Galactic Center $\simeq 7 \times 10^{13}$ cm may be highly significant. If this were the ultimate size of the "machine," it would tell us a great deal. However, not only is this a very weak source, which may be due to a single star, it is also possible that the more powerful nuclei may contain machines which are intrinsically much bigger.

Ideally, in studying the nuclei of nearby galaxies, we would like to test the predictions of the various theoretical models against the observations, or failing this, attempt to rule out some theories on the basis of observations. Unfortunately we are not anywhere near the stage where this can be done. The procedure that is still being adopted by theoreticians is to adopt a model, e.g. a massive black hole surrounded by an accretion disk, and then attempt to speculate on a scenario which will give some of the observed properties. Since there is usually a wealth of free parameters, there is no quantitative way to estimate the plausibility of a chosen model.

In this lecture I cannot improve on this situation. I will simply briefly discuss a few observational discoveries which relate to the nuclei of galaxies.

SOME OBSERVED PHENOMENA

Obscured Nuclei

There are two well known nearby galaxies which have been thought to have active nuclei which are heavily obscured by dust. They are NGC 5128 and M82.

NGC 5128 was long ago identified as a powerful extended radio source, and the inner double lobe structure showed that more than one outburst was involved. However only recently has the nuclear structure been studied at infrared, X-ray, and γ-ray wavelengths. The object is powerful and rapidly variable, but since it will be discussed by M. Rees I shall not mention it further.

M82 has had a chequered history. It was one of the original galaxies in which it was believed that a violent explosion had occurred, the evidence coming from the radio properties, the velocity field which suggests ejection along the rotation axis, and high polarization of the continuum radiation from the optical filaments extending above and below the plane which was interpreted as optical synchrotron radiation. The discovery that there was a high degree of polarization in the Hα emission lines from the filaments meant that another explanation for the polarization was required. Solinger, Morrison, and Markert (1977) have now concluded that all of the evidence for an explosion has effectively been removed, and their arguments at this point should be taken very seriously. Studies of the central region of M82 in the near and far infrared and also in radio wavelengths (Raff 1969; Kleinmann and Low 1970a, b; van den Bergh 1971; Hargrave 1974; Kronberg and Wilkinson 1975) show that there is no evidence for a powerful nonthermal nuclear source. The radiation from the central region at these wavelengths is resolved into a number of discrete sources with a highly complex pattern. They may simply be due to heavily obscured O and B associations similar in some respects to the pattern in our own Galactic Center. Of course the existence of a very weak nonthermal source is not excluded.

Probably the most difficult problem which we encounter in reinterpreting the evidence for a galactic explosion in M82 is to find an alternative explanation for the velocity field. Solinger et al. have argued that it can be explained by supposing that the galaxy is drifting through a large cloud of intergalactic dust with the dust grains acting as moving mirrors. While this explanation is ingenious, it is still not very satisfactory, since a very large cloud of dust of unspecified origin is required.

But on balance we must probably now exclude M82 from the class of galaxies in which violent activity is seen.

Powerful Infrared Sources in Galaxies

Several years ago Low and his associates (Kleinmann and Low 1970a, b; Aumann and Low 1970; Low 1970) measured the infrared fluxes out to $\sim 25\,\mu$ in a number of bright galaxies and obtained very large luminosities which they concluded were due to nonthermal processes in the nuclei. These observations were not all confirmed. However, while the initial measurements may have had problems, studies carried out since 1970 by Rieke et al. (1973), Clegg et al. (1976), Low and several other groups (Rieke and Low 1972; Penston et al. 1971, 1974; Hildebrand et al. 1977) show that large infrared fluxes are indeed present. The observations have now been extended in a few cases out to wavelengths as long as 1 mm. It is clear that in many cases the luminosities out to $\sim 25\,\mu$ are comparable to or greater than the total optical luminosities of the galaxies.

Among the galaxies which have been detected in this way are the classical Seyfert galaxies NGC 1068 and NGC 4151, M82, NGC 253, NGC 5236, and others.

There are two possible mechanisms operating to give the infrared flux:

(i) A nonthermal process, meaning that it is likely to be incoherent synchrotron or Compton scattered radiation,
(ii) Thermal radiation from dust.

If (ii) is operating, the grain temperatures must be low ($\lesssim 100°$ K) and thus the sources must be extended. Consequently we would not expect to see variations in flux in the infrared. On the other hand, if the radiation is nonthermal, we expect that it does arise in the machine in the nucleus. Even if the radiation is thermal, however, it may well be that the source which is heating the dust is nonthermal. As was pointed out by Rieke and Low (1972) very considerable problems are encountered if we attribute the energy source to stars. For large luminosities $\sim 10^{44}-10^{45}$ erg sec^{-1} very large numbers ($\sim 10^8$) of high luminosity O and B stars would be required. They would comprise a large fraction of the mass in the central region of the galaxy. However, their evolutionary lifetimes are very short ($\sim 10^6$ years). Such luminosities could not be maintained for more than a small fraction of the lifetime of the galaxy. Thus, the existence of such high infrared luminosities would only be expected in rare circumstances. Thus the fact that nearby galaxies are commonly found to have large infrared luminosities, suggests that the radiation is ultimately of nonthermal origin.

As the evidence stands at present, the only good case for variability in the infrared is NGC 4151. It appears that NGC 1068 has shown no variations and this source is extended.

In some cases the form of the infrared spectrum strongly suggests a thermal origin.

A serious problem associated with the large far-infrared luminosities is the large amount of dust and interstellar matter which is apparently required. In a recent study, Hildebrand et al. (1977) have concluded that to explain far infrared fluxes of 1.2×10^{45} erg sec^{-1} and 6×10^{43} erg sec^{-1} in NGC 1068 and NGC 253 respectively; the mass of dust required is 10^8 $M_\odot$ in NGC 1068 and 8×10^6 $M_\odot$ in NGC 253. Assuming a gas-to-dust ratio = 100, this gives total amounts of diffuse matter of 10^{10} $M_\odot$ in NGC 1068 and 8×10^8 $M_\odot$ in NGC 253. Now rotation curves are available for both of these galaxies (Burbidge, Burbidge and Prendergast 1959; Burbidge et al. 1962) and thus an upper limit to the mass contained in the same volume can be obtained. In both cases the total mass is comparable with, or considerably less than, the mass apparently required to explain the infrared observations. There are three possible ways to resolve this dilemma.

(1) To argue that the radiation is nonthermal. However, the spectra appear to have rough blackbody forms, so that this is unlikely.
(2) To argue that the gas-to-dust ratio is much less than 100.
(3) To suppose that the dust is a much more efficient radiator than is generally supposed, i.e. $Q(\nu)$, the emissivity of the grains, is very different from the values assumed.

Probably both (2) and (3) are important. Almost certainly the emissivity of the grains is more efficient, than has been assumed so far. But at present the observations present a puzzle.

X-Ray Emission from Seyfert Galaxies

Until very recently very few of the classical (nearby) Seyfert galaxies were known to be X-ray emitters. However, recent results from the Ariel V Sky Survey instrument (Elvis et al. 1977) show that many Seyfert galaxies are powerful X-ray sources. Classical Seyferts which are now identified include NGC 4151, NGC 3227, NGC 5548, NGC 6814, and NGC 1275. In total 15 Seyferts are now reported as X-ray sources. However, the others are further away and do not fall into the loose category of nearby Seyfert galaxies.

The luminosities lie in the range 10^{43}-10^{44} erg sec^{-1} in the photon energy range 2-10 keV (NGC 4151 is a strong source at $\sim$ 100 keV), and are thus 10 to 100 times more energetic than the optical fluxes from Seyfert nuclei.

This puts further demands on the energetic properties of the central machine.

It is much too early to say anything definitive about the mechanism of X-ray generation. The discoverers argue that the X-ray power is correlated with the IR and continuum optical flux and with the luminosity in Hα. They conclude that it arises in a region < 0.1 pc from the center. Possible mechanisms are bremsstrahlung, the hot gas arising from shock heating in the highly turbulent center where the broad lines arise, or Compton radiation from optical or IR photons generated by the synchrotron process. They favor the latter process.

Ejection of Large Gas Masses from Nuclei

One of the earliest indicators of violent activity in galactic nuclei was the evidence that matter in considerable amounts is being ejected at high velocities. Different kinds of observations suggest this. Here are some examples.

(a) High velocities in Seyfert nuclei (1000-10 000 km sec^{-1}) which are far greater than the escape velocities.

(b) The apparent explosive ejection in M82 which may now need to be reinterpreted.

(c) Ejection of a large mass of gas in NGC 1275 with a line of sight velocity difference of $\sim$ 3000 km sec^{-1} with respect to the center. The total mass is of order 10^8 $M_\odot$. Attempts to reinterpret the observation in terms of a colliding or intervening galaxy are, in my opinion, unconvincing.

(d) Similar evidence for a large ejection in the radio galaxy DA 240 which has a large jet. Gas in the jet has a line-of-sight velocity some 3000 km sec^{-1} less than the velocity of recession. DA 240 does not lie in a cluster of galaxies.

(e) The double structures in the emission lines in N systems like 3C 390.3, 3C 227, etc. are probably due to phenomena similar to those seen in NGC 1275 and DA 240. The separation of the emission line peaks is $\sim$ 3000-4000 km sec^{-1}.

(f) If the absorption-line systems in the spectra of QSOs are intrinsic to the objects, as seems likely, they indicate that gas shells are being ejected with velocities typically of order 0.1c.

(g) Ejecta from M87. The jet appears to be made up of a series of highly compact synchrotron sources. Velocities and masses are not known.

(h) Shreds of gas ejected from NGC 5128.

(i) Non-circular motions observed in the central regions of many spiral galaxies are most likely to be due to explosive ejection from the nuclei.

CONCLUSION

No attempt has been made here to review all of the many recent observations of the activity in the nuclei of nearby galaxies. Instead we have chosen to discuss a few phenomena which have recently been discovered. In each case the new observations show that nuclei are even more energetic than we have believed before.

The connections between theory and observation are still very tenuous. However, it does appear likely that almost every galaxy contains a "machine" in its nucleus which is able to release energy in many exotic forms, and which is active for a large fraction of the life of the galaxy, though it may, for long periods, operate at a low level of activity.

Work reported here has been supported in part by the National Science Foundation and in part by NASA through grant no. NGL 05-005-004.

REFERENCES

Aumann, H. H., and Low, F. J.: 1976, Astrophys. J. Letters 159, L159.
Burbidge, E. M., Burbidge, G. R., and Prendergast, K. H.: 1959, Astrophys. J. 130, 26.
Burbidge, E. M., Burbidge, G. R., and Prendergast, K. H.: 1962, Astrophys. J. 136, 339.
Burbidge, G. R., Burbidge, E. M., and Sandage, A. R.: 1963, Rev. Mod. Phys. 35, 947.
Clegg, P. E., Ade, P. A. R., and Rowan-Robinson, M.: 1976, in M. Rowan-Robinson (ed.), Far Infrared Astronomy, Pergamon Press, London, p. 209.
Elvis, M., Maccacaro, T., Wilson, A. S., Ward, M. J., Penston, M. V., Fosbury, R. A. E., and Perola, G. C.: 1977, preprint.
Hargrave, P. J.: 1974, Monthly Notices Roy. Astron. Soc. 168, 491.
Hildebrand, R. H., Whitcomb, S. E., Winston, R., Stiening, R. F., Harper, D. A., and Moseley, S. H.: 1977, Astrophys. J., in press.
Kleinmann, D. E., and Low, F. J.: 1970a, Astrophys. J. Letters 159, L165.
Kleinmann, D. E., and Low, F. J.: 1970b, Astrophys. J. Letters 161, L203.
Kronberg, P. P., and Wilkinson, P. N.: 1975, Astrophys. J. 200, 430.
Low, F. J.: 1970, Astrophys. J. Letters 159, L173.
Oort, J. H.: 1977, Ann. Rev. Astron. Astrophys., in press.
Penston, M. V., Penston, M. J., Neugebauer, G., Tritton, K. P., Becklin, E. E., and Visvanathan, N.: 1971, Monthly Notices Roy. Astron. Soc. 153, 29.
Penston, M. V., Penston, M. J., Selmes, R. A., Becklin, E. E., and Neugebauer, G.: 1974, Monthly Notices Roy. Astron. Soc. 169, 357.
Raff, M. I.: 1969, Astrophys. J. Letters 157, L27.
Rieke, G. H., Harper, D. A., Low, F. J., and Armstrong, K. R.: 1973, Astrophys. J. Letters 183, L67.
Rieke, G. H., and Low, F. J.: 1972, Astrophys. J. Letters 176, L95.
Solinger, A., Morrison, P., and Markert, T.: 1977, Astrophys. J. 211, 707.
van den Bergh, S.: 1971, Astron. Astrophys. 12, 474.

DISCUSSION FOLLOWING REVIEW IV.3 GIVEN BY G. BURBIDGE

OORT: In connection with your enumeration of the few cases of ordinary galaxies with direct evidence for expulsion of gas from the nuclear region one should mention the phenomena in our own Galaxy, in which among other phenomena indicating such expulsion there is the massive ring of molecular clouds at about 190 parsecs from the center expanding at a velocity of 150 km/s. In this case there is not much room for doubt that we are witnessing gas expelled from a small nuclear region, and in such quantity that quite high energies must have been involved.

EMISSION FROM THE NUCLEI OF NEARBY GALAXIES: EVIDENCE FOR MASSIVE
BLACK HOLES?

M.J. Rees
Institute of Astronomy,
Cambridge,
England.

1. INTRODUCTION

The non-stellar activity in the nuclei of nearby galaxies poses problems of its own. But it gains added interest insofar as it may provide clues to the nature of quasars and the unusually energetic nuclei of some more remote galaxies. The evident qualitative resemblance between these spectacular phenomena and some of the nearby galactic nuclei discussed at this symposium suggests that we may be witnessing a scaled-down or slow-motion version of the same physical mechanism; and the likelihood that dead quasars vastly outnumber living ones suggests that defunct remnants - perhaps displaying some low-level residual activity - may lurk in the centres of most large galaxies.

In this contribution I shall first venture some conjectures on scenarios whereby a massive black hole could form in a galactic nucleus, and the characteristic features of accretion onto it. I shall then suggest how some observed phenomena - particularly the radio and x-ray continuum - can be interpreted in terms of such processes. And in conclusion I shall mention some other tests for (or limits on) massive black holes in nearby galaxies.

Once the star or gas density in a galactic nucleus exceeds some threshold value, well-known arguments suggest that some kind of 'runaway' evolution ensues. This evolution can plausibly lead to the formation of a massive black hole, and the quasar phenomenon can be interpreted as a result of rapid accretion onto such a body. This process seems able to convert rest-mass energy into appropriate non-thermal forms with higher efficiency than any likely precursor stages. The scenarios that terminate in massive black holes are numerous (see Figure 1) and various lower-level manifestations of activity in nuclei may be attributable to the preliminary stages. (e.g. star-gas interactions or multiple supernovae in dense star clusters, supermassive objects or 'magnetoids', etc).

The evolution sketched in Figure 1 suggests that 'dead' quasars - interpreted as massive black holes that are now being starved of fuel-outnumber active ones by a factor which, though uncertain, may lie in the range 10^4-10^5. This figure allows for the number of generations of quasars in a Hubble time (10^3-10^2 if the characteristic lifetime is 10^7-10^8 years) and for the evidence that their density per comoving volume was perhaps 10^2-10^3 times higher at the epoch corresponding to $z \simeq 2$ than at the present time. It is thus possible that dead quasars could be as common as large galaxies.

These massive black holes could be detected by their influence on the density distribution or velocity dispersion of the stars around them; or there may be some luminosity arising from slow accretion of stars and gas. This latter process could be the explanation for some of the compact sources observed in some nuclei (though we should of course be open-minded about the possibility that some of the precursor stages shown in Figure 1 might also be relevant).

2. LOW-LEVEL ACCRETION ONTO MASSIVE BLACK HOLES

The masses involved in the production of a quasar or strong radio source are likely to be $\gtrsim 10^7$ solar masses, so it is convenient to introduce the parameter M_7. The Schwarzschild radius is then $r_s \simeq 3.10^{12} M_7$ cm. It is interesting to ask whether the limited data are consistent with a model in which the compact continuum emission from nuclei results from accretion onto such objects. The theory of this process, with special application to quasars, has been discussed at length elsewhere (Rees 1977, and references cited therein). The rate of accretion required in the present context is low, in the sense that $\dot{M} \ll \dot{M}_{crit}$. The 'critical' accretion rate $\dot{M}_{crit} \simeq 10^{-2} \varepsilon^{-1} M_7$ solar masses per year is that needed to generate the Eddington luminosity, assuming an efficiency factor of ε. When $\dot{M} \ll \dot{M}_{crit}$, the densities around the hole are lower than in the 'critical' case, and the cooling timescale tends to be longer than the infall time, unless the temperature gets so high that relativistic cooling processes (pair production, synchrotron and inverse Compton emission, etc) can come into play. This means that the infalling gas will heat up to a very high temperature, and the fraction of radiation emitted non-thermally from near the hole would be higher than for quasars. If there is a magnetic field in the gas, it may be amplified to a value comparable to the equipartition strength, which is approximately

$$B_{eq} \simeq 10^6 M_7^{-\frac{1}{2}} (\dot{M}/\dot{M}_{crit})^{\frac{1}{2}} (r/r_s)^{-5/4} (v_{infall}/v_{free\ fall})^{-\frac{1}{2}} G.$$

For disc-type accretion, v_{infall} may be much less than $v_{free\ fall}$, but when the inflow is 'quasi-spherical' the two would be comparable. When $\dot{M} \ll \dot{M}_{crit}$ the opacity of the infalling material is negligible except at radio wavelengths, and the flow pattern is unlikely to be confused by such complications as radiation-driven winds.

IV.4 NUCLEI OF NEARBY GALAXIES: EVIDENCE FOR MASSIVE BLACK HOLES? 239

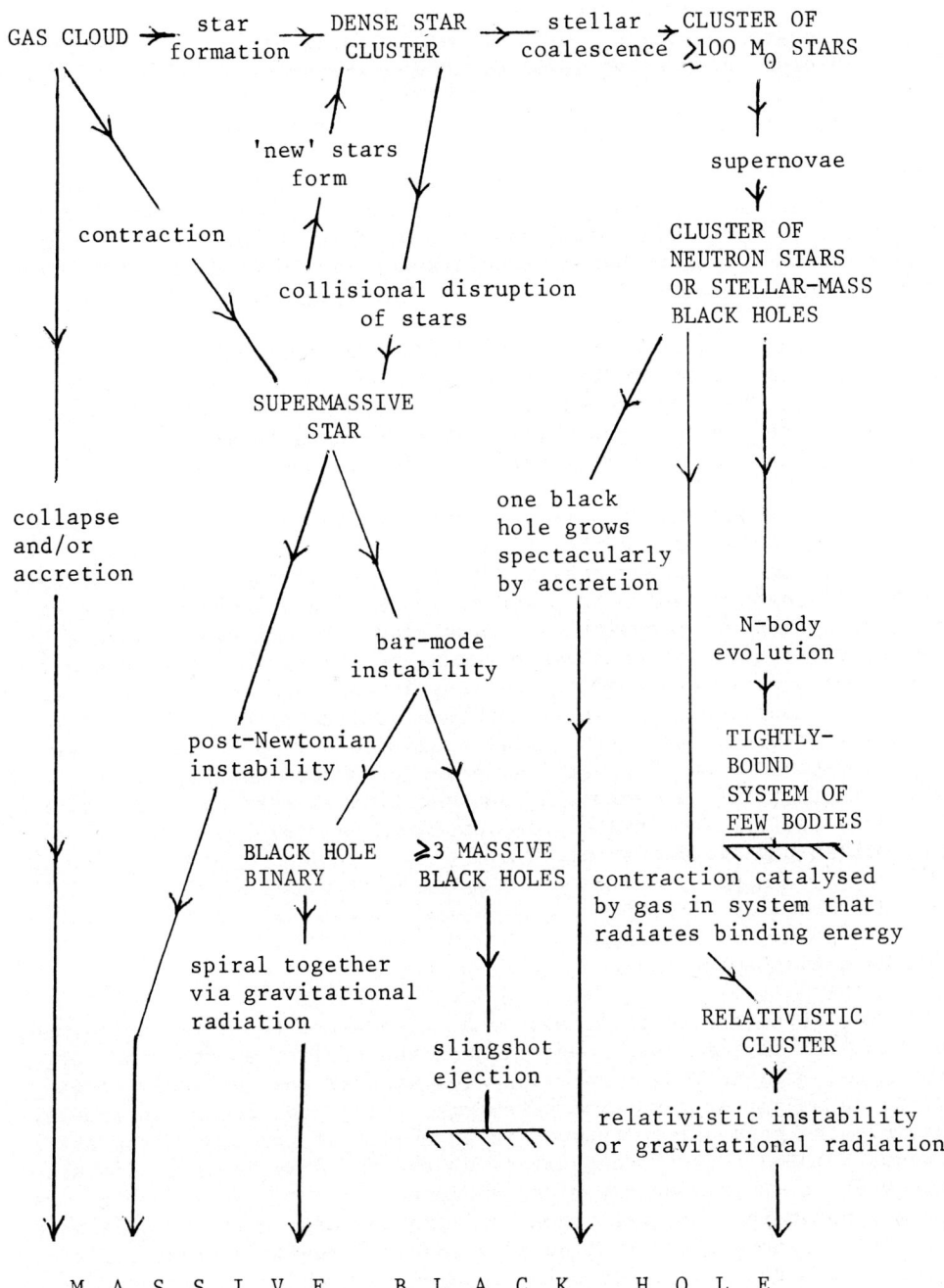

Figure 1. Scenarios for the formation of a massive black hole in an active galactic nucleus.

There are three general types of mechanism which could generate a non-thermal spectrum of relativistic electrons:

(i) Processes associated with magnetic flares or reconnection in a disc (Shields and Wheeler 1976) or in quasi-spherical inflow (Meszaros 1975).

(ii) Shock waves occurring near the hole, where the velocities are a good fraction of c (Fabian et al 1976). (It is known that shocks in supernova remnants, where the velocities are only 0.02c, are capable of accelerating relativistic electrons; and whatever process operates there should work even better when the velocities involved are higher.)

(iii) Exotic electromagnetic processes leading to production of electron-positron-pairs near a Kerr black hole (Blandford and Znajek 1977).

Electrons resulting from any of the above mechanisms would emit radiation via the synchrotron or inverse Compton process. Note, as regards (i), that the demands made on such a process are less severe when even the 'thermal' particles may themselves be mildly relativistic than in (e.g.) solar flares where the temperatures are much lower.

If the radio emission were synchrotron radiation, then self-absorption would require the effective emitting surface to have dimensions $\gg r_s$ (except possibly in the case of the Galactic Centre, where the power is only $\sim 10^{33}$ erg/sec). There are no cases where the current brightness temperature limits from VLBI are embarassingly high. On the other hand, it would by no means be surprising if some coherent radiation process operated in the very strong magnetic fields expected near accreting black holes. Synchrotron or inverse Compton radiation in the infra-red, optical or x-ray bands could in principle emerge from a region as small as $\sim r_s$, displaying correspondingly rapid variability. It still seems an open question whether the x-ray flux is more likely to be inverse Compton (Grindlay 1975) or thermal bremsstrahlung (Fabian et al 1976).

3. SOME SPECIFIC OBJECTS

The Galactic Centre

The very compact radio source at the Galactic Centre (Kellermann et al 1977) is much less powerful than the nuclear sources in, for instance, M81 and M82. Its properties make it most unlikely to be either a pulsar or a supernova remnant, though one cannot so readily exclude the possibility that it is an object of the same class as Cygnus X3. The recent observations of the infrared Ne line, discussed elsewhere at this meeting, place an upper limit of 5.10^6 solar masses on a hypothetical central black hole (which means that our Galaxy is unlikely to have ever experienced a phase of really violent radio or quasar-like activity). Accretion onto such an object could however provide a natural explanation for the radio component, and maybe also for some of the other phenomena occurring in the same region (Lynden-Bell and Rees 1971). If the radio source were the only manifestation of accretion, the required inflow rate would be only $\sim 10^{-14} \varepsilon^{-1}$ solar masses per year. The same source could perhaps be contributing to the infra-red flux from the Centre, in which case variability on timescales down to minutes could occur.

Centaurus A

Although the total radio power output of Cen A is only $\sim 10^{42}$ erg/sec, the energy stored in the very extended radio lobes is $\gtrsim 10^{60}$ ergs. This suggests that, when in its prime, Cen A might have been as spectacular an object as Cygnus A or 3C 273. If this energy were generated by a black hole (or its progenitor) the mass would be 10^8 solar masses, assuming reasonable efficiencies. The very compact central radio source, and the $\sim 10^{44}$ erg/sec variable x-ray source, can be explained in terms of accretion onto such an object (Fabian et al 1976). If this interpretation is correct, Cen A is the nearest galaxy that displays evidence of accretion onto a black hole at an abnormally high level.

M 87

Recent photometry of the central region of M87, together with evidence that the stellar velocity dispersion rises towards the centre, is suggestive of the presence of a central 'dark mass' of 10^{10} solar masses (Young 1977). Several hypotheses are consistent with the present tentative data, but one obvious possibility is that this mass constitutes a black hole, which could be the remnant of a violent outburst in the past. A hole as large as this would swallow stars whole, without tidally disrupting them; but it would be worthwhile to attempt to interpret the peculiarities of M 87's nucleus on the basis of the black hole hypothesis (c.f. Lynden-Bell and Rees 1971).

4. SOME OBSERVATIONAL QUESTIONS

(i) The magnetic fields in these compact radio components may be higher than in the central components of quasars and strong radio galaxies. The lower limits to surface brightness (proportional to $B^{-\frac{1}{2}}$ for a self-absorbed source) which can be set by VLBI measurements may therefore set significant constraints on models.

(ii) In the case of the weak source at the Galactic Centre, the magnetic field may be so high in the emitting region that circular polarization could be detected.

(iii) At the moment it is unclear whether the variable x-rays from Cen A (and also from other extragalactic sources such as 3C 120 and 3C 273) is thermal or non-thermal. Detection of a broad, variable (and possibly gravitationally-redshifted) x-ray Fe line would clinch this question, and also provide firmer evidence for the existence of a relativistically deep potential well.

(iv) The detection of variability on timescales down to r_s/c would be expected according to the black hole hypothesis in all bands except the radio.

(v) Statistical studies of variability could test whether there is evidence for any outburst of standardized energy, such as might be attributed to supernovae, or to stars being disrupted and swallowed by a black hole.

(vi) Correlations between spectrum, luminosity, variability, etc. could in principle help to decide how the observed properties of the nuclei depend on the hole mass, on $\dot{M}$, and on the angular momentum or other properties of the infalling material.

REFERENCES

Blandford, R.D., Znajek, R.L.: 1977, Monthly Notices Roy.Astron.Soc. 179, 433.
Fabian, A.C., Maccagni, D., Rees, M.J., Stoeger, W.R.: 1976, Nature 260, 683.
Grindlay, J.E.: 1975, Astrophys.J. 199, 49.
Kellermann, K.I., Shaffer, D.B., Clark, B.G., Geldzahler, B.J.: 1977, Astrophys.J.Letters 214, L61.
Lynden-Bell, D., Rees, M.J.: 1971, Monthly Notices Roy.Astron.Soc. 152, 461.
Meszaros, P.: 1975, Astron.Astrophys. 44, 59.
Rees, M.J.: 1977, Ann.N.Y.Acad.Sci. (in press).
Shields, G., Wheeler, J.C.: 1976, Astrophys.Letters 17, 69.
Young, P.J.: August 1977, talk given at Cambridge quasar conference.

DISCUSSION FOLLOWING REVIEW IV.4 GIVEN BY M.J. REES

APPENZELLER: I would like to comment on one of the processes for producing black holes indicated in your "flow diagram": As you know Dr. Klaus Fricke and myself have carried out detailed model computations for the consequences of the relativistic instability of massive objects. The results do not agree with your diagram. If you start with an equilibrium supermassive star ($10^5 \lesssim M/M_\odot \lesssim 10^9$) and make it unstable by either adding mass or by removing angular momentum, the object will explode rather than collapse to a black hole. Thus, the relativistic instability will normally lead to explosive events (which are observed in active galaxies) but not to the formation of massive black holes.

REES: Yes, I agree. This possibility should be represented by an extra arrow in my (already overcomplicated!) "flow diagram".

SHU: I have a question in connection with your comments on mass ejection being a natural expectation of this model. How does mass ejection occur simultaneously with mass accretion, especially in a spherical geometry? Could you elaborate how you visualize matter both to come and go?

REES: There must obviously be some deviation from strict spherical symmetry. One possibility is that inflow occurs in an equatorial plane, with outflow along a rotation axis (as in the "turn exhaust" radio source model); another possibility is that one has a turbulent or "two phase" mixture of rising and falling elements. Of course, the net flow must always be inward if accretion is the power supply.

WAXMAN: Concerning the simultaneous accretion and ejection of matter from an initially spherically symmetric accreting hole, we should realize that as the luminosity approaches the Eddington limit, the radiation field might be thought of as a "fluid". Then the radiation – gas interface may go unstable in analogy to a Rayleigh-Taylor instability yielding both channels of infalling matter and channels of outpouring radiation which may carry some matter with it.

REES: Yes. We would expect fluid elements where p_{rad}/p_{matter} exceeds the average value to develop into jets or bubbles. When this happens, there is of course no reason why the Eddington limit should not be exceeded.

OORT: You have made suggestions for the interpretation of many phenomena connected with nuclear activity, and it seems ungrateful to ask whether you could suggest an explanation for one more phenomenon, viz. the tendency for intermittancy in the expulsions observed. In the motions of the large molecular complexes near the Galactic center time scales of the order of a million years are suggested. And similar time scales occur in the emission of radio blobs in some radio galaxies, in particular some head-tail ones.

REES: Possibilities might include: (1) the interval between the disruption of successive stars by a massive black hole, (2) the time scale of gas flows in the Galactic nucleus, or (3) the time scale for gas accumulating in a potential well to become so dense that its cooling time becomes less than its free-fall time, leading to sudden infall. But none of these mechanisms leads to a natural time scale of a million years.

VAN DER LAAN: Your lecture casts doubt on the appropriateness of the ratio of "living/dead" nuclei of galaxies and quasars with which it began. A large range of temporal variations for any one nucleus' power seems implied by the physical scenarios you sketched, with potential activity depending only on appropriate fuel. If you must retain a biotic metaphor, perhaps the ratio "awake/asleep" is preferable.

REES: It is unclear whether low-luminosity galactic nuclei represent slowly-dying quasars, or whether they have never been luminous. It is also possible that some "dormant" black holes may be "revived" if they receive a renewed gas supply. (The work of Condon and Dressel suggests that this may happen in disturbed interacting systems.) I agree that "awakening" is a more appropriate term for this star "resurrection"!

CAPACCIOLI: In addition to what you said about M87, I like to mention that near the center of NGC 3379 de Vaucouleurs and I found a strong excess of light with respect to the requirements of the isothermal model. This excess has been analyzed using the model worked out by P. Young and is consistent with the presence of a nuclear "black hole" having a mass of 4×10^8 $M_\odot$.

V THE OUTSKIRTS OF GALAXIES

"My life is made miserably difficult by these facts which seem to be getting better and better."

A. Toomre in Discussion V.2

GALAXY HALOES AND THE MISSING MASS PROBLEM

Sidney van den Bergh
David Dunlap Observatory, University of Toronto

> *I am afraid my subject is rather an exciting one and as I don't like excitement, I shall approach it in a gentle, timid, roundabout way.*
> - Max Beerbohm

ABSTRACT

Available evidence on the mass-to-light ratios in binary galaxies, small clusters and in rich clusters is reviewed. The interpretation of the binary data remains ambiguous. The relative velocity of M81 and M82 implies that $\mathcal{M}/L_V > 12$. The Local Group data also appear to favour the existence of some hidden mass associated with M31 and the Galaxy. In the Virgo cluster a significant fraction of the virial mass may be concentrated in the halo of M87. In Coma the missing mass was probably stripped from the haloes of the two dominant E galaxies and is now distributed throughout the cluster.

It is suggested that metal-poor stars in the galactic disc have a higher CNO abundance than do halo stars of similar metallicity. This may explain why globular cluster stars lie on taller red giant branches than do moderately high-velocity stars near the Sun. The dominant old red giant population in the disc of M31 appears to be similar to that in the Galaxy. Most of the red giants in M33, NGC 6822 and in the LMC lie on taller red giant branches than do those in the Galaxy and M31. This suggests that old evolved stars in these relatively low-luminosity galaxies have a lower heavy element abundance than do their counterparts in the galactic disc. The red giants in M32 appear to be intermediate between those in M31 and M33. This supports the idea that M32 was once much more luminous than it is at present.

1. STELLAR POPULATIONS IN THE HALO AND DISC

1.1 Old red giants in the Galaxy

According to ideas that were first introduced by Baade (1944) a spiral galaxy consists of a disc of (metal-rich) Population I stars that is embedded within a halo consisting of (metal-poor) Population II stars. The most striking difference between these two populations is that halo Population II stars have tall red giant branches with $-3 < M_V < -2$ whereas the old giants of Population I are much fainter and have $-1 < M_V < 0$.

The first indication that something might be wrong with this beautifully simple picture was obtained by Keenan and Keller (1953) who showed (cf. Figure 1) that high-velocity stars in the solar vicinity have a colour-magnitude diagram resembling that of old Population I. This result was totally unexpected because high-velocity stars in the disc were supposed to be representative members of the galactic halo population.

The puzzling observations of Keenan and Keller were ignored for the next two decades. Baade believed (Sandage 1977) that the spectroscopic luminosity estimates by Keenan and Keller must have been falsified by metallicity effects.

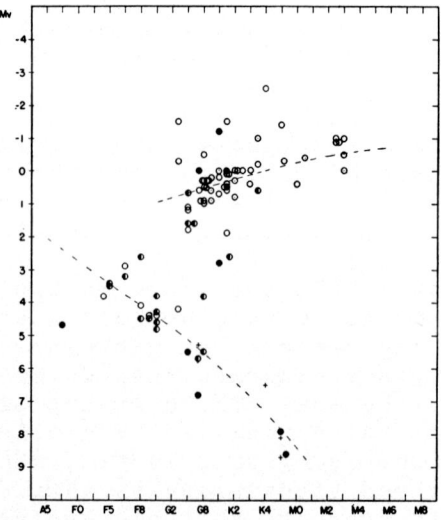

Fig. 1. Hertzsprung-Russell diagram for high-velocity ($V > 85$ km s^{-1}) stars according to Keenan and Keller (1953). The figure shows that most nearby high-velocity giants lie well below the giant branch of a typical halo globular cluster.

Striking confirmation of the results obtained by Keenan and Keller was, however, obtained by Hartwick and Hesser (1972). These authors showed (see Figure 2) that high-velocity field stars with ultraviolet excesses (which measures [Fe/H]) $\delta(U-B) \simeq +0.11$ and globular cluster giants in 47 Tucanae, for which $<\delta(U-B)> = +0.10$, have differing red giant luminosities. This conclusion is strengthened and confirmed by the work of Demarque and McClure (1977) who show (cf. Figure 3) that the old open cluster NGC 2420, for which $\delta(U-B) = +0.11$, has a fainter red giant branch than does the relatively metal-rich globular cluster 47 Tuc. Calculations by these authors show that the observed differences between the giants in 47 Tuc and NGC 2420 might be explained if either (1) 47 Tuc is richer in helium than NGC 2420 by $\Delta Y \sim 0.1$ or (2) if 47 Tuc has a ten times lower Z(CNO) than does NGC 2420. Substantial support for the latter alternative is provided by the work of Hearnshaw (1972) who concludes that "*iron-deficient old disc stars are generally no more than marginally carbon-deficient, a result which is probably not valid for halo stars.*" Additional evidence for this conclusion is

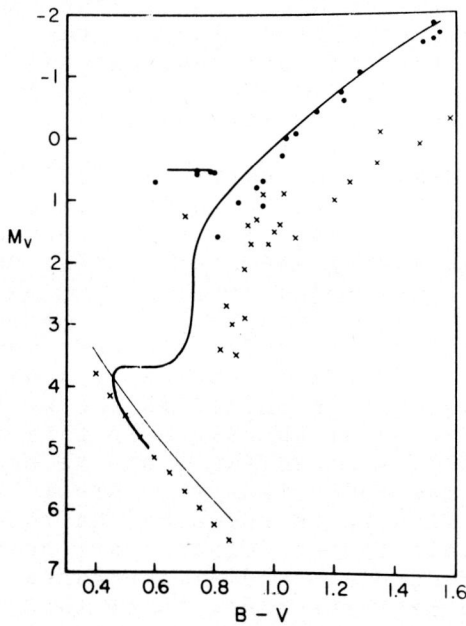

Fig. 2. Colour-magnitude diagram (Hartwick and Hesser 1972) for nearby stars (×) with $\delta(U-B) \simeq +0.11$ and for stars (•) in the globular cluster 47 Tucanae which have $\delta(U-B) = +0.10$. The figure shows that metal-poor disc stars are fainter than halo stars of similar metallicity.

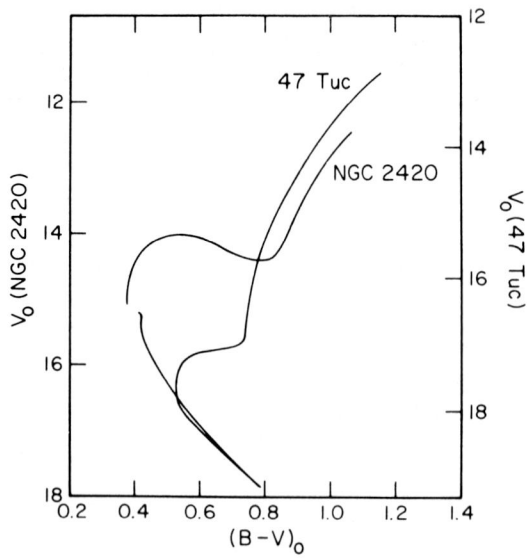

Fig. 3. Comparison of the colour-magnitude diagrams of the metal-poor open cluster NGC 2420 and the globular 47 Tucanae. The fact that 47 Tuc has the tallest giant branch implies (cf. Demarque and Geisler 1963) that Z(47 Tuc) < Z(NGC 2420) despite the fact that both clusters have similar Fe/H values.

provided by the work of Hesser, Hartwick and McClure (1976) who find that the average CN strength in the giant stars of the relatively metal-rich globular cluster 47 Tucanae is lower than that in field high-velocity stars of similar metallicity.*

---

*A possible difficulty with this point of view is that Peimbert (1973) finds that CNO/Fe is unusually *high* in the planetary nebula K 648 which is situated in the metal-poor globular cluster M15. The metal abundance in this cluster is down by a factor of $\sim$100 whereas the O and Ne abundances in K 648 are only $\sim$10 times lower than they are in the Sun. An even more extreme difficulty is posed by the recently-discovered planetary nebula in the Fornax dwarf galaxy. Danziger et al. (1977) find the O and Ar abundances in this planetary to be only two or three times lower than they are in the Orion Nebula! This result is particularly surprising because the globular clusters associated with the Fornax system (van den Bergh 1969, Danziger 1973) are exceedingly metal-poor. In view of these results it might be necessary to reconsider Peimbert's (1973) conclusion that the material in the envelopes of planetary nebulae cannot have been enriched in heavy elements by their central stars.

Additional evidence in favour of the view that halo stars have a low CNO/Fe abundance ratio is provided by chemical abundance analyses of halo stars. One of the best known examples of a nearby star which kinematically belongs to the galactic halo is the sub-dwarf Groombridge 1830. In this unevolved star carbon and nitrogen are found to be underabundant relative to iron (Tomkin and Bell 1973). Analysis of the ^{12}CH and ^{13}CH absorption features in Gmb 1830 by Lambert and Sneden (1977) shows that $^{12}C/^{13}C \geqslant 90$ which is similar to the $^{12}C/^{13}C$ ratio in the Sun. Since nucleosynthesis in very massive objects (Wagoner, Fowler and Hoyle 1967) is believed to produce roughly equal amounts of ^{12}C and ^{13}C this suggests that very massive objects did not dominate the heavy element enrichment during the halo phase of galactic evolution. It therefore seems improbable that the remnants of such massive objects could contribute significantly to the "missing" halo mass.

1.2 Red giants in other Local Group galaxies

The bright red giants that Baade (1944) discovered in the disc of M31 are embedded in a much richer population of faint giants which contribute most of the light. A similar situation prevails in the galactic disc near the Sun and in the galactic nuclear bulge (Arp 1959, van den Bergh 1971a). Baade (1963) refers to the bright Population II giants in the disc of the Andromeda Nebula as "the frosting on the cake".

Recently Sandage (1977) has emphasised the fact that the red giants in the disc of M33, in NGC 6822 and in IC 1613 (Sandage 1971) are *all* bright (V $\sim$ 21.5) and easily resolvable. Sandage points out that the surface density of these bright red stars in M33 is $\sim$100 times greater than it is in the galactic disc near the Sun. It follows that these bright red giants in M33 constitute the dominant old population in the Triangulum Nebula; not a "frosting on the cake" as they do in M31 and the Galaxy.

Probably the most straightforward interpretation of this result is that the old red giants in M33, NGC 147, NGC 185, NGC 205 and NGC 6822 lie on tall red giant branches because they have low Z values. The alternative hypothesis that these stars lie on tall branches because they have high Y values appears to be ruled out by the observation that the *present* helium abundance in M33 and in NGC 6822 (Peimbert and Spinrad 1970a, b) is slightly below that prevailing in galactic HII regions.

In IC 1613 and in the Small Magellanic Cloud the interpretation of the observations is rendered uncertain by the possibility that a major fraction of the bright red giants might resemble those in the intermediate-age SMC cluster NGC 419.

A comparison between recent abundance determinations for HII regions in the SMC, the LMC and the Galaxy is given in Table I. These data suggest that Z(Galaxy)/Z(LMC) $\simeq$ 3 and Z(Galaxy)/Z(SMC) $\simeq$ 10. These results are consistent with the weakening of the lines of N, O, Si (and probably of Mg and C as well) that has been reported by Osmer (1973) in the spectra of LMC and SMC stars. Little direct evidence is available on the Fe abundance of stars in the Magellanic Clouds. From DDO photometry of the giant stars in the intermediate-age LMC cluster NGC 2209 Gascoigne et al. (1976) obtain [Fe/H] $\sim$ 1.0. From synthetic spectra Bell and Parsons (1972) find that stars with $T_{eff} \sim$ 6000 K will become $\sim$0.1 mag bluer in B-V if their metal abundance is reduced by a factor of 4. From a compilation of data on all Cepheids with photoelectric B-V observations (van den Bergh 1977a) it appears that [Fe/H] $\gtrsim$ -0.6 for the LMC and [Fe/H] $\lesssim$ -0.6 for the SMC. These results indicate that both the abundances of the CNO group and of the metals are presently substantially lower in the Magellanic Clouds than they are in solar-type stars. This suggests that the old stars in the Magellanic Clouds must also have significantly lower Z values than do similar stars in the Galaxy. This conclusion is supported by Hardy's (1977) observations of stars in the disc of the LMC. These data, which are plotted in Figure 4, show that most of the old disc giants in the LMC are situated on evolutionary tracks that lie between the giant branches of the SMC globular NGC 121 and the relatively metal-rich galactic globular 47 Tuc. Hardy's data suggest, but do not yet prove conclusively, that the disc of the LMC contains few if any of the high Z M67 and NGC 188 type of red giants with $M_V \sim$ 0.

TABLE I

HEAVY ELEMENT ABUNDANCES*

Region	He	N	O	Ne
SMC HII	10.89	6.4	8.0	7.3
LMC HII	10.92	7.0	8.5	7.8
Orion Nebula	11.00	7.7	8.89	8.15

*Entries in the table are log [N(X)/N(H)] + 12.00 taken from a compilation by Dufour and Killen (1977).

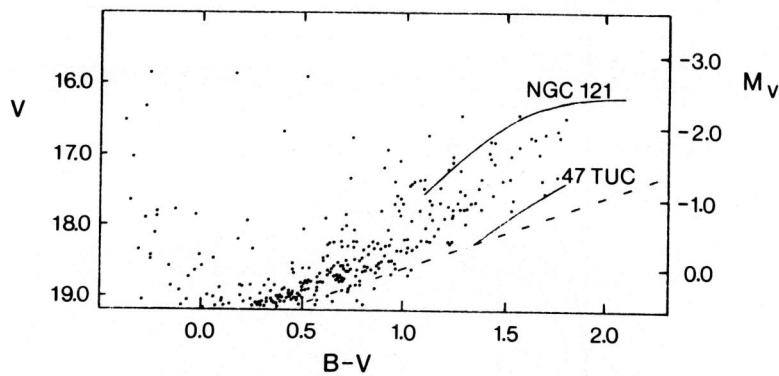

Fig. 4. Colour-magnitude diagram for a region in the disc of the LMC (Hardy 1977). The figure shows that most giants lie on giant branches that are intermediate between those of NGC 121 and 47 Tucanae. The dashed line shows the limit of completeness of the data.

The results obtained above may be summarised as follows: The galaxies of the Local Group may be divided into two groups according to the properties of their dominant red giant populations. In M33, the LMC, NGC 6822, NGC 147, NGC 185, NGC 205 and probably in the SMC and IC 1613 as well the red giants are bright indicating a relatively low heavy element abundance. In the Galaxy and M31 the dominant red giant population consists of old faint giants that are composed of material with a relatively high Z value. The red giants in M32 seem to have characteristics that are intermediate between those of the high-luminosity and low-luminosity members of the Local Group. This result is consistent with the work by Faber (1973a, b) who finds that M32 has lost a considerable amount of mass by tidal stripping resulting from encounters with M31. Its CN + Mg index suggests that M32 may now be ∼5 times less luminous than it was originally.

In the dwarf galaxies of the Local Group and in the halo of the Galaxy low metallicity and low CNO (and hence low Z) go together. This contrasts with the situation in the galactic disc in which Z(CNO) appears to have remained constant while Z(Fe) continued to rise (*cf*. Figure 5). These results show that the traditional practice of using the terms metallicity (Fe/H) and heavy element abundance (Z) interchangeably is both misleading and incorrect.

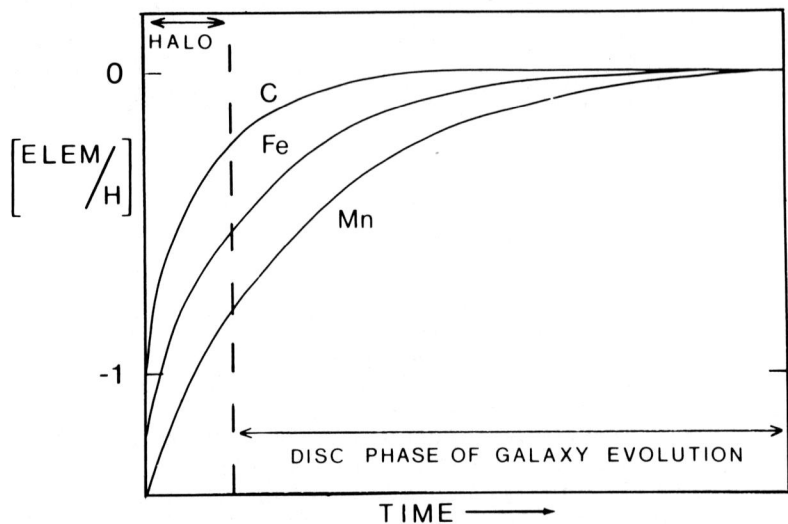

Fig. 5. Schematic plot illustrating the time-dependent changes of the C, Fe and Mn abundances in the galactic disc according to Hearnshaw (1972). The diagram shows that moderately metal-poor disc stars might have almost normal carbon abundances.

2. EVIDENCE FOR MISSING MASS

2.1 Missing mass near the Sun

In a recent review Gliese (1976) shows that the total density of known stars and interstellar material near the Sun amounts to $0.07 \lesssim \rho \lesssim 0.09$ $m_\odot pc^{-3}$. This value appears to be significantly smaller than the dynamically-determined local mass density of 0.15 $m_\odot pc^{-3}$ (Oort 1965). It is profoundly disturbing to find that as much as one third to one half of the mass in the nearest region of space may exist in some unknown form. According to Schmidt (1975) the local density of halo stars, defined as objects with tangential velocities larger than 250 km s^{-1}, is 1.7×10^{-4} $m_\odot pc^{-3}$. This represents only $\sim 1 \times 10^{-3}$ of the total local density. It follows that the missing mass near the Sun cannot be in the form of normal $(0.1 < m/m_\odot < 1)$ halo stars.

Theoretical work by Kalnajs (1972) and by Ostriker and Peebles (1973) appears to show that "cold disc" galaxies are unstable unless they are embedded in massive haloes. To date direct optical evidence for the existence of such massive haloes is still lacking. Available colour measurements show that the haloes of most galaxies are bluer than are their main bodies (*cf*. Simkin 1975, Strom and Strom 1977). The most straightforward interpretation of this observation is that globular cluster-like stars do and that faint M-type dwarfs do not contribute significantly to the integrated light of galaxy haloes.

2.2 Velocity dispersion of galactic globular clusters

Hartwick and Sargent (1977) have obtained new radial velocities for distant Palomar-type globulars and for dwarf spheroidal companions to the Galaxy. For eleven such objects, which are situated at an effective distance of ~ 75 kpc, Hartwick and Sargent get a total galactic mass of $(9 \pm 2) \times 10^{11}\, m_\odot$ if the cluster velocities are assumed isotropic and $(3.2 \pm 0.6) \times 10^{11}\, m_\odot$ if the cluster orbits are purely radial. These values appear to be significantly larger than those obtained from conventional galactic models (*eg*. Schmidt 1956, Innanen 1973) which yield masses in the range $1 - 2 \times 10^{11}\, m_\odot$. This conclusion is, however, quite uncertain because the number of outlying objects for which radial velocities are available is quite small. Radial velocities of outlying clusters in M31 may soon provide a value for the mass of the halo of the Andromeda Nebula.

2.3 Rotation curves of spirals

21-cm observations of neutral hydrogen gas have made it possible to extend the rotation curves of galaxies to much greater radial distances than was possible previously. Such observations (Roberts 1975, Krumm and Salpeter 1977) appear to show that a substantial number of spirals have rotation curves that are essentially flat out to the largest distances to which neutral hydrogen can be detected. Additional observations are, however, needed to prove that the observed rotation curves are not significantly affected by non-circular motions or by "spillover". Such spillover might be important in some cases where an antenna points toward the faint outer regions of a galaxy while receiving emission from the bright central area of that galaxy into a side lobe.

2.4 Binary galaxies

In principle observations of the velocity differences of pairs of galaxies (Page 1952) should be able to provide information on the total masses of galaxies. Due to complex selection effects and differing methods of analysis different authors still obtain quite widely divergent results. Assuming $H = 50$ km s^{-1} Mpc^{-1} Karachentsev (1977) obtains $m/L \sim 7$ for spiral-spiral pairs whereas Turner (1977) gets $m/L \sim 65$. The former result is quite compatible with that obtained from rotation curves whereas the latter value is consistent with massive halo models. It is unlikely that the difference between these results is due to a difference in the mean separations of the pairs studied by these two authors. For Karachentsev's spiral-spiral pairs the mean separation of the components is 50 kpc, compared to a mean separation of 64 kpc for Turner's field galaxy sample. In the case of elliptical galaxies (which are usually located in clusters), the determination of m/L ratios is particularly tricky because it is often difficult to distinguish between optical and physical pairs.

2.5 Small clusters of galaxies

The density contrast between small clusters and the general background field of galaxies is low. As a result small cluster samples are always in danger of contamination by field galaxies. Simple-minded attempts to reduce such contamination by using galaxy velocities as a selection criterion are self-defeating since they result in truncation of the Maxwellian velocity distribution. This leads to an artificial reduction in the velocity dispersion and hence to m/L ratios that are too low. Such a bias may, at least in part, account for the low mass-to-light ratios obtained by Materne and Tammann (1974). An additional complication is that the observed velocity dispersion in some density enhancements on the sky can be reduced (Tully and Fisher 1976) by assuming that they consist of two or more clusters at different distances that happen to be superimposed on each other. Finally a number of small clusters have "crossing times" that are longer than the Hubble time (Jackson 1975). Such groups need not be gravitationally bound.

From a recent analysis of data on 39 groups of galaxies Gott and Turner (1977) obtain $m/L \sim 140$. For the richest subset in their sample these authors obtain $m/L \approx 200$ whereas they find that $m/L \sim 65$ is more appropriate for small groups. High mass-to-light ratios for many (but not all) small groups are also derived by Rood et al. (1970) and by Rood and Dickel (1976). Einasto et al. (1975) have argued that these high m/L ratios are due to hidden mass that is located in the haloes ("coronas") of the giant galaxies that dominate some of these groups. Additional information on the m/L ratios in small clusters may be obtained from observations of the Local Group and the M81 group.

Perhaps the most remarkable feature of the Local Group (Kahn and Woltjer 1959) is that M31 and the Galaxy (which probably contain most of the mass) are presently *approaching* each other. If M31 and the Galaxy were formed close together 1.3×10^{10} yr ago, reached a maximum separation of 850 kpc 8.5×10^9 yr ago and are now falling towards each other at

70 km s^{-1} then their combined mass must be $\sim 2.9 \times 10^{12}$ $\mathcal{M}_\odot$ from which $\mathcal{M}/L \sim 125$ (Gunn 1974). The conclusion that the dynamics of the Local Group is dominated by missing mass can only be avoided (Herbst 1975) if the galactic rotational velocity of the Sun is $\gtrsim 290$ km s^{-1}, which appears to be unacceptably high. This problem is, however, slightly alleviated by the fact that all of the companions to the Andromeda Nebula (see Table II) for which radial velocities are available are redshifted relative to M31 itself.

On the basis of the frequency with which supergiant galaxies occur in nearby space van den Bergh (1971b) finds that the *a priori* probability that M31 and the Galaxy are unrelated galaxies that are passing each other like ships in the night is only ~ 4 percent. Furthermore, such a situation is physically implausible (Silk and Lea 1973) because decay of primaeval random motions in an expanding Universe would be expected to reduce the relative velocities of clusters to at most a few km s^{-1} at the present epoch.

The M81 group is dominated by M81 and M82. M81 has a radial velocity $V = -44$ km s^{-1} ($V_0 = +95$ km s^{-1}) (de Vaucouleurs, de Vaucouleurs and Corwin 1976). For M82 neutral hydrogen observations yield $V = +205$ km s^{-1} (Crutcher, Rogstad and Chu 1977) and optical observations (O'Connell and Mangano 1977) give $+200.6 \pm 2.4$ km s^{-1}. Adopting $V = +203$ km s^{-1} ($V_0 = +345$ km s^{-1}) for M82 yields a velocity difference $\Delta V_0 = 250$ km s^{-1}. For a closed orbit

$$\Delta V_0^2 \, P < 2G \, (M_{81} + M_{82}). \tag{1}$$

At a distance of 3.25 Mpc (Tammann and Sandage 1968) the projected separation P of M81 and M82 is 36 kpc so that $M_{81} + M_{82} > 2.6 \times 10^{11}$ $\mathcal{M}_\odot$. Since $L_{81} + L_{82} = 2.2 \times 10^{10}$ $L_\odot$ (Holmberg 1958) it follows that $\mathcal{M}/L_V > 12$.

TABLE II

RADIAL VELOCITIES OF COMPANIONS TO M31

Galaxy	V(km s^{-1})	V$_0$(km s^{-1})	ΔV$_0$(km s^{-1})
M31	-299	-61	0
M33	-180	+5	+66
NGC 221	-195	+43	+104
NGC 205	-239	+1	+62
NGC 185	-208	+41	+102
NGC 147	-157	+95	+156

2.6 Rich clusters of galaxies

Application of the Virial theorem to rich clusters of galaxies appears to give overwhelming support to the notion that the dynamics of such clusters is dominated by hidden mass. Recent determinations of the m/L ratios in the nearest giant clusters (van den Bergh 1977b) are summarised in Table III.

X-ray observations of the Virgo cluster (Gorenstein et al. 1976) show that the diffuse emission from this cluster is centred on the giant elliptical M87 to an accuracy of $\sim 3'$. The x-ray emission comes from a region with a radius of $\sim 1°$, which is significantly smaller than the $\sim 6°$ radius of the Virgo cluster core. If the observed x-rays are generated by hot gas that is confined by the potential well of M87 then the total mass of this galaxy must be a significant fraction of the Virial mass of the Virgo cluster (Bahcall and Sarazin 1977, Mathews 1977). This observation constitutes one of the strongest pieces of evidence in favour of the view that some galaxies may have very massive haloes (Ostriker, Peebles and Yahil 1974). The situation in Virgo appears to differ from that prevailing in the Coma cluster in which the low degree of concentration of massive galaxies towards the cluster centre (Rood 1965, White 1977) shows that most of the cluster mass is not bound to individual galaxies. For the intergalactic material in the Coma cluster Melnik, White and Hoessel (1977) find $m/L \gtrsim 1000$. Possibly this material was originally located in the haloes of the two dominant galaxies in the Coma cluster and was subsequently stripped off by tidal interactions (cf. Gallagher and Ostriker 1972). Richstone (1976) has shown that as much as ninety percent of the total mass of galaxies in a typical rich cluster might be stripped off by tidal shocks during one Hubble time.

TABLE III

MASS-TO-LIGHT RATIOS IN THE NEAREST RICH CLUSTERS

Cluster	m/L_V*
Perseus	350
Coma	200
Virgo	250

*$H = 50$ km s^{-1}Mpc^{-1} assumed.

The importance of tidal interactions on the outer structure of E galaxies is beautifully illustrated by Kormendy's (1977) photometry. His results show that elliptical galaxies with companions have much more extended haloes than do isolated ellipticals. Additional support for the view that the outer structure of E galaxies is affected by tidal interactions is provided by recent CTIO 4-m observations (van den Bergh 1978). This work shows that the envelopes of "cD" galaxies in *poor* clusters (Morgan, Kayser and White 1975) almost all fall into two classes: (A) objects with faint symmetrical envelopes that were presumably produced during tidal encounters that took place long ago and (B) galaxies embedded in bright asymmetrical envelopes that were probably formed by recent tidal encounters.

Finally deep photographs of radio sources associated with nearby elliptical galaxies such as Centaurus A (NGC 5128 = Figure 6) and Fornax A (NGC 1316 = Figure 7) show that these galaxies are embedded in extended haloes that exhibit a rather chaotic structure. Neither the physical nature of these haloes nor their origins are presently understood.

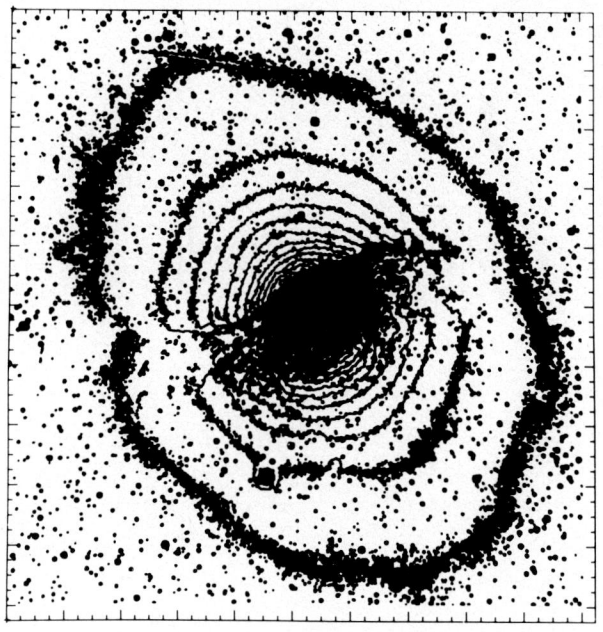

Fig. 6. Outer isophotes of NGC 5128 in yellow (103aD + GG495) light according to Dufour and van den Bergh (1978).

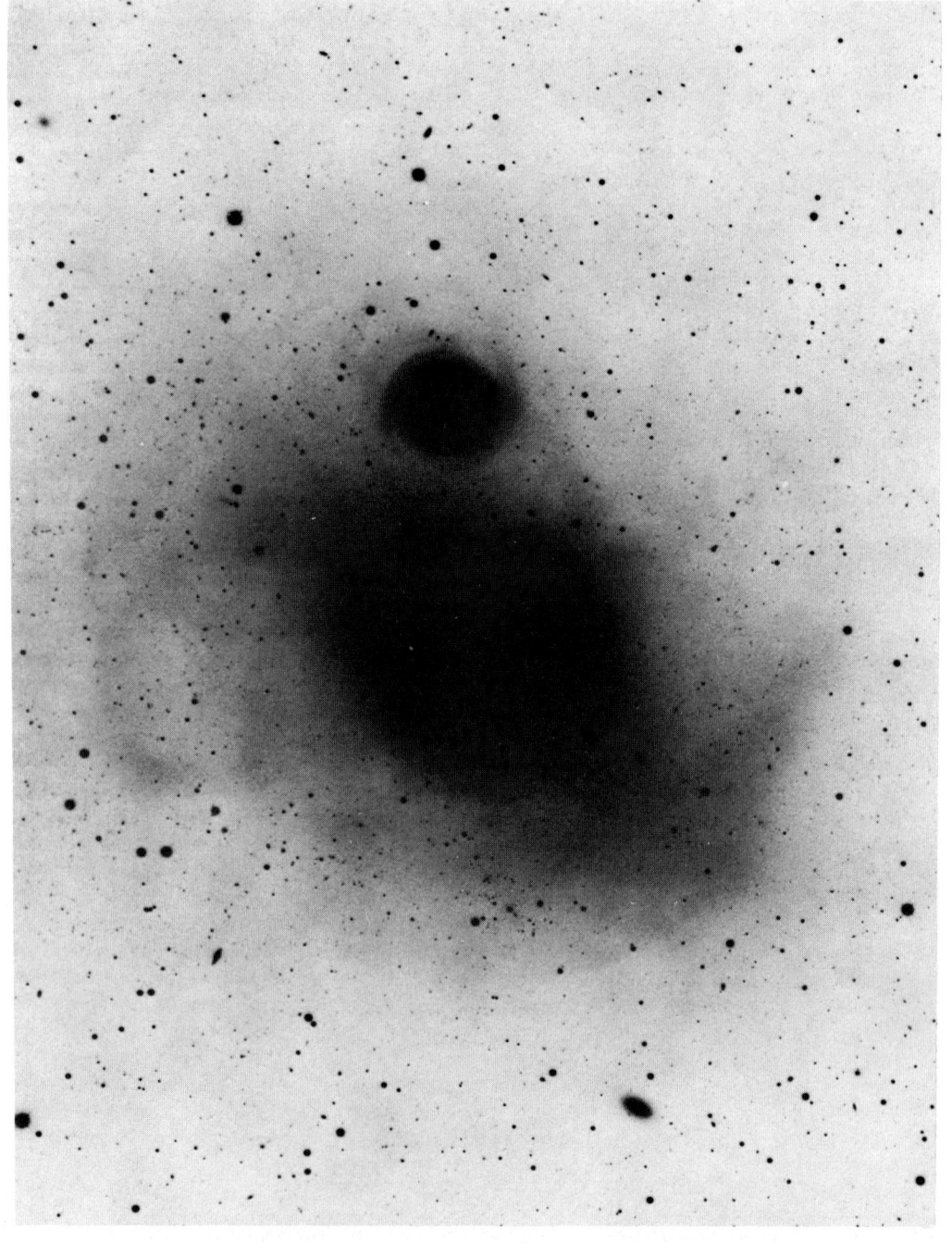

Fig. 7. Photograph of the halo of NGC 1316 (Fornax A) obtained in red light (127-04 + GG495) with the CTIO 4-m reflector.

REFERENCES

Arp, H. C.: 1959, *Astron. J.* 64, 33.
Baade, W.: 1944, *Astrophys. J.* 100, 137.
Baade, W.: 1963, *Evolution of Stars and Galaxies*, Cambridge, Harvard University Press.
Bahcall, J. N. and Sarazin, C. L.: 1977, *Astrophys. J. Letters* 213, L99.
Bell, R. A. and Parsons, S. B.: 1972, *Astrophys. Letters* 12, 5.
Crutcher, R. M., Rogstad, D. H., and Chu, K.: 1977 (preprint).
Danziger, I. J.: 1973, *Astrophys. J.* 181, 641.
Danziger, I. J., Dopita, M. A., Hawarden, T. G., and Webster, B. L.: 1977, *Astrophys. J.* (in press).
Demarque, P. and Geisler, J. E.: 1963, *Astrophys. J.* 137, 1102.
Demarque, P. and McClure, R. D.: 1977, *Astrophys. J.* 213, 716.
de Vaucouleurs, G., de Vaucouleurs, A., and Corwin, H. G.: 1976, *Second Reference Catalogue of Bright Galaxies*, Austin, University of Texas Press.
Dufour, R. J. and Killen, R. M.: 1977, *Astrophys. J.* 211, 68.
Dufour, R. J. and van den Bergh, S.: 1978 (in preparation).
Einasto, J., Kaasik, A., Kalamees, P., and Vennik, J.: 1975, *Astron. Astrophys.* 40, 161.
Faber, S. M.: 1973a, *Astrophys. J.* 179, 423.
Faber, S. M.: 1973b, *Astrophys. J.* 179, 731.
Gallagher, J. S. and Ostriker, J. P.: 1972, *Astron. J.* 77, 288.
Gascoigne, S. C. B., Norris, J., Bessell, M. S., Hyland, A., R., and Visvanathan, N.: 1976 (preprint).
Gliese, W.: 1976, in E. K. Kharadze (ed.), *Proceedings of the Third European Astronomical Meeting of the I.A.U.*, Abastumani, Georgian Academy of Sciences, p. 463.
Gorenstein, P., Topka, K., Fabricant, D., and Harnden, F. R.: 1976, *Bull. Am. Astron. Soc.* 8, 553.
Gott, J. R. and Turner, E. L.: 1977, *Astrophys. J.* 213, 309.
Gunn, J. E.: 1974, Comments on *Astrophys. Space Sci.* 6, 7.
Hardy, E.: 1977, *Astrophys. J.* 211, 718.
Hartwick, F. D. A. and Hesser, J. E.: 1972, *Publ. Astron. Soc. Pacific* 84, 813.
Hartwick, F. D. A. and Sargent, W. L. W.: 1977 (preprint).
Hearnshaw, J. B.: 1972, *Mem. Roy. Astron. Soc.* 77, 55.
Herbst, W.: 1975, *Publ. Astron. Soc. Pacific* 87, 827.
Hesser, J. E., Hartwick, F. D. A., and McClure, R. D.: 1976, *Astrophys. J. Letters* 207, L113.
Holmberg, E.: 1958, *Lund Medd.* Ser. 2, no. 136.
Innanen, K. A.: 1973, *Astrophys. Space Sci.* 22, 39.

Jackson, J. C.: 1975, *Monthly Notices Roy. Astron. Soc.* 173, 41p.
Kahn, F. D. and Woltjer, L.: 1959, *Astrophys. J.* 130, 705.
Kalnajs, A. J.: 1972, *Astrophys. J.* 175, 63.
Karachentsev, I. D.: 1977, in C. Balkowski and B. E. Westerlund (eds.), *Décalages Vers le Rouge et Expansion de l'Univers*, Paris, C.N.R.S., p. 321.
Keenan, P. C. and Keller, G.: 1953, *Astrophys. J.* 117, 241.
Kormendy, J.: 1977, *Astrophys. J.* (in press).
Krumm, N. and Salpeter, E. E.: 1977, *Astron. Astrophys.* 56, 465.
Lambert, D. L. and Sneden, C.: 1977, *Astrophys. J.* 215, 597.
Materne, J. and Tammann, G. A.: 1974, *Astron. Astrophys.* 37, 383.
Mathews, W. G.: 1977, *Lick Obs. Bull.* no. 766.
Melnik, J., White, S. D. M., and Hoessel, J.: 1977, *Monthly Notices Roy. Astron. Soc.* 180, 207.
Morgan, W. W., Kayser, S., and White, R. A.: 1975, *Astrophys. J.* 199, 545.
O'Connell, R. W. and Mangano, J. J.: 1977 (preprint).
Oort, J. H.: 1965, in A. Blaauw and M. Schmidt (eds.), *Galactic Structure*, Chicago, University of Chicago Press, p. 455.
Osmer, P. S.: 1973, *Astrophys. J. Letters* 184, L127.
Ostriker, J. P. and Peebles, P. J. E.: 1973, *Astrophys. J.* 186, 467.
Ostriker, J. P., Peebles, P. J. E., and Yahil, A.: 1974, *Astrophys. J. Letters* 193, L1.
Page, T.: 1952, *Astrophys. J.* 116, 63.
Peimbert, M.: 1973, *Mem. Soc. Roy. des Sciences de Liège* 6th Ser. 5, 307.
Peimbert, M. and Spinrad, H.: 1970a, *Astrophys. J.* 159, 809.
Peimbert, M. and Spinrad, H.: 1970b, *Astron. Astrophys.* 7, 311.
Richstone, D. O.: 1976, *Astrophys. J.* 204, 642.
Roberts, M. S.: 1975, in A. Hayli (ed.), *Dynamics of Stellar Systems*, Dordrecht, Reidel Publ. Co., p. 331.
Rood, H. J.: 1965, University of Michigan (unpublished Ph.D. Thesis).
Rood, H. J. and Dickel, J. R.: 1976, *Astrophys. J.* 205, 346.
Rood, H. J., Rothman, V. C. A., and Turnrose, B. E.: 1970, *Astrophys. J.* 162, 411.
Sandage, A. R.: 1971, *Astrophys. J.* 166, 13.
Sandage, A. R.: 1977, paper presented at the Pomona Meetings of the Astronomical Society of the Pacific, June 21, 1977.
Schmidt, M.: 1956, *Bull. Astron. Inst. Neth.* 13, 15.
Schmidt, M.: 1975, *Astrophys. J.* 202, 22.
Silk, J. and Lea, S.: 1973, *Astrophys. J.* 180, 669.
Simkin, S. M.: 1975, *Dudley Obs. Repr.* no. 9, p. 401.

Strom, K. M. and Strom, S. E.: 1977, in B. Tinsley (ed.), *The Evolution of Galaxies and Stellar Populations* (in press).
Tammann, G. A. and Sandage, A. R.: 1968, *Astrophys. J.* 151, 825.
Tomkin, J. and Bell, R. A.: 1973, *Monthly Notices Roy. Astron. Soc.* 163, 117.
Tully, R. B. and Fisher, J. R.: 1976, in E. K. Kharadze (ed.), *Proceedings of the Third European Astronomical Meeting of the I.A.U.*, Abastumani, Georgian Academy of Sciences, p. 481.
Turner, E. L.: 1977, in C. Balkowski and B. E. Westerlund (eds.), *Décalages Vers le Rouge et Expansion de l'Univers*, Paris, C.N.R.S., p. 337.
van den Bergh, S.: 1969, *Astrophys. J. Suppl.* 19, 145.
van den Bergh, S.: 1971a, *Astron. J.* 76, 1082.
van den Bergh, S.: 1971b, *Astron. Astrophys.* 11, 154.
van den Bergh, S.: 1977a, in C. Balkowski and B. E. Westerlund (eds.), *Décalages Vers le Rouge et Expansion de l'Univers*, Paris, C.N.R.S., p. 13.
van den Bergh, S.: 1977b, *Vistas Astron.* 21, 71.
van den Bergh, S.: 1978 (in preparation).
Wagoner, R. V., Fowler, W. A., and Hoyle, F.: 1967, *Astrophys. J.* 148, 3.
White, S. D. M.: 1977, *Monthly Notices Roy. Astron. Soc.* 179, 33.

DISCUSSION FOLLOWING REVIEW V.1 GIVEN BY S. VAN DEN BERGH

OORT: You mentioned the possibility that the relative velocity of M31 and the Galaxy might indicate a chance encounter. However, there are other, independent galaxies in the Local Group, for instance NGC 6822, which make a chance encounter unlikely.

FREEMAN: Perhaps I could add here that the whole question of what actually is in the Local Group, and what sort of velocity dispersion you derive from this mass, is a controversial one. Sandage has recently made a solar motion solution where he feels justified in throwing out all galaxies at a velocity that deviates more than 3σ from the solar motion solution. Then, of course, the whole mass of our Local Group can shrink abruptly. This point is far from solved.

Furthermore, I want to reinforce Dr. van den Bergh's remarks about problems in deriving M/L from binary galaxies. Turner analysed his sample by two methods: Page's which gave M/L $\sim$ 3, and his own which gave M/L $\sim$ 40. The problem here is in the procedure, primarily. A definitive paper, on how to do this properly, will appear soon.

M.S. ROBERTS: S. Peterson has recently completed a study of binary galaxies combining Turner's observations with additional new data he obtained from 21-cm line observations. He finds M/L $\sim$ 25 (H = 50 km s^{-1}

Mpc^{-1}) for spiral pairs. This value is essentially constant with angular separation. The total mass of a pair increases with separation up to $\sim$ 100 kpc and then remains constant at $\sim 2 \times 10^{12}$ M$_\odot$ for larger separations. Thus these data do not indicate an extensive massive halo.

TULLY: We have done a virial analysis on a dozen or so nearby small groups, making use of a large body of new 21-cm velocities. In a large majority of cases the virial masses are comparable to masses found from rotation curves, although we concede the possibility they might be a factor of two greater. The most remarkable exception was the M81 group where the virial theorem suggests a very large mass. However, almost all the kinetic and potential energy are in the pair M81-M82. If these two are taken as a single entity and the virial analysis is repeated then there is no difficulty with the proposition that the group is stable with conventional masses.

REES: Recent numerical simulations by Simon White at Cambridge indicate that, if the galaxies in binaries had halos extensive enough to be almost merging, they would spiral together in little more than one orbital timescale. This suggests that - for galaxies in binaries at least - the "halos" do not extend beyond about 50 kpc.

HUCHTMEIER: SENSITIVE OBSERVATIONS OF HI ENVELOPES OF LATE-TYPE GALAXIES

Neutral hydrogen has been observed with a sensitivity of a few times 10^{18} cm^{-2} with the 100-m radio telescope (Effelsberg) in the environment of the nearby late-type galaxies Holmberg I, IC 1613, M33, Holmberg II, WLM, M101, NGC 2366, NGC 3109, Sextans A, M83, IC 10, partially in NGC 2403 and 6822. The ratios of HI-radius to Holmberg-radius are 1.3, 1.6, 2.2, 2.4, 3.6, 3.6, 3.7, 4.5, 5.8, 6.2, and 9, respectively. The diameters of the HI envelopes correspond to linear distances of 5, 11, 38, 200, and 230 kpc in the case of Holmberg I, WLM, M33, M101, and M83, respectively.

For M101 about 30% of the HI mass of 24×10^9 M$_\odot$ is found outside of the Holmberg diameter. 1.5×10^9 M$_\odot$ are associated with the southern companion NGC 5474. The velocity field in the HI envelope is in agreement with the assumption that most of the material in the envelope is in corotation with the galaxy.

There are a few indications for a possible gravitational interaction between M101 and its companions. Between M101 and NGC 5474 an HI link is seen in two velocity channels (separated by 16.5 km/s) at a surface density of 5×10^{18} cm^{-2} (beam averaged), which is well above possible side lobe contributions. A few of the M101 velocity channels show asymmetries of the kind expected from interactions.

The area of this link and an area close to the companion dwarf galaxy NGC 5477 show considerably broader profiles ($\Delta v > 60$ km/s instead of < 40 km/s) than nearby regions. This is probably the effect of different HI complexes along the line of sight and suggests bridge-like features between these galaxies. However, the outer contours of the HI distribution of M101 are considerably more symmetric than the central part which is strongly disturbed.

LANDECKER: Considering the known sidelobe levels of the telescope, what is the dynamic range attainable in such measurements?

HUCHTMEIER: The outermost detected HI has a column density of about 3×10^{18} cm^{-2}, and the accumulating effects of sidelobe errors for the channels where the strongest HI signals have been observed is about half that value.

GIOVANELLI: Does the velocity field of M101 behave regularly throughout the HI distribution?

HUCHTMEIER: The velocity field is very regular. With the present resolution it does not show any deviations from an extrapolated field of circular rotation.

ON PERIPHERAL DYNAMICS

A. Toomre
Massachusetts Institute of Technology, Cambridge

A review of the observed facts and the observed theories of distortions of the outer parts of galaxies.

[Manuscript not received]

DISCUSSION FOLLOWING REVIEW V.2 GIVEN BY A. TOOMRE

ALLEN: Some years ago we used to think that such peculiar galaxies were only a small fraction of all galaxies in the sky. The new deep photographs seem to be changing this picture. Are there any galaxies left which remain simple and symmetric as one goes to even fainter isophotes?

TOOMRE: Perhaps not. And my life is made miserably difficult by these facts which seem to be getting better and better.

PFLEIDERER: I refer to the velocity distribution in the long whisp of NGC 3628. Do you have a simple explanation for the double peak in the outer part of the whisp?

TOOMRE: No, impossible. I would guess it might just be a remnant of there having been spiral structure in the galaxy, or may be it was lumpy like in arms already and somehow it got squashed.

WRIGHT: Could you remind us of the range of collision parameters which give strong tidal interactions and what fraction of encounters produce strong disturbances?

TOOMRE: The interacting systems should have roughly comparable mass, must be originally bound, and should pass at about an arm's distance.

GALLAGHER: The NGC 5128 situation may be even more complex than you have suggested. NGC 5102 is a member of the same group and is peculiar in that a high HI content and young stellar population appear superimposed on a normal S0. Van den Bergh has interpreted his photometry as evidence for a burst of star formation a few times 10^8 years ago. Thus whatever caused NGC 5128 may have been a group-wide phenomenon.

FREEMAN: I might add that the same group contains NGC 5253, an Irr II in which the supernovae rate is about three times higher than expected.

M.S. ROBERTS: The velocity field along the minor axis of NGC 2685 is, not surprisingly, peculiar; in addition to radiation at the systemic velocity, long velocity wings, both blue and red, are present. It is remarkable that almost exactly the same minor axis velocity field occurs in the presumably normal galaxy M31!

FREEMAN: People use the Rogstad ring procedure to estimate circular velocities for warped galaxies. This procedure is obviously not dynamically selfconsistent and the inferred circular velocities could easily be <u>systematically</u> wrong by 10 or 20 km/s; this is just the size error that could be a problem for calculating M/L in the outer parts of galaxies. What is your advice to people who measure rotation curves in warped galaxies?

TOOMRE: Just one word: wait!

M.S. ROBERTS: Could I comment on this? If one gives a rotation curve it should be clearly stated as to how this was obtained. The safest thing is that you don't even correct for the inclination. If you have a modest bend the effects on the rotation curve are not very great. Only when you have a great bend in a peculiar position you can grossly distort your data so that your rotation curve can fool you.

INTERACTING GALAXIES: THE KINEMATICS OF NGC 4038/39 AND THE HI BRIDGE
BETWEEN M81 AND NGC 3077

J.M. van der Hulst
Kapteyn Astronomical Institute
University of Groningen

1. THE KINEMATICS OF NGC 4038/39

Computer simulations of galaxy encounters (Toomre 1974 and references therein) have strongly supported the idea that the luminous bridges and tails in many multiple galaxies are a result of gravitational interaction. The need for an observational test of the kinematics of these models has prompted observations of NGC 4038/39 (the "antennae") in the 21 cm line of neutral hydrogen (HI) with the Westerbork Synthesis Radio Telescope (WSRT) at an angular resolution of 100"×150" ($\alpha \times \delta$). A detailed description of the observations and data reduction is given elsewhere (van der Hulst, 1977). Neutral hydrogen is definitely associated with the tails as was already indicated by the filled aperture observations of Huchtmeier and Bohnenstengel (1975) at the lower resolution of 8.'7. The distribution of HI column density is shown in Figure 1 superposed on a print of a plate obtained by F. Schweizer with the 4 m telescope at Cerro Tololo. Owing to the low declination of the object ($-19°$) the WSRT map is affected by several sidelobes; the main effect is an attenuation of extended east-west structures, for which reason the HI along the southern tail in Fig. 1 shows a gap. Computer experiments and a comparison with the filled aperture observations show that the WSRT data are entirely consistent with an HI distribution which is uniform along this tail and increases with about a factor 2 at the extreme end, where the tail is very blue and small HII regions are present (Schweizer, elsewhere in this volume). The tails contain 2.9×10^9 $M_\odot$ of HI (assuming a distance of 20 Mpc), 85% of which is in the southern and more prominent tail. About 1.1×10^9 $M_\odot$ of HI is associated with the galaxies which are not spatially resolved at the present angular resolution.

The comparison of the velocity information with the model of Toomre and Toomre (1972) is shown in Figure 2. The left diagram shows the geometry; the spatial coordinates are given in units of the perigalactic distance. The middle diagram shows the radial velocities of the test particles in the left diagram. Velocities to the left and the right of the vertical axis are approaching and receding; the unit is 415 km s^{-1}.

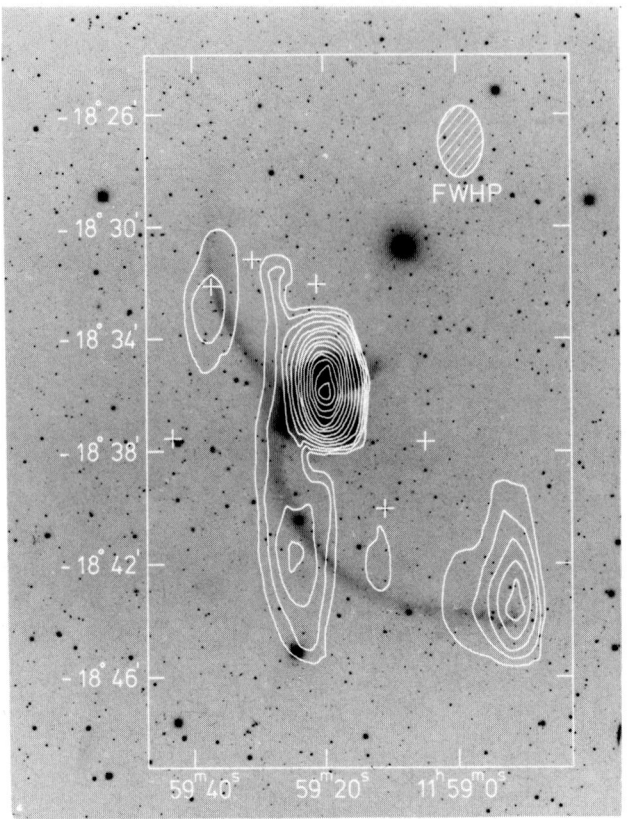

Figure 1. Distribution of HI column density in NGC 4038/39 as measured with the WSRT superposed on a 4 m CTIO photograph. The contour interval is 4.5×10^{19} cm^{-2}. The ellipse denotes the half power beam width.

Three features are essential to the model velocities in the tails: (i) the velocity along a tail is approximately constant with a slow decline towards the mean velocity of the galaxies at the tip; (ii) profiles along the tails are narrow; (iii) the tail velocities have opposite sense with respect to the mean velocity. The right diagram shows a position-velocity map centered at $\alpha = 11^h 59^m 21^s$, $\delta = -18°36'$ constructed by interpolating the observations onto a line of position angle 40°, chosen such to obtain close agreement with the coordinate system of the left diagram for the southern tail. Superposed are the model velocities (black dots) after scaling of the dimensions of the model by a factor 0.3 in order to match the observed length of the southern tail. Also the sense of rotation of the original model has been inverted. The model masses are now 2.7×10^{10} $M_\odot$, because the model velocities did not require substantial scaling. The quantitative agreement is quite good and the three crucial kinematical features are well represented by the observations. The broad galaxy profiles of the model are still well separated, because even after the scaling the separation of the model galaxies is a factor 2 larger than is observed. Only at the tip of the southern tail the observed velocities

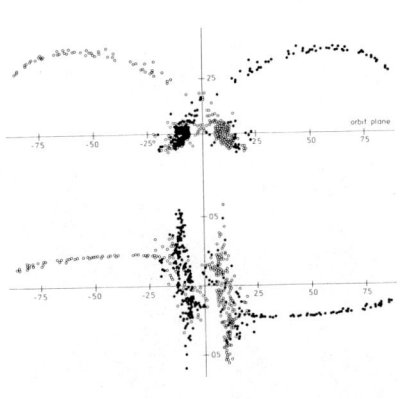 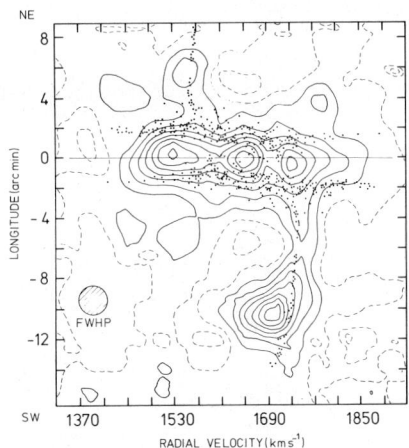

Figure 2. The geometry and velocities (left and middle diagram) of the model of Toomre and Toomre for NGC 4038/39 and the comparison of this model with the WSRT observations (right diagram). The contour interval for the position-velocity map is 0.1 K; the zero and negative contours have been dashed. The circle denotes the position-velocity half power "beam". I thank A. Toomre for kindly providing his model data.

are lower and the profiles broader than the model predicts. This may be a result of integration along the line of sight, the southern tail being seen in parallel at its extreme.

2. THE NGC 3077 HI BRIDGE

The early type spiral galaxy M81 and its two IO companions, M82 and NGC 3077 are known to have a common HI envelope (Roberts 1972, Davies 1974). The southern part of this triplet, including NGC 3077, has been observed with the WSRT (van der Hulst, 1977) at 50" resolution in order to determine the detailed structure and kinematics of the HI in this region. The WSRT data have been combined with observations from the 100-m Effelsberg telescope (Harten, Mebold and Shane, priv. comm.) in order to incorporate the extended ($\gtrsim 15'$) HI emission which is missing from the interferometric observations. The results are shown in Figure 3, combined with the WSRT observations of M81 (Rots and Shane, 1975) smoothed to the same angular resolution. Superposed are contours of equal radial velocity. The nuclei of M81 and NGC 3077 are indicated by a white triangle and filled circle respectively. The areas in the top left and bottom right corner are blank, because no WSRT observations of these regions are yet available.

The most remarkable feature is an HI bridge between M81 and NGC 3077 which appears to be a regular eastward continuation of the outer HI spiral structures at the west side of M81. Farther east this continuation is less evident, though one may tentatively try to identify the double

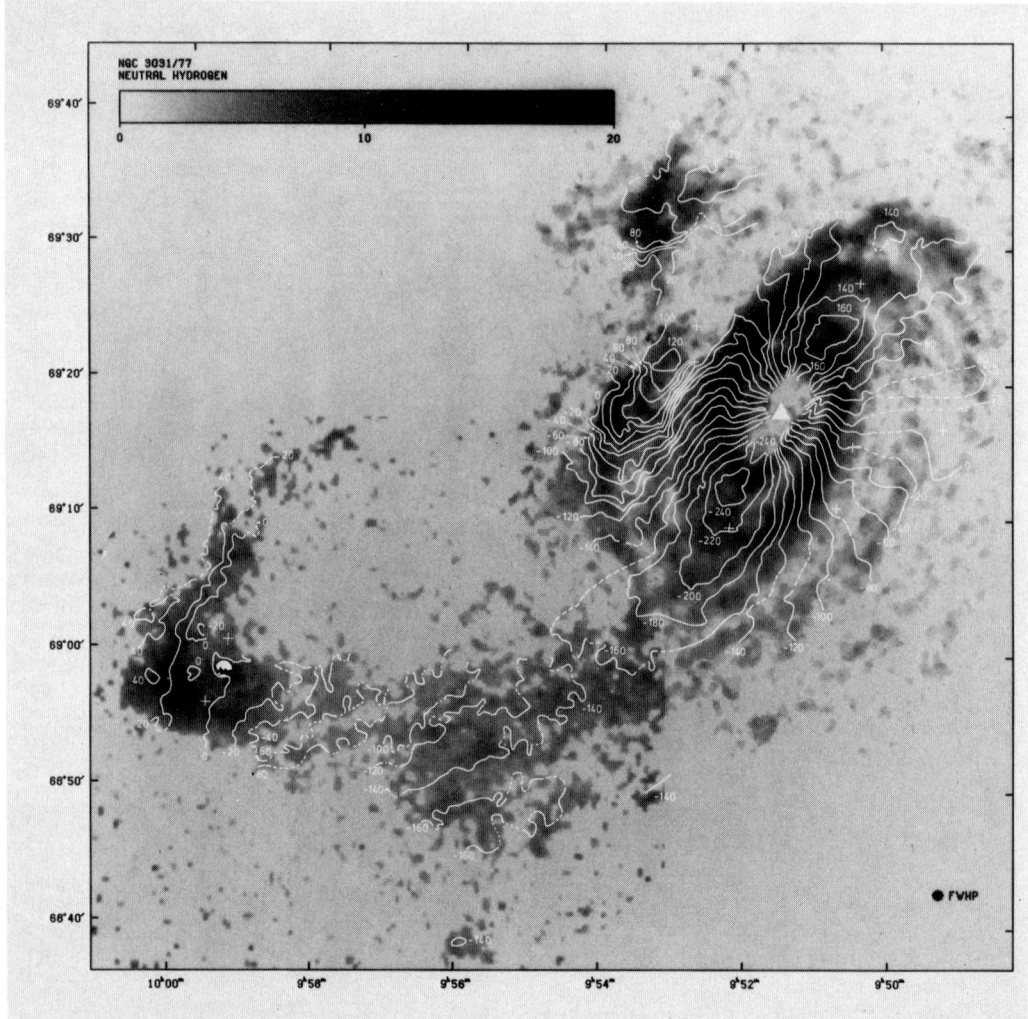

Figure 3. The distribution and kinematics of the HI in M81 and NGC 3077. The grey scale (top left) is in units of 10^{20} cm^{-2}. This diagram has been prepared by A.H. Rots with the Dicomed filmrecording device of the NRAO.

structure in the bridge close to NGC 3077 with the same spiral structure. The HI distribution around NGC 3077 is asymmetric; more than 50% of the HI lies beyond the Holmberg radius, partly in a bright concentration to the south-east of the nucleus and partly in a tail which extends northward. No optical counterpart is found for these features on a print from a deep IIIa-J plate (courtesy H.C. Arp), except for small condensations (Barbieri, Bertola and di Tullio, 1974) coinciding with the bright southeastern concentration. The amount of HI in the bridge is 5×10^8 M$_\odot$; the tail contains 2.3×10^8 M$_\odot$ of HI whereas 5×10^8 M$_\odot$ is associated with NGC 3077. The HI velocities in the bridge show a regular velocity gradient which persists across the HI around NGC 3077 and across the tail. The bridge velocities blend smoothly into the differential rotation velocity field of M81.

The properties of the bridge and the tail (width: 4-8 kpc, length: 25 kpc, peak column densities: $3-5 \times 10^{20}$ cm^{-2} and profile halfwidth: 20-40 km s^{-1}) are very similar to those of the Magellanic Stream (Mathewson, Cleary and Murray, 1974) and the HI filaments in NGC 4631/56 (Weliachew, Sancisi and Guélin, 1977). For the Magellanic Stream three hypotheses have been put forward: (i) the gas is primordial (Mathewson 1976, Mathewson and Schwarz 1976); (ii) the gas traces the wake of the Magellanic Clouds in the hot halo of the Galaxy (Mathewson, Schwarz and Murray, 1977); (iii) the gas has been drawn out of the Clouds by the Galaxy during a past encounter (Mathewson et al.1974, Fujimoto and Sofue 1976, Lynden-Bell 1976, Davies and Wright 1977). Of these hypotheses only the latter is likely to apply to the NGC 3077 bridge. A primordial hypothesis raises a stability problem, because the bridge will disperse within about 6×10^8 years as estimated from its velocity gradient. In a galactic wake hypothesis the bridge and the tail cannot be explained as such at the same time. The good match of the structure and velocities of the bridge with M81 strongly suggests that the gas has been drawn out of M81. Gravitational model calculations (van der Hulst, 1977) for the interaction between M81 and NGC 3077 and between M81 and M82 indicate that either NGC 3077 or M82 can possibly produce the bridge with the correct velocities. A model involving all three galaxies may be required to fully describe the interaction.

REFERENCES

Barbieri, C., Bertola, F., di Tullio, G.: 1974, Astron.Astrophys.35,463
Davies, R.D.: 1974, in The Formation and Dynamics of Galaxies, IAU Symp. no. 58, ed. J.R. Shakeshaft, p.119
Davies, R.D., Wright, A.E.: 1977, Monthly Not.Roy.Astron.Soc. 180, 71
Fujimoto, M., Sofue, Y.: 1976, Astron. Astrophys. 47, 263
Huchtmeier, W.K., Bohnenstengel, H.D.: 1975, Astron.Astrophys. 41, 477
Hulst, J.M. van der: 1977, Dissertation, University of Groningen
Lynden-Bell, D.: 1976, Monthly Not.Roy.Astron.Soc. 174, 695
Mathewson, D.S.: 1976, Roy. Greenwich Obs. Bull. No. 182
Mathewson, D.S., Schwarz, M.P.: 1976, Monthly Not.Roy.Astron.Soc.176,47p
Mathewson, D.S., Cleary, M.N., Murray, J.D.: 1974, Astrophys. J. 190,291
Mathewson, D.S., Schwarz, M.P., Murray, J.D.: 1977, Astrophys. J. Letters 217, L5
Roberts, M.S.: 1972, in External Galaxies and Quasi Stellar Objects, IAU Symp. no.44, ed. D.S. Evans, p.12
Rots, A.H., Shane, W.W.: 1975, Astron. Astrophys. 45, 25
Toomre, A.: 1974, in The Formation and Dynamics of Galaxies, IAU Symp. no.58, ed. J.R. Shakeshaft, p.347
Toomre, A., Toomre, J.: 1972, Astrophys. J. 178, 623
Weliachew, L., Sancisi, R., Guélin, M.: 1977, Astron. Astrophys., submitted

DISCUSSION FOLLOWING PAPER V.3 GIVEN BY J.M. VAN DER HULST

VAN WOERDEN: Does the tidal HI filament around NGC 3077 also have a dwarf-irregular (possibly DDO 66) at its tip?

VAN DER HULST: I would definitely say not; it is way off.

DAVIES: DISTRIBUTED HI IN SMALL GROUPS OF GALAXIES

In cooperation with G.P. Davidson, L. Hart, S.C. Johnson and P.N. Appleton an extensive programme of mapping the 21-cm neutral hydrogen emission in and around small groups of galaxies has been undertaken at Jodrell Bank using the MK IA radio telescope which has a beamwidth of 13'. The group with the most distributed neutral hydrogen emission is the M81/M82/NGC 3077 group. A detailed examination of the data shows two neutral hydrogen "bridges" between NGC 3077 and M81, a bridge between M81 and M82 and a bridge between M81 and the SW cloud. In addition there is a neutral hydrogen tail projecting from M82 in the direction opposite to M81. Velocity gradients are evident along some of these features which indicates that they may have a tidal origin resulting from the gravitational interaction between the galaxies. Some of the gas, the SW cloud for example, may be primordial and may not have ever been bound to any of the galaxies.

Two pairs of galaxies show a simpler intra-cluster neutral hydrogen distribution. NGC 4151 and NGC 4145 show a bridge connecting the two galaxies. The NGC 4725/4727 pair appears to have generated a neutral hydrogen tidal bridge and tail extending from the less massive galaxy.

NGC 1023, an S0 galaxy, is in a different category. It has an adjacent neutral hydrogen companion situated at a projected distance of $\sim$ 60 kpc and having a velocity difference of approximately 200 km/s. This appears to be a purely hydrogen galaxy with no detectable optical emission on the Palomar Sky Survey prints. The other members of the NGC 1023 group are at such large distances that they cannot have pulled this gas from NGC 1023.

ALLEN: Concerning the detectability of faint extended features in the region of the bridge between M81 and NGC 3077 on the Westerbork maps: I should like to point out that the maps of this area shown by Mr. van der Hulst are in fact a combination of WSRT interferometer observations and single dish maps made with the 100-meter Effelsberg telescope.

VAN WOERDEN: TIDAL INTERACTION AND ACCRETION IN THE GALAXY PAIR NGC 1512 AND 1510

The southern ringed, barred lenticular galaxy NGC 1512 is immersed in a vast amount of hydrogen (van Woerden et al. 1976, P.A.S.A. 3, 68). Its E0-type companion, NGC 1510, at 5' = 20 kpc distance, is 2 mag fainter and has quite blue colours. Disney and Pottasch (1977, A.A. 60, 43) observe an A-type absorption spectrum and strong emission lines; they interpret NGC 1510's colours and spectrum as evidence of recent formation, $\sim$ 500 Myr ago.

Our Parkes observations (Hawarden et al. 1977, M.N.R.A.S., in press) show that the 11 x 10^9 $M_\odot$ of neutral hydrogen is extended over > 100 kpc diameter; from its rotation follows a mass of 2×10^{11} $M_\odot$ for NGC 1512.

Deep photographs with the Siding Spring Schmidt telescope reveal an irregular pattern of arms and filaments, reaching 33 kpc from NGC 1512, and a distortion of the ring. These, apparently tidal, effects indicate for NGC 1510 a mass of 1 - 10% that of NGC 1512. However, if NGC 1510 is young, its luminosity and colours indicate a mass $\lesssim 1 \times 10^8$ $M_\odot$. Also, the Disney-Pottasch coeval model fails to account for the ionization.

Our model proposes that NGC 1510 is a dwarf-elliptical galaxy of 2×10^9 $M_\odot$, which in the last 100 Myr has accreted enough gas to form a young population of $\sim 50 \times 10^6$ $M_\odot$. This model is consistent with the observed colours, absorption and emission spectra, and tidal interaction. The interaction suggests that the inclination of NGC 1510's orbit about NGC 1512 is low. Assuming a velocity of NGC 1510, relative to the surrounding gas, of 100 km/s (the observed radial velocity difference is < 50 km/s), we estimate that $(10 - 100) \times 10^6$ $M_\odot$ of gas would have been accreted.

The low metal abundance observed in the emission lines must, on either model, be due to the composition of the gas disk around NGC 1512.

VAN DEN BERGH: I suspect that many of the problems to which you referred will resolve themselves when good optical classifications based on homogeneous plate material become available.

GALLAGHER: For NGC 6902, if you include the extra light in the outer part of the galaxy, the ratio M_{HI}/L_B is reduced by about a factor of 2 which is normal for an Sbc.

SCHWEIZER: If we accept the idea that NGC 1512 and 1510 interacted tidally rather recently, the observation of a strong burst of star formation in the smaller galaxy agrees nicely with Arp's (1969, A.A. 3, 418) finding that such bursts are a frequent phenomenon. In his sample of six spiral galaxies with companions at the end of one spiral arm, four of the companions show blue absorption-type spectra indicative of strong recent star formation (i.e. spectra with Balmer lines in absorption).

COMBES: TIDAL INTERACTIONS WITHIN THE NGC 4631 GROUP OF GALAXIES

The two galaxies NGC 4631/4656 have been observed in the 21-cm line of atomic hydrogen by Weliachew et al. (A.A., submitted). The isophotes of neutral hydrogen emission show very important distortions; mainly around NGC 4631 we note at least four gas features around this galaxy. A numerical model considering only a restricted three-body problem, has been tried to interpret the distortions in terms of a tidal interaction.

This model shows that at least two of the features, the bridge between NGC 4631 and NGC 4656, and the counter arm almost parallel to the plane of NGC 4631, can be interpreted in terms of a parabolic passage of the two galaxies. The other two features, almost perpendicular to the plane of NGC 4631, are interpreted as being related to a third neighbouring galaxy NGC 4627. The latter does belong to the

NGC 4631 and 4656 group according to its systemic velocity and its optical appearance, showing filaments towards NGC 4631. Due to its much smaller total mass, NGC 4627 has a negligible action on NGC 4631 and 4656 but is greatly being damaged itself. Though NGC 4627 is classified now as a dwarf-elliptical galaxy, it is assumed to have been an irregular or spiral galaxy in the past and to have lost its neutral gas in the recent encounter. This assumption is supported by the presence of rather young stars in this small galaxy.

GIOVANELLI: HI OBSERVATIONS OF THE M51 SYSTEM

Observations in the 21-cm line (in cooperation with M.P. Haynes and M.S. Burkhead), made with the 100-m telescope of the MPIfR and the 92.6-m telescope of the NRAO, have revealed an extended distribution of peripheral HI in the M51 system, well beyond that reported by previous observers. Its characteristics may be summarized as follows:
(1) The projected diameter is 125 kpc for an assumed distance to M51 of 9.7 Mpc.
(2) The velocity field is quite irregular, extending almost 200 km/s beyond the cutoff of the profile of NGC 5194, and the gas does not co-rotate with the material in the disk of NGC 5194.
(3) The HI mass exceeds 10^9 $M_\odot$.
(4) The contour map of the HI column density at velocities larger than 600 km/s yields a suggestive match with a peculiar optical extension of the system toward the NW, recently traced by Burkhead (1977, preprint).
(5) No positive detection of NGC 5195 may be claimed, in contrast with the result of Rood and Dickel (1976, Ap.J. 205, 346).

Although the M51 system is by no means unique in being surrounded by an extended distribution of material luminous both optically and in the HI line, the obvious peculiarities of the system in combination with the details of the circumgalactic HI make it quite special. Either a primordial or tidal origin is possible. Our low resolution observations do not yield information on the structure of the gas at a scale smaller than about 40 kpc; the distribution may then be an assembly of Sculptor-like clouds as found by Mathewson et al. (1975, Ap.J. 195, L97) and Haynes and Roberts (in preparation). The association between optical and HI luminosity, on the other hand, recalls the case of the Leo triplet (Haynes et al. 1977, B.A.A.S. 9, 361), although the large velocities of the gas and its extension make a tidal origin after a recent pericenter transit, as in the picture proposed by Toomre and Toomre (1972, Ap.J. 178, 623), rather improbable. The possibility that this material may have been swept out from the galaxies in the system during earlier close passages of the history of their affair cannot however be ruled out.

SANCISI: GAS DISTRIBUTION AND VELOCITY FIELD OF THE BARRED SPIRAL GALAXY NGC 5383

In cooperation with R.J. Allen a new, more sensitive 21-cm line study of NGC 5383 has recently been completed with the Westerbork Synthesis Radio Telescope. The results confirm and strengthen the main conclusions from the preliminary observations reported by Allen et al. (1974, The Formation of Galaxies, ed. J.R. Shakeshaft, p. 425) and by

Sancisi (1975, La Dynamique des Galaxies Spirales, ed. L. Weliachew, p. 403).

Hydrogen distribution. Most of the hydrogen is found in the optically bright parts of the galaxy and is concentrated especially in the central region and in the spiral arms. There is apparently little or no HI emission from the bar. HI is detected out to about 3' ($\sim$ 40 kpc) or twice the de Vaucouleurs radius, where the column densities fall to values of about 1×10^{20} cm^{-2}. Clear evidence of related faint optical emission of similar extent in the outer regions of NGC 5383 is shown by a deep IIIa-J plate recently taken by van der Kruit and Bosma (1976, private communication) with the 48-inch Schmidt telescope at Mt. Palomar.

Velocity field. The overall shape of the velocity field is typical of an axisymmetric disk in circular differential motion. The rotation curve is approximately flat within 20 km/s out to the outer parts of the galaxy, indicating an extended mass distribution. In the inner regions near the bar the velocity field clearly shows large scale deviations of order 100 km/s from circular motion: in particular, the iso-velocity contours tend to become parallel to the direction of the bar.

A comparison of the HI velocities with those derived from optical emission lines in the central regions by Peterson et al. (1977, preprint) shows good agreement: the same velocity pattern indicating non-circular motion in the optically derived velocity field is easily recognizable also in the radio map even though the radio observations have much lower resolution.

It is most likely that these velocity perturbations are associated with the presence of the bar.

Southern companion. HI emission has also been detected in the small 16.5 mag SBdm galaxy (U 8877; Nilson 1973, Uppsala Gen. Cat. Gal.) located about 3' to the south of NGC 5383 (12.5 mag). Its systemic velocity V_{Hel} = 2370 km/s is close to that of NGC 5383 (V_{Hel} = 2250 km/s), providing further evidence for a mutual association.

VAN DER KRUIT: Surface photometry by myself and Bosma on the deep plate just showed indicates that the inner region is an ovally distorted lens with the bar along the major axis. The outer structure seems to be the response of the outer disk to the ovally distorted lens. Also, its flat rotation curve indicates that M/L in the outer region must be 10 times higher than in the lens.

"Something strange is going on."

M.S. Roberts in Discussion V.2

GALAXIES WITH LONG TAILS

François Schweizer
Cerro Tololo Inter-American Observatory*

Today I would like to show you some animals from the zoo of Alar Toomre. These animals are all characterized by long tails and belong to two subspecies: The first consists of animals which always huddle in pairs, each individual having one long tail. The second, rarer subspecies comprises those strange animals which live alone but have two tails!

1. THE ANTENNAE GALAXIES

A prime example of a pair of the first subspecies are the well-known Antennae, NGC 4038/9. Their tails are long indeed, with an overall projected size (from tip to tip) of 160 kpc (20') at the redshift distance corresponding to $H_o = 50$ km s^{-1} Mpc^{-1}. The projected separation of the two main bodies is about 9 kpc and the absolute magnitude of each galaxy $M_V = -21$. The Toomres (1972, TT) proposed, through their tidal-interaction model for these galaxies, that during a close encounter some 7×10^8 yr ago material from the outer fringes of the disks was ejected into the present-day tails. The narrow width of these tails is most remarkable. On Figure 1a the southern tail appears only 4-5 kpc wide, or about 1/20th of its length. The TT model reproduces this narrow width only by assuming a zero velocity dispersion for the test particles orbiting in the initial galactic disks. A reasonable velocity dispersion of, say, 10-20 km s^{-1} would broaden the tails substantially. A very deep photograph obtained with the CTIO 4-m telescope (Fig. 1b) does indeed show the tails to be about three times wider than previously seen; the projected width of the southern tail now appears to be ~15 kpc. By dividing half that width by the above interaction age, I estimate a velocity dispersion of 10-15 km s^{-1} for stars now at the edge of the tail, depending on the projection factor.

An especially interesting feature is the patch of luminous material near the tip of the southern tail. This patch appears to be a stellar system of very low surface brightness (V $\gtrsim$ 25 mag arcsec^{-2}), similar to

*Supported by the National Science Foundation under contract NSF-C866.

the dwarf irregular galaxy IC 1613 in the Local Group, but about twice as large (11x19 kpc). The integrated absolute magnitude is $M_V \approx -16.5$. There is good evidence (see below) that this dwarf system is physically associated with the southern tail.

What is the stellar and gaseous content of the tails? Photoelectrically measured UBV colors of five selected areas in the tails indicate stellar populations similar to those typically found in late-type spiral galaxies. If the stellar populations were coeval, the measured U-V indices would indicate ages of $1-3 \times 10^9$ yr, if one uses the cluster models by Searle et al. (1973). Since these ages are of the order of two to four times the interaction age, we conclude that the tail populations probably consist mainly of stars pulled out of the main bodies with only a sprinkle of stars possibly formed more recently out of the gas. One notable exception is an area measured towards the tip of the southern tail; its color is significantly bluer than the others, bluer in fact than the color of an average irregular galaxy. The associated age of 3×10^8 yr for a coeval population indicates that at the tip of the tail significant numbers of stars must have formed recently, *after* the closest approach of the two galaxies. This conclusion is supported by the

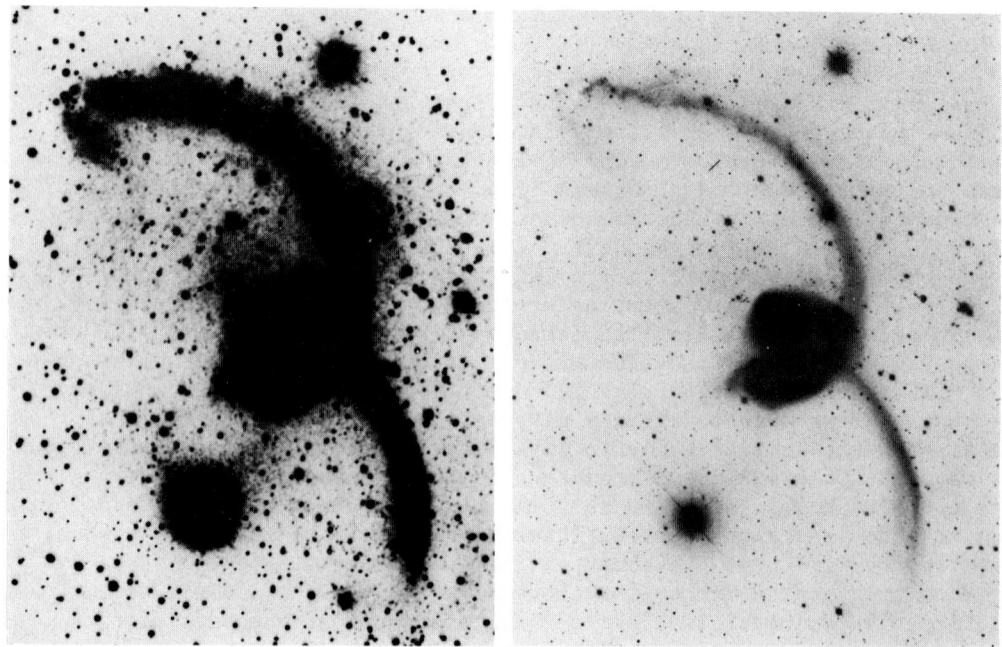

Figure 1. NGC 4038/9 on IIIa-J plates obtained with the CTIO 4-m telescope. North is at the top and east is at the left. (a) (left) Exposure of 50 minutes. (b) (right) Superposition print of two plates totalling 3.5 hours of exposure time. Note the dwarf stellar system near the tip of the southern tail.

appearance of the tails on the best CTIO 4-m plates and by spectroscopic observations. Whereas the whole northern tail and most of the southern tail appear smooth and unresolved even under 0".8 seeing, implying the absence of objects brighter than $M_B \approx -8.5$, the tip of the southern tail is resolved into bright stars (clusters?) and fuzzy knots. These knots are clearly H II regions, as shown by their emission-line spectra (Hα, [N II], [S II], and [O II] λ3727 lines on my spectrograms). The Hα fluxes are of the order of 100 times the Hα flux of the entire Orion nebula, and the line ratios indicate an excitation similar to that of, e. g., H II regions at intermediate radii in M 101. The remarkable fact which I wish to emphasize here is that *at about 100 kpc projected distance from the main bodies of NGC 4038/9, stars are still actively being formed as a consequence of some tidal interaction which took place 7×10^8 yr ago!* Furthermore, the gas out of which these stars are forming is not as metal-poor as one might expect from its remote location, but rather seems of a metallicity found typically at the outskirts of a giant disk galaxy. This, of course, fits in nicely with the tidal-interaction model for the Antennae.

The radial velocities of the four H II regions intersected by the spectrograph slit (1690, 1708, 1710, and 1711 km s^{-1}) agree closely with the 1710 km s^{-1} velocity of the H I gas observed by van der Hulst (this meeting). However, note that the H II regions are clearly located within the tip of the tail, whereas the H I gas seems to be more concentrated in that dwarf stellar system near the tip of the tail. The dwarf is therefore likely to be physically associated with the tail. Although nothing is known as yet about its stability, we should envisage the possibility that *tidal interactions may create dwarf galaxies* and that *these dwarfs may contain more metals than we would expect* if we somehow think of primordial material out there. This possible mechanism for the formation of dwarfs was emphasized already long ago by Zwicky (1956).

2. INTERGALACTIC RECYCLING

Large and even giant H II regions at the ends of tidal tails are a rather general phenomenon in interacting galaxies, providing direct evidence that (metal-enriched) gas is being returned to intergalactic space. To quote just a few examples: In NGC 4676 (the "Mice"), the northern tail consists of a young stellar population with a Balmer absorption-line spectrum, and H II regions occur out to 58 kpc (90") projected distance (Stockton 1974). The tidal model (TT) shows that most of this tail material will escape from the system. In NGC 2623 (= Arp 243), I have found, on spectra obtained with the Hale telescope, H II regions out to halfway along the northern tail (30", or only 16 kpc in that dwarfish system). In NGC 6621/2 (= Arp 81), several giant H II regions sit right at the tip of the tail, at 21 kpc (33") projected distance from the main galaxy, or 79 kpc (125") if one measures the projected distance along the tail. In NGC 3256, a southern system consisting of two partially merged galaxies, a complex of H II regions occurs at 39 kpc (154") projected distance along the eastern tail. Finally, in

NGC 2535/6 (= Arp 82), I have found H II regions at 58 kpc (147") projected distance from the main spiral, near the tip of the long tail to the Northwest. These examples show that we can expect to find in increasing numbers young stars of all degrees of metallicity in intergalactic space. These stars may well appear isolated, since the tidal filaments are often significantly fainter than the H II regions at their tips and are bound to dissipate into invisibility on a time-scale of $\sim 10^9$ yr. Why young stars should form preferentially at the ends of tidal tails is not clear at present, but perhaps this is due to a concentration of H I gas there, as observed in the Antennae by van der Hulst. The concentration of gas, in turn, might be explained by its exterior location in the pre-encounter galactic disks, since it is clear from the TT models that exterior particles fly farthest in any given time interval.

3. ON MERGING AND MERGED PAIRS OF GALAXIES

In 1972, the Toomres proposed a sequence of increasingly merged pairs of galaxies (see also Toomre 1977). This incited me to study observationally whether I could find an object in the last throws of merging, or perhaps even one which would recently have completed the merging process. As a start, I surveyed the Toomre sequence and some additional objects with near-infrared plates to see whether one or two nuclei were present. The technique is quite powerful, as illustrated in the case of NGC 4038/9. A near-infrared plate ($\lambda\lambda$ 7000-9000 Å) shows distinctly the two nuclei at 9 kpc projected separation, whereas plates obtained in the usual bandpasses show two messy bodies full of young stars and emission-line regions. This survey of "mergers" produced two results: (1) Either there were two well-separated nuclei (projected separations of 5-10 kpc and more), or there was only one nucleus to the resolution limit set by the seeing (corresponding to 0.25-1 kpc at the distances of the galaxies). This finding is compatible with the theoretical notion that the merging process should be rapid in its final stages. (2) There were five galaxies with single nuclei, but for a variety of reasons only two of these seemed good candidates for recent mergers: NGC 3921 and NGC 7252.

At this point, we should ask ourselves: What are the characteristics to be expected of a recently merged pair of galaxies? I can think of five characteristics to look for: (1) A pair of long tidal tails is the safest indicator of two participants. (Here "tidal" means that an explosive origin can be excluded.) (2) The candidate merger should be isolated to exclude the possibility of tidal damage from neighbors. (3) One might expect to see a single nucleus. (4) Relative to this nucleus, the two tails should move in opposite directions, since, after all, tides are symmetrical. (5) Motions in the main body might still be rather chaotic, if the merging took place recently.

NGC 3921 (= Arp 224), although located in a small group of galaxies, shows at least three of these characteristics: It has two massive tails, a single nucleus (with an unusually strong Balmer absorption-line spectrum indicative of young stars), and a chaotic body, whose internal motions I am still studying.

NGC 7252 shows just about all of the above characteristics to a striking degree (Fig. 2a). The galaxy has a beautiful pair of tails, each of which appears to be smoothly connected to one of a set of loops. (My wife says the object reminds her of a crumpled spider!) The loops probably consist mainly of stars, since I have been unable to pick up any emission lines (with one exception) with the powerful spectrograph of the CTIO 4-m telescope. The connection of stellar tails to stellar loops strongly implies that the tails are tidal. Explosive ejection of a large stellar aggregate seems difficult enough, but explosive ejection "around the corner" and into different orbital planes seems virtually impossible. Luckily, NGC 7252 is highly isolated. The first major galaxy of similar redshift lies at a projected distance of 2.5 Mpc away, whereas several nearby faint galaxies all turn out to be distant background objects. Quite apart from the fact that two long tails normally point to two interacting bodies of about equal mass (TT) rather than to one galaxy tidally damaged by another, there simply is no such other galaxy around! There is also only one nucleus, as sequences of exposures of decreasing length show. The resolution limit in 1" seeing corresponds to about 500 pc at the galaxy. Figure 2b shows that the tails move in opposite directions relative to the nucleus, as expected under the tidal hypothesis. In this galaxy as well as in the others mentioned earlier, tail velocities relative to the nucleus are of the order of 100 km s^{-1}. These low velocities, too, imply a tidal origin of the tails rather than an explosive origin. Note that the tail velocities of NGC 7252 could not have been determined if it weren't again for the presence of giant H II regions at the tip of each tail, at projected distances of 118 kpc and 70 kpc from the nucleus. By dividing these distances by the relative velocities (on the assumption that the transverse velocities

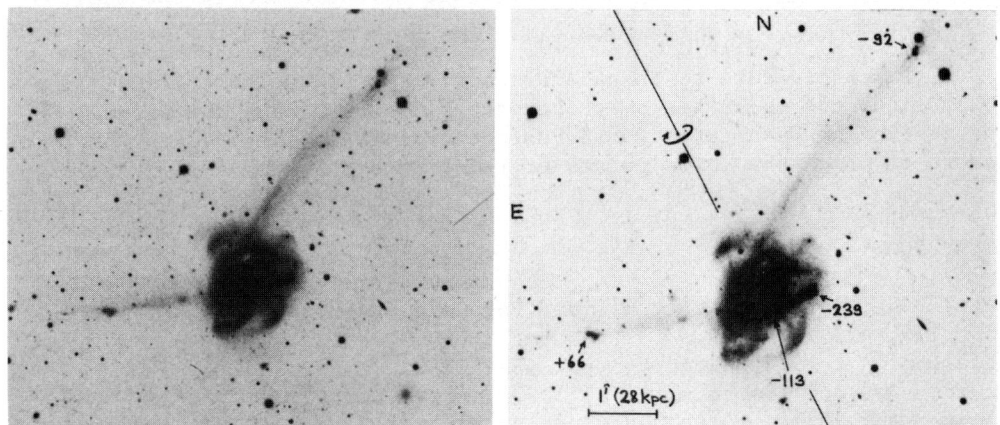

Figure 2. NGC 7252, a recent merger. (a) (left) IIIa-J plate obtained with the CTIO 4-m telescope. (b) (right) Rotation axis of inner gas disk, and tail and loop velocities (in km s^{-1}) relative to the nucleus.

are of the same order as the radial velocities), we estimate an interaction age of 1×10^9 yr. Finally, the spectroscopic data indicate that the motions in the main body are chaotic indeed. The nucleus appears to be surrounded by a disk of gas, which rotates with $v \sin i \approx 100$ km s^{-1} around the axis shown in Figure 2b. Between this inner disk of about 4 kpc radius and the outer loops, the sense of rotation of the gas seems to reverse, if rotation is a proper description at all! However, farther out, in the one loop in which H II regions were detected (see Fig. 2b), the sense of motion has reversed again compared to the intermediate region and the material moves in the same sense as the inner disk.

In summary, NGC 7252 shows several characteristics which one would expect of a merger: a pair of tidal tails despite the splendid isolation, a single nucleus, tail motions in opposite directions relative to the nucleus, and chaotic motions in a strangely looped main body. This main body plus loops, by the way, has gigantic dimensions ($\sim$55 kpc) and a $M_V = -22.8$, which is about 1 mag brighter than an average Sc I galaxy. Certainly, this high luminosity is compatible with the idea that this heap of stars contains two former galaxies.

Finally, let me be so rash as to ask three provocative questions. Suppose that this IAU meeting were held not in Bad Münstereifel, but instead in NGC 7252 at some 10 kpc distance from the nucleus. Suppose also that we had studied the nearby galaxies in much the same way as we have discussed at this meeting. *Question 1:* What Hubble type would we give the galaxy we would live in? S0, as NGC 7252 has been classified, or E, Sa, Sb, or Sc? *Question 2:* What would our views be on galaxy evolution in general and, in particular, on chemical stratification as a consequence of that evolution? *Question 3:* How would we interpret the complicated velocity field in our galaxy? Especially, would considerations about the energetics coupled with observations of external, active galaxies lead us to believe that there was an explosion in the nucleus?

It is a pleasure to thank Alar Toomre for innumerable stimulating discussions and exchanges over the past five years, and the Carnegie Institution of Washington for a Postdoctoral Fellowship, during the tenure of which the early phases of this research were carried out.

REFERENCES

Searle, L., Sargent, W. L. W., and Bagnuolo, W. G.: 1973, *Astrophys. J.* **179**, 427.
Stockton, A.: 1974, *Astrophys. J.* **187**, 219.
Toomre, A., and Toomre, J. (TT): 1972, *Astrophys. J.* **178**, 623.
Toomre, A.: 1977, in *The Evolution of Galaxies and Stellar Populations* (New Haven: Yale University Obs.), 401.
Zwicky, F.: 1956, *Ergebnisse der Exakten Naturwissenschaften* **29**, 344.

DISCUSSION FOLLOWING PAPER V.4 GIVEN BY F. SCHWEIZER

FREEMAN: The ages inferred from the Searle et al. colors are probably somewhat large; the stellar populations in the tails of the Antennae might have ages comparable to the interaction age.

The dwarf galaxy at the end of NGC 4038/9 tail would probably have a mass $\gtrsim 10^8$ $M_\odot$. Does that seem reasonable if this dwarf comes from the tidally produced tail?

TOOMRE: 10^8 $M_\odot$ is only about one thousandth of the total mass; that doesn't seem unreasonable.

SHU: Does your determination of a velocity dispersion of $\sim$ 15 km/s in the "disk" stars of the tail in the Antennae mean that this is a first encounter for the system?

TOOMRE: Presumably the motions slowed down since the previous encounter.

MATERIAL IN THE VICINITY OF GALAXIES

Bernard F. Burke
Massachusetts Institute of Technology

It is increasingly certain that the principal concentrations of neutral hydrogen gas are in spiral and irregular galaxies. The intergalactic medium (IGM) appears to be either a near-void with $<\rho> \lesssim 1.5 \times 10^{-11}$ cm^{-3} from the 3C9 measurements of Gunn and Peterson (1965) or at best a hot, highly ionized gas with very little neutral hydrogen present (Field and Perrenod, 1977). In clusters of galaxies, the intracluster medium (ICM) also appears to be mostly hot and ionized, as suggested by X-ray observations (Jones et al. 1977) and by observations of head-tail radio galaxies. Even if most of the IGM and ICM were ionized, however, one might imagine that there exist high-density concentrations where the cooling rate is sufficient to allow recombination, particularly in the vicinity of galaxies. One might expect such condensations to be revealed either by 21-cm emission, or by absorption against a bright background continuum source.

A study of galaxy-quasar pairs by Haschick, Baan, and Burke has revealed one such example. A total of 8 close pairs were examined, Table I, with negative results in 7 cases but a clearly positive result for the pair NGC 3067/4C32.33. The absorption line was narrow (5 km/s) and has a radial velocity well within the range of velocity exhibited by the hydrogen emission of NGC 3067. The absorbing cloud clearly is associated with the galaxy, yet must lie about 60 kpc distant from the center if it belongs to the disc population, a distance of several Holmberg radii. If the cloud is in the halo, it is remarkable that there is so little internal motion. The observed column density is 2.7×10^{17} T$_s$ cm^{-2}, or 2.7×10^{19} cm^{-2} if the state temperature is 100 K, a value comparable to that observed for the high-latitude, high-velocity gas observed in our own galaxy.

The same method was used by Baan, Haschick, and Burke to search for neutral hydrogen in clusters of galaxies, using continuum radio sources lying in or beyond rich clusters. 15 clusters of galaxies were observed, with positive results in one case. The one positive case was Markarian 6 (IC450), in the cluster Zw 0642.0 + 7334, but since this is a Seyfert galaxy of type II, and hence an active galaxy,

Table I

QSO - Galaxy pairs

Quasar/Galaxy	Optical Depth $\Delta\tau_{rms}$	Radial Velocity V, ΔV (km sec^{-1})	Disc Distance of Quasar (kpc)	Angular Separation (arc min)	Observed Column Density N_H/T_s (cm^{-2} K^{-1})
4C32.33/NGC 3067	.027±.005	1494,76	58	1.9	2.7 (17)
3C268.4/NGC 4138	<.005	1020	29	2.9	
3C275.1/NGC 4651	<.006	795	16	3.5	
3C309.1/NGC 5832	<.008	457	28	6.2	
3C455/NGC 7413	<.005	10050	31	0.39	
1749+701/NGC 6503	<.003	70	30	5.4	<2.8 (17)
CTA102/NGC 7305	<.001	4100	-	5.1	
3C345/NGC 6212	<.002	-	-	4.6	

it was considered premature to classify this as a positive detection of an ICM. The absorbing gas cannot be very close to the nucleus of the galaxy, however, since the resulting ionization state and state temperature would require a very large cloud indeed. The cloud has a column density of 6×10^{18} T_s cm^{-2}, and has a velocity outward from the nucleus of about 170 km/s, with a line-width of 32±6 km/s.

This detection prompted an investigation of other active galaxies, and a similar line was found in Mark 1 (NGC 447). In this case, the apparent velocity is 300 km/s <u>inward</u>. Since the optical velocities may well have systematic deviations from the true galactocentric velocity, these apparent relative velocities should be interpreted with some caution. Recently, Balick (1977) has reported a strong absorption line in Mark 231, and both Balick and we have seen a complicated profile in Mark 3 that may be caused by absorption also. In general the observed absorption lines imply column densities of $10^{18} - 10^{19}$ T_s cm^{-2}, or $10^{20} - 10^{21}$ cm^{-2} for $T_s = 100$ K, and probably are situated at some distance from the nucleus of the galaxies.

A third class of absorption line was detected by Steigerwald and Roberts (1977) and by Haschick, Baan, and Burke (1977), in the galaxy

Table II

Catalog of Material in Vicinity of Galaxies

System	Reference	Column Density $T_s = 100$ K (cm^{-2})	Mass $(M_\odot)$
HI Certainly External			
M81/82/NGC 3077	Em (1)	2×10^{20}	3×10^9
NGC 55/300	Em (2)	$2-10 \times 10^{20}$	1.6×10^9
NGC 3623/3627/3628	Em (3)	10^{20}	10^9
NGC 3067	Abs (4)	2.7×10^{19}	
HI Probably External			
Markarian 6 (= IC 450)	Abs (5)	6×10^{20}	
Markarian 1 (= NGC 447)	Abs (5)	10^{21}	
Markarian 231	Abs (6)		
3C178 (= NGC 2377)	Abs (7)	3.6×10^{20}	
Magellanic Stream	Em (8)	$1-10 \times 10^{20}$	
Complicated Cases			
NGC 1275	Abs (9)	3.5×10^{20}	
1506 + 37	Abs (10)	1.1×10^{22}	
M82	Abs (11)	6.6×10^{21}	
NGC 5128	Abs (12)	$> .7 \times 10^{20}$	
NGC 253	Abs (13)	4.3×10^{21}	
NGC 4945	Abs (14)	9×10^{21}	

(1) Cottrell, G.A.: 1977, Monthly Notices Roy. Astron. Soc. 178, 577.
(2) Mathewson, D.S., Cleary, M.N., Murray, J.D.: 1975, Astrophys. J. 195, L97.
(3) Haynes, M.P., Giovanelli, R., Roberts, M.S.: 1977, Bull. Am. Astron. Soc. 9, 361.
(4) Haschick, A.D. and Burke, B.F.: Astrophys. J.
(5) Haschick, A.D., Baan, W.A., Burke, B.F.: 1977.
(6) Balick, B.: private communication.
(7) Roberts, M.S. and Steigerwald, D.G.: 1977, in press.
  Baan, W.A., Haschick, A.D. and Burke, B.F.: 1977, in press.
(8) Mathewson, D.S., Cleary, M.N., Murray, J.D.: 1974, Astrophys. J. 190, 291.
(9) DeYoung, D.S., Roberts, M.S., Saslaw, W.C.: 1973, Astrophys. J. 185, 809.
(10) Haschick, A.D.: private communication.
(11) Guélin, M. and Weliachew, L.: 1970, Astron. Astrophys. 9, 155.
(12) Roberts, M.S.: 1970, Astrophys. J. 161, L9.
(13) Weliachew, L.: 1971, Astrophys. J. 167, L47.
(14) Whiteoak, J.B. and Gardner, F.F.: 1976, Proc. Astron. Soc. Austr. 3, 71.

3C178. Here the strongest absorption is very nearly at rest with respect to the center of the system, a circumstance very difficult to understand dynamically if the gas is associated with the nucleus.

A summary of the known examples of hydrogen clouds associated with galaxies is given in Table II.

Under "complicated cases" are listed a number of systems in which the interpretation is not clear, and in fact in several of these the HI may be inside the galactic system, or systems. The upper two classes are much more definite. In the emission systems, and NGC 3067, the physical separation is directly observed. The Markarian galaxies, 3C178, and the Magellanic Stream are less definite, but indirect arguments are strong. The Magellanic Stream has, in part, such an obvious positional and dynamic relation to the Magellanic Clouds that it is almost certainly external. Mathewson, Schwarz, and Murray (1977) have argued that these clouds extend throughout the local group, so the uncertainty in distance prevents making a mass estimate. The column densities all appear to be within a remarkably small range, considering the diversity of forms.

The origin of these outlying hydrogen clouds is not clear. In one case, NGC 3067, Boksenberg and Sargent (1977, in press) have shown, by observing the quasar spectrum, that CaII appears in absorption at the radial velocity of the HI absorbing cloud. The observed quantity of CaII is consistent with a normal stellar abundance, implying that the material is not primeval, but has been processed by stellar nucleosynthesis. At least three modes of genesis are feasible: tidal interaction, expulsion by activity in a galactic nucleus, and accretion from the IGM/ICM.

The relative frequency of each mode cannot yet be determined from the limited samples available. The isolated galaxies, and especially the Seyfert galaxies are good candidates for expulsion, with the gas condensing in the halo and perhaps falling back in some instances. The extensive M81-M82-NGC 3077 complex may well be an instance of tidal interaction, although the possibility of the material being left over from an earlier epoch cannot be excluded. A search for other clouds in the M81 group, and for clouds in the CVnI and NGC 1023 groups has been completed by Sargent and Lo (1977, unpublished). A few small (.5 to 3×10^6 $M_\odot/Mpc^2$) clouds were detected, but no further large systems. The large ($> 10^8$ $M_\odot$) systems appear to be uncommon.

REFERENCES

Field, G.B. and Perrenod, S.C.: 1977, Astrophys. J. 215, 717.
Jones et al.: 1977, in press.
Mathewson, D.S., Schwarz, M.P., Murray, J.D.: 1977, Astrophys. J. 217, L5.
Other References appear in Table II

DISCUSSION FOLLOWING PAPER V.5 GIVEN BY B.F. BURKE

VAN DER LAAN: Your negative results for the quasar/galaxy pairs may be disappointing, but also very important. From results presented earlier in this symposium one may retain the impression that column densities $N_{HI} > 3 \times 10^8$ cm^{-2} at $R \lesssim 100$ kpc from the galaxy nucleus are the rule for spirals. What are your distances and upper limits?

BURKE: The upper limit is a column density of a few times 10^{18} cm^{-2} at typically 30 kpc from the nucleus.

VAN DER LAAN: So for the impression from the results just referred to, to remain credible in the face of your results, we must presume that the hydrogen in galaxy outskirts is very clumpy, but smoothed completely by beams $\sim 10'$.

ALLEN: As to the interpretation of your inability to detect HI absorption in 7 out of 8 quasar/galaxy pairs, one should remember that we have no reason to believe that the HI gas in the outer regions of galaxies should be smoothly distributed. In fact if we look at the faint outer parts of the synthesis map of M101 for example, we see a very clumpy distribution which must have a rather low filling factor. It seems unlikely that the even fainter gas observed at 2 or 3 galaxy diameters away would have a very different morphology. So the chance that the line of sight from the distant quasar would go through a hole is probably rather large.

BOKSENBERG: I would like to make two comments: (1) I suggest that the very few detections of absorption by HI at 21 cm in the case of QSO/galaxy pairs is merely due to the low sensitivity of the radio measurements, for which the detection limit is comparable with the expected signals for rather dense columns. If you could improve your sensitivity by a factor 10^6, such as is given by Lyman α observations, you might find HI to be widely present. (2) Markarian 231 (which is probably a QSO with a high degree of reddening) shows optical absorption lines of CaII, NaI and HeI $\lambda 3888$ at velocities of several thousand km/s outwards from the nucleus. NGC 4151 shows HeI $\lambda 3888$ and Balmer lines which are seen to vary in times of about 1 month. In these cases, where we observe variability, lines from other than ground states, or high relative velocities, we are probably seeing material physically in close relation to the nuclear activity and not at large extensions to the galaxies.

TULLY: At Green Bank Fisher and I have observed some 1500 positions associated with galaxies and an equal number of off positions. In addition we have looked at about 1000 positions in a large region in the vicinity of the M81 group. So in total we looked at 4000 positions. Our bandpass covers velocities up to 3000 km/s and our sensitivity is 1 or 2×10^6 M$_\odot$ at 1 Mpc. In all this volume of space, there are a couple of interesting cases which might represent detection of intergalactic HI clouds well removed from visible galaxies. However our results should be interpreted as upper limits. These limits are a factor of four or so more severe than the limits published recently by Shostak.

HIGH VELOCITY CLOUDS : GALACTIC OR EXTRAGALACTIC ?

Riccardo Giovanelli
Istituto d'Astronomia, Università di Bologna

While most of the sky has been surveyed in the 21 cm line at velocities within a few hundred km/s with respect to the LSR, the coverage is highly inhomogeneous in sensitivity, angular and spectral resolution and velocity extent. As interesting information has recently become available on the structure of the clouds, however, inferences on the dynamical and thermal state of the gas have become possible. Here, we shall concentrate on the direction in which this structural information influences the dichotomy stated in the title of this paper.

EXTRAGALACTIC MODELS

Verschuur(1969) first proposed that High velocity clouds (HVC's) may be primordial intergalactic clouds. Since then, two facts have contributed most to reactivate interest in extragalactic interpretations of the HVC phenomenon: (a) the discovery of the Magellanic Stream (MS)(Wannier and Wrixon 1972; Mathewson et al. 1974), and (b) the discovery of the clouds apparently associated with the Sculptor group of galaxies (Mathewson et al. 1975). A number of authors (Lynden-Bell 1976; Kunkel and Demers 1976; Mathewson 1976; de Vaucouleurs and Corwin 1975; Einasto et al. 1976) have claimed that HVC's, distant globular clusters and nearby dwarf galaxies describe some or another great circle. In some models, the implication of such an alignment is that the HVC's form tails of tidal debris in the orbits of nearby dwarf galaxies undergoing tight pericenter transits with the Galaxy; in others, the clouds are primordial inhabitants of intergalactic space. While the MS stretches close to a great circle, the other HVC's do not. Actually, the various great circles proposed by those authors are quite different. The statistical samples used in the above-mentioned studies were not complete: (a) no a priori reasons exist for excluding from the sample the so-called "intermediate" velocity clouds (IVC's) at high galactic latitude (particularly since the differentiation of the IVC's from the HVC's was not based on any physical grounds, and in fact, arose from their LSR rather than GSR velocities), while (b) a number of clouds with very high velocities have been recently found far from the proposed great circles (Shostak 1977; Greisen, private communication; Giovanelli, in preparation; Hulsbosch, communication by Prof. Oort).

The parameters of most HVC's when assumed to be at distances comparable with those of other galaxies in the Local Group, are similar to the ones of the Sculptor clouds (if those are indeed within the Sculptor group) : sizes of several tens of kpc, HI masses between 10^8 and 10^9 M_o, HI densities on the order of 5 10^{-4} cm^{-3}. However, as pointed out by Giovanelli (1977), there is substantial evidence indicating that the northern galactic hemisphere HVC's are rather different objects than either the Sculptor clouds or the clouds in the MS. Fine spectral and morphological structure as well as very strong velocity and brightness gradients characterize the northern HVC complexes. The large and disordered velocity gradients coupled with the angular extent of the features infer that the HVC's are transient. The narrow components in the line profiles indicate that a large fraction of the gas is at low kinetic temperature and cannot be supported in steady-state conditions in intergalactic space. The severe brightness gradients along cloud peripheries suggest that effective compression mechanisms are at work. Moreover, very low brightness emission has been detected in some fields between complexes, implying that the gas distribution may be far from patchy. A distribution of primordial intergalactic clouds, as proposed by Mathewson (1976), de Vaucouleurs and Corwin (1975) and Eichler (1976), cannot explain the dynamical unrest of the clouds, their low brightness connections, or the low temperature inferred from their line profiles. The idea that HVC's form tidal streams in the wakes of nearby dwarf galactic systems, as proposed by Lynden-Bell (1976), leaves, on the other hand, the majority of the HV material unexplained, since no obvious association with globular clusters or dwarf galaxies exists for most complexes.

GALACTIC MODELS

Two scenarios that put the HVC's within the Galaxy have dominated: Oort's infall model and the Verschuur-Davies distant spiral arm model. The merits and pitfalls of both have been illustrated in review papers (Hulsbosch 1975; Verschuur 1975); both meet with considerable difficulties in trying to incorporate parts of the observational data. Clouds at high galactic latitude, near $l = 180°$, cannot be interpreted as parts of distant spiral arms; the general velocity field of the HVC's shows a large anisotropy, if they are to be interpreted as an infall phenomenon; while the very high velocity clouds recently discovered present additional problems. And the MS cannot of course be accounted for as either infalling or spiral arm material. On the other hand, galactic interpretations can more flexibly adapt themselves to the transient character and the occurrence of condensations of cold material in the northern HVC's.

SO, WHAT THEN ?

The panoramic view of models is then not particularly satisfying.

V.6 HIGH-VELOCITY CLOUDS: GALACTIC OR EXTRAGALACTIC?

A combination of several of them could possibly leave no major piece of information unexplained, at the price, however, of a very poor balance in the economy of the hypotheses. For example, the gas at not too high galactic latitude with intermediate or moderately high velocities may be distant spiral arms; the high northern latitude gas, the shell of an old nearby supernova remnant; the MS, the tidal debris of the Magellanic Clouds (MC); and the very high velocity clouds, primordial intergalactic objects. The number of degrees of freedom of such a "model" is large enough that it can probably accomodate all of the known phenomenology and, presumably, observational data acquired for some time ahead. However, to understate the problem, this approach utterly lacks simplicity. Hopefully, it should be possible to do better than explaining the whole picture by means of proposing one model per phenomenological item.

Possibly, all the observations can be incorporated in a unified picture. Tidal interaction between neighboring galaxies is known to severely distort the gas within the system, as in the cases of the Leo triplet and NGC 4631/56. The Galaxy/MC system may suffer analogous distortion, yielding the MS as the most prominent evidence. Earlier numerical models of the tidal interaction between the Galaxy and the MC met difficulties in describing the large negative velocities at the northernmost tip of the MS. They may be overcome if orbits with relatively small pericenter distances are assumed, which allow for capture of Magellanic material by the Galaxy. Angular momentum conservation will then yield high orbital velocities. Davies and Wright (1977) obtain a numerical solution of that type, which reproduces the velocity field along the MS by locating its tip at less than 10 kpc from the galactic center. It is clear that orbiting material with a galactocentric distance smaller than the radius of the galactic disk will eventually collide with galactic gas in a time smaller than its orbital period. Such collision processes may indeed have already started; we propose that the northern HVC's may be their conspicuous result. As the infalling Magellanic material approaches the galactic disk, its evolution will be similar to the one predicted within the framework of Oort's model, with the notable advantage that here a large "fly-by" component of motion and the high velocities beyond the tip of the MS come naturally, in addition to the infall. Savedoff et al. (1967) and Chow and Savedoff (1972) elaborated the hydrodynamical and thermal evolution details of the infall model. Their guidelines are still applicable: northern HVC's would then be high z galactic material, accelerated by the infalling flux, as seen after undergoing shock compression and subsequent cooling. The fine structure and the disordered velocity field present in this material may be easily accounted to density or phase inhomogeneities of the high z galactic gas. The large, quasi-continuous distribution of HV material in the northern hemisphere reflects the nearness of this material, probably located in the high z extension of our spiral arm. Supersonic impact between infalling and galactic gas will produce temperatures in excess of 10^6 K. For temperatures of that order, thermal emission peaks in the 0.25 keV (44-70 Å) X-ray domain. An external test of the scheme being proposed is possible, by comparing the expected soft X-ray flux that would be produced by the infall, with the observed high latitude background. This test applies equally well to Oort's model. The emission measure EM (cm^{-6} pc), the emissivity J (erg cm^3 s^{-1}) and the photon flux F (cm^2 s sterad keV)$^{-1}$ in the 44 to 70 Å energy range, are related according to F = 2 10^{27} J EM (Silk 1973). Using Chow and Savedoff's (1972) density and temperature profiles of the post-shock gas, the EM of the infalling gas in the specified energy range is on the order of 0.013. By assuming an emissivity of 10^{-23} (Tucker and Koren 1971), one gets F = 260, which compares well with the value of 300 observed at high galactic latitudes. This hypothesis also accomodates well the observed enhancement in flux and the softening of the spectrum observed toward high northern galactic latitudes (Levine et al. 1976). Although this hypothesis seems capable of encompassing all the known observational information, further numerical tests of Magellanic passages and revision of the calculations of Chow and Savedoff are needed.

In conclusion, observational evidence indicates structural differences between the MS and the rest of the HVC gas. Existing extragalactic or galactic models cannot interpret consistently all of the data. It is proposed that the northern hemisphere HVC's are the result of Magellanic material infalling on the galactic disk after a close pericenter passage of the MC. The high latitude enhancement of the soft X-ray flux and the softening of its spectrum may be partly produced as a consequence of this infall.

Chow,T.L. and Savedoff,M.P.: 1972, Nuovo Cimento 8B, 130
Davies,R.D. and Wright,A.E.: 1977, Mon. Not. R. Astr. Soc. 180, 71
de Vaucouleurs,G. and Corwin,H.G.: 1975, Astrophys. J. 202, 327
Eichler, D.: 1976, Astrophys. J. 208, 694
Einasto,J.,Haud,U., Joeveer,M. and Kaasik,A.: 1976, Mon. Not. R. Astr. Soc. 177, 357
Giovanelli,R.: 1977, Astron. and Astrophys. 55, 395
Hulsbosch,A.N.M.: 1975, Astron. and Astrophys. 40, 1
Kunkel,W.E. and Demers,S.: 1976, preprint
Levine,A., Rappaport, S., Doxsey,R. and Jernigan,G.: 1976, Astrophys. J. 205, 226
Lynden-Bell,D.: 1976, Mon. Not. R. Astr. Soc. 174, 695
Mathewson,D.S.: 1976, preprint, Herstmonceux conference
Mathewson,D.S., Cleary,M.N. and Murray,J.D.: 1974, Astrophys. J. 190, 291
Mathewson,D.S., Cleary,M.N. and Murray,J.D.: 1975, Astrophys. J. 195, L97
Savedoff,M.P., Hovenier,J.W. and van Leer,B.: 1967, Bull. Astr. Inst. Neth. 19, 107
Shostak,G.S.: 1977, Astron. and Astrophys. 54, 919
Silk,J.: 1973, Ann. Rev. Astron. Astrophys. 11, 269
Tucker,W.H. and Koren,M.: 1971, Astrophys. J. 170, 621
Verschuur,G.L.: 1969, Astrophys. J. 156, 771
Verschuur,G.L.: 1975, Ann. Rev. Astron. Astrophys. 13, 257
Wannier,P. and Wrixon,G.T.: 1972, Astrophys. J. 173, L119

V.6 HIGH-VELOCITY CLOUDS: GALACTIC OR EXTRAGALACTIC?   297

DISCUSSION FOLLOWING PAPER V.6 GIVEN BY R. GIOVANELLI

WRIGHT: Isn't the infall theory for HVCs in serious trouble for velocities as high as -400 to -500 km/s?

GIOVANELLI: According to the infall theory of Professor Oort such clouds would still be Galactic clouds. If they were material broken off the Magellanic Stream then it has not collided with Galactic material.

WRIGHT: But you would then expect the velocity profiles to look similar in the Magellanic Stream and in the HVCs — which they don't!

OORT: It is somewhat surprising that one does find such a preponderance of negative velocities in these high velocity objects. In part, of course, these can be explained by the fact that velocities are relative to the Local Standard of Rest, and part of them can be ascribed to rotation of the Galaxy. But even if you subtract that you are left with a great majority of negative velocities, also in the other regions near the anticenter where rotation of the Galaxy does not play any role. It is probably too early to speculate about this until we get data on the whole sky with the same completeness.

EINASTO: HIGH VELOCITY CLOUDS AND WARPING OF THE GALACTIC PLANE
    We have collected available data on the kinematics and distribution of high velocity clouds. The data show that HVCs can be divided into two populations. One population is concentrated towards the Galactic plane and takes part in Galactic rotation. The other population is concentrated towards a plane perpendicular to the Galactic plane. Southern HVCs of this population form the Magellanic Stream, northern HVCs form three shorter streams. The kinematics of both stream systems is similar: near $\ell = 90°$ the southern stream is approaching the Galaxy and the northern stream is receding it, while near $\ell = 270°$ the situation is vice versa.
    Numerical calculations carried out by Mr. Haud (Astr. Circ. USSR No. 958, 1977) indicate that the infall of HVCs to the Galactic gas from opposite directions to opposite Galactic sides can give rise to the warping of gas in the outskirts of the Galaxy.
    A similar mechanism may act also in companions of massive galaxies. Suppose that a small galaxy like M33 is circulating around a massive one (M31) and that there is some extragalactic gas in the system. The circulation of the small galaxy in the extragalactic medium is equivalent to the rotation of this medium around the galaxy and the warping of the gas disk of the galaxy is a natural consequence.

DAVIES: What intergalactic density do you require for the intergalactic wind?

EINASTO: A density of about 0.001 cm^{-3}.

OORT: Your plot showing the far "streams" of HVCs seems to me oversimplified. There are important groups of clouds lying outside these streams. I find it difficult to imagine a rotation between the two streams. The one lies outside the Galaxy and has clearly been tidally drawn out from the Magellanic Clouds. The other lies in a rather different part of space and seems actually to be falling into the halo of the Galaxy.

MIRABEL: FINE STRUCTURE IN THE MAGELLANIC STREAM

In order to study the small-scale structure of the Magellanic Stream, high resolution and high sensitivity neutral hydrogen observations have been carried out on selected areas of the sky (in cooperation with R.J. Cohen and R.D. Davies). The 250 foot Mk IA radio telescope (beamwidth 12' x 12' at 21 cm) has been used with receiver channel spacings of 2.06 and 1.03 km/s.

The gas has a hierarchical structure in which small bright clouds are distributed along narrow filaments embedded in larger areas of low emissivity. These bright clouds have elongated shapes with major axes generally aligned parallel to the great circle described by the large scale distribution of the Magellanic Stream. Their typical minor axis diameters are $0°.4$ to $0°.6$, with masses in the range $10 - 20$ D^2 (kpc^2) solar masses. Although there are velocity differences up to 30 km/s between adjacent clouds, there are no velocity gradients within the clouds themselves. The spectra show no evidence for two phases in the gas. Typical half-power widths are in the range 25-30 km/s; any component narrower than 16 km/s is less than 5 per cent of the HI column density observed.

The differences in spectral properties between bright clouds in the Magellanic Stream and high velocity clouds in the northern galactic hemisphere (where two phases were found by Cram and Giovanelli (1976, A.A. 48, 39)) provide a clue for the understanding of their different origin and nature.

The lifetime of the large-scale features in the Stream is consistent with tidal models. However, the present observations show that small-scale features have much shorter lifetimes and require a containment mechanism to explain their continued existence.

DISCUSSION V.7 ON IRREGULAR GALAXIES

KINMAN: HI OBSERVATIONS OF DWARF GALAXIES

Few dwarf galaxies ($M_{pg} > -15$) are known outside our Local Group. Although Fisher and Tully (1974, A.A. 44, 151) detected HI by the 21-cm line in 179 out of the 243 objects in the DDO catalog of dwarf galaxies, only 8 percent of those detected have $M_{pg} > -15$, and only 15 percent have $M_{pg} > -16$. The Arecibo radio telescope was therefore used by E.K. Conklin and myself at 21 cm in the velocity range -500 to 1500 km/s to observe dwarf galaxy candidates to a hydrogen flux limit of $\sim 4 \times 10^4$ $M_\odot$ Mpc^{-2} (velocity resolution 4 km/s); photoelectric magnitudes and colors were obtained at KPNO. Among candidates chosen for low surface brightness, HI was found in three out of seven objects taken from Karachentsava's (1968, Comm. Byurakan Obs. 39, 61) catalog of Sculptor systems. These are dwarf irregular galaxies with M_{pg} = -12.8, -14.4, and -15.5 (H_0 = 50 km sec^{-1} Mpc^{-1}), which suggests that Karachentsava's catalog (like the DDO catalog) consists primarily of late-type systems. HI was also found in one DDO galaxy (No. 72, M_{pg} = -15.6) out of four searched, in which none had been detected by Fisher and Tully. Among compact, blue emission-line galaxies, neutral hydrogen was found in seven out of eight low-velocity galaxies discovered by Rubin et al. (1967, A.J. 72, 59) in the direction of the Virgo cluster. One of these (RMB 132) is a satellite of M87. A provisional comparison of the radio and optical properties of these compact galaxies with some low-velocity Markarian galaxies supports the idea that these RMB objects are members of the Virgo cluster with M_{pg} in the range -14.1 to -16.6. Spectroscopic observations are in progress in collaboration with K. Davidson and R.M. Humphreys.

TULLY: GAS DISTRIBUTION, MOTIONS AND DYNAMICS FOR SOME DWARF IRREGULAR GALAXIES

The magellanic irregular galaxies DDO 125 (M_{pg} = $-15^m.9$) and Ho I (M_{pg} = $-14^m.4$) have been observed with the Westerbork Synthesis Radio Telescope in the 21-cm HI line. These systems are intrinsically the faintest such objects for which we now have high-resolution kinematic information. Both galaxies are found to have well-ordered velocity fields which can be described as solid-body rotation over most of the disk, with a hint of a velocity turnover toward the extremities. For both, mass models have been fitted, based on gaussian density distributions characterized by a central density and scale lengths in and perpendicular to the disk. In both galaxies the random gas motions are comparable to the rotation velocities and provide a significant pressure term in the equations of motion. The masses derived are of order 5×10^8 $M_\odot$, and the M/L ratios are of order 3 in both systems. These values should be considered as lower limits since considerable mass might be located at radii exceeding the scale length. Similar gaussian models have been fitted to the previously observed irregulars: the Small Magellanic Cloud, NGC 3109 and NGC 6822. The large-scale features of all five galaxies are compared. A more extensive discussion is in press (1978, A.A.).

HEIDMANN: MASS OF CLUMPS IN THE IRREGULAR GALAXY MARKARIAN 296

In a previous work we drew attention to a new class of irregular galaxies which are larger, brighter and with larger internal motions than classical irregulars, have UV emission and are characterized by a clumpy structure made up of half a dozen of clumps scattered in a common envelope. Only six cases are known so they appear to be rare. It was suggested that these giant clumpy irregular galaxies could still be in a transient state of evolution with large cells where the rate of star formation is high (Casini and Heidmann 1976, A.A. 47, 371).

C. Casini and I obtained a high dispersion Hα spectrum and the 21-cm line profile of one of them, Markarian 296. The spectrum shows a linear velocity gradient of 120 km/s along the line of 8 clumps, which is 5 kpc long. If stable, and with a reasonable inclination, the system of clumps has a total mass 1.2×10^9 $M_\odot$. This gives for each clump and its neighbourhood a mean mass 1.5×10^8 $M_\odot$.

The HI mass of the galaxy is three times larger, 3.4×10^9 $M_\odot$. Thus neutral hydrogen is probably scattered in a larger volume and it would be interesting to map it with the Westerbork radiotelescope.

A more complete study of this galaxy with low dispersion spectra obtained by M. Tarenghi is in preparation in collaboration with him.

BURBIDGE: What is the M/L ratio?

HEIDMANN: For the system of clumps, the M/L ratio is very small. If one uses what we call the indicative total mass for the total mass of the galaxy, as evaluated from the width of the 21-cm line and the optical dimensions, the M/L ratio is 10. But of course it would be nicer to evaluate the total mass from a kinematical mapping of the neutral hydrogen with the Westerbork telescope.

LANDECKER: HI OBSERVATIONS OF THE IRREGULAR GALAXY NGC 4214

Aperture synthesis observations have been made of neutral hydrogen and continuum emission from the type I irregular galaxy NGC 4214 by P.E. Dewdney and myself. Angular resolution is 2 x 4 arcmin and velocity resolution 5 km/s.

The hydrogen distribution shows a marked concentration to the center of the galaxy, coinciding with the bulk of the optical emission and a weak continuum source. There is clear evidence for rotation, and a well defined axis and center of rotation have been found. The profiles obtained suggest a thick spheroidal distribution as an appropriate model. However a Brandt rotation curve is asymmetrical and cannot easily be fitted to the data. Only a lower limit to the mass has been obtained.

Optical parameters
    Distance   6.3 Mpc
    Holmberg Size   10.6 x 10.6 arcmin
Derived parameters are
    Neutral Hydrogen Extent   18 x 13 arcmin
    Systemic Velocity   302 $\pm$ 5 km/s
    Rotation Axis   15° W of North $\pm$ 5°
    Inclination to Line of Sight   39° $\pm$ 5°

V.7 DISCUSSION ON IRREGULAR GALAXIES

Continuum Flux   50 ± 10 mJy
Neutral Hydrogen Mass 3×10^9 solar masses
Total Mass Lower Limit 5×10^9 solar masses

SEAQUIST: RADIO RECOMBINATION LINES IN M82

M.B. Bell and I have detected the H102α (6.1 GHz) and H85α (10.5 GHz) lines in M82 with the Algonquin 46-meter telescope. These results confirm the existence of observable recombination lines in M82, in agreement with our earlier work (Bell and Seaquist 1977, A.A. 56, L461) at H102α and the work of Shaver, Churchwell, and Walmsley (1977, preprint) at H166α (1.4 GHz) and H110α (4.9 GHz). Figure 1 shows our results at both 900 kHz and 300 kHz resolution.

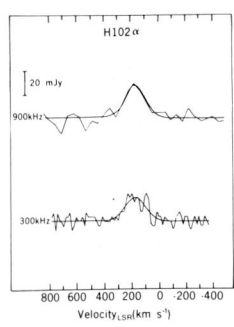
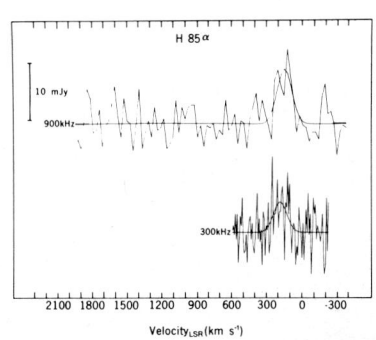

When the observed peak intensities (24 ± 4 mJy at H102α; 7.5 ± 2 mJy at H85α) are combined with the existing data on M82 at H166α and H110α made with a single paraboloid, the peak intensities show a decrease with increasing frequency. This decrease is inconsistent with an origin for the lines by spontaneous emission, but is consistent with stimulated emission involving ionized gas in front of the nonthermal radio source in the nucleus of M82, as originally suggested by Shaver, Churchwell and Rots (1977, A.A. 55, 435).

The mean velocity width (160 ± 40 km/s), and centroid (V_{LSR} = 177 ± 20 km/s) from Figure 1 are consistent with the known properties of ionized gas in the nucleus. We are currently developing a simple model to account for the recombination line data.

VAN DER HULST: The mean velocity of the emission lines (H110α, H102α, H85α) observed with a single dish is about 192 km/s. You suggest that the emission is arising from a small central region in M82 (i.e. the region outlined by the radio continuum map of Kronberg et al.). The H166α emission line, measured with the Westerbork array (P. Shaver et al. 1976) comes from about the same region. Yet the mean velocity of the H166α line is about 240 km/s. Do you have an explanation for this difference?

SEAQUIST: Within the nuclear region there is a velocity spread of at least 200 km/s, and it is possible that different regions contribute at different frequencies. On the other hand, the difference may not be real; because of the low signal-to-noise ratio of the profiles the velocities may not be determined as well as we think.

GIOVANELLI: You showed us several profiles in two different resolutions. Were they independent measurements, confirming each other, or the same signal observed with different bandwidths?

SEAQUIST: No, they are not dependent data. We used one front-end connected to two spectrometers in parallel.

GIOVANELLI: You quoted errors of 10 or 20 km/s for the lines' central velocity, and the widths are as large as 300 km/s. Since several lines were only a few channels wide, how can you determine central velocities with such accuracy?

SEAQUIST: The error quoted is half the velocity resolution. I don't think this is unreasonable.

COMTE: PRELIMINARY RESULTS FROM AN Hα-[NII] LINE SPECTROSCOPIC STUDY OF M82

Spectra of the system of emissive filaments in the galaxy M82 have been taken with the image-tube nebular spectrograph attached to the Cassegrain focus of the 193-cm telescope in Haute Provence. The preliminary results of this survey are:

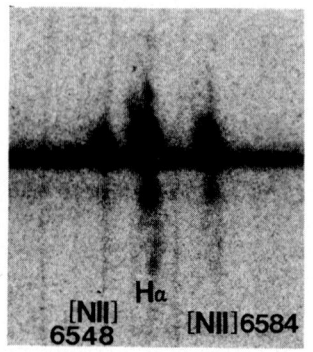

(1) Up to ±2' from the adopted "nucleus", the Hα/[NII] line intensity ratio is not significantly different from Peimbert and Spinrad's value.
(2) The velocity field of the filaments up to ±3' is very complex but the main outline is consistent with previous measurements by other authors, and recent 21-cm line data, i.e. blueshift with respect to the systemic velocity on the southern side and redshift on the northern side.
(3) Detailed analysis of the Hα and [NII] line profiles in the filaments show several (3 to 4) velocity components within a line, becoming more and more resolved with increasing distance from the center. Each component shows a half-maximum linewidth superior to the instrumental response (75 km/s); the internal velocity dispersion of components varies, but does not exceed 150 km/s.

OORT: What are the velocity differences between the components?

COMTE: About 300 km/s.

SCARROTT: A DUST SCATTERING MODEL OF M82

The optical linear polarisation map of M82 shows a centrosymmetric pattern of polarisation orientation, and a polarisation intensity that increases up to ∼ 30% at a distance of 200 arc seconds from the nucleus.

A model of the galaxy has been constructed by assuming the body of the galaxy to be an edge-on luminous disk with a radial luminosity fall off given by exp(-kr) with k = 2.0 kpc^{-1}. The disk is assumed to be situated in a large uniform dust cloud of dimensions 20 kpc. Further dust associated more closely with the galaxy is also assumed and this is taken to fall off as k Z^{-n} where Z is the height above the galactic plane and k and n are model parameters.

Using this model and assuming Rayleigh scattering by the surrounding dust of light originating in the galactic disk we have determined the Stokes parameters (Q, U, I) for various positions along the minor axis of the galaxy. We have compared the model predictions with the experimental data as shown in figure 1 and optimised the values of k and n. We find a difference between the north and south directions in the galaxy and these are represented by the different values of the model parameters as indicated in figure 1. (This work was done in co-operation with H.G. Perkins, W.S. Pallister and R.G. Bingham.)

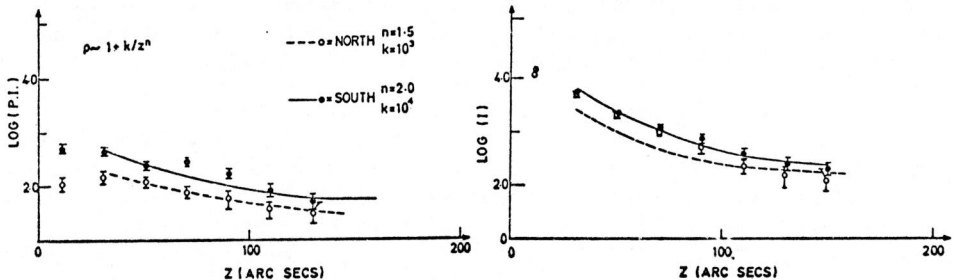

Fig. 1. Model predictions to polarised intensity and total intensity for the galaxy M82.

VAN DER HULST: Does your model give a number for the density of the dust around M82?

SCARROTT: No. We normalise the predicted polarised intensity to our data at one point 100 arcseconds from the center of M82. To obtain an absolute number density of the dust we would need at least the luminosity of the nucleus and unseen face of M82; this is not known.

BALDWIN: Can you guess whether your data would fit models in which the dust, supposedly associated with M82 itself, is in fact at some other point in the line of sight?

SCARROTT: If the dust were between M82 and ourselves the scattering that gives rise to the polarisations would be at small angles and consequently any polarisation would be very much less than the 20-30% observed. The dust therefore must be local to M82 to give a scattering geometry compatible with the observed polarisation.

GIOVANELLI: There is gas in the system of which M82 is a member and there are published synthesis maps of the gas distribution. How does it match the dust contribution that you assumed?

SCARROTT: We have yet to compare our dust density distribution with the gas distribution derivable from published radio maps. It has to be seen whether the radio maps are of sufficient detail to enable the comparison to be made; remember that our plots only extend to 150 arcseconds from the plane of the galaxy.

OORT: Isn't it difficult to imagine that such thin filaments as are seen in M82 could have persisted in the intergalactic medium?

COURTÈS: It is interesting to note that on the deep $H\alpha$ photograph shown by Comte, the filaments exhibit a loop-like structure. This morphology should be in favour of a physical relation between M82 and the filaments and not of the intergalactic clouds interpretation.

SCARROTT: For our polarisation map an integration bin of 7 by 7 arcseconds was used and this smeared out any effects of the filaments in the presented polarisation pattern. Our model does not implicitly include the filaments and represents the general dust distribution around M82. Deep electronographs would best indicate the radial extent of the filaments.

INDEX OF GALAXIES AND CLUSTERS OF GALAXIES

<u>n</u>: object is mentioned on more than 2 consecutive pages of the Paper and Discussion containing page n.

ABELL 1060 - 94
ABELL 1367 - 84, 85
CENTAURUS A - see NGC 5128
COMA CLUSTER - 4, 75, 82, <u>85</u>, 258
CYGNUS A - 241
CV n I - 290
3C 9 - 287
3C 31 (= NGC 383) - 52
3C 120 - 241
3C 178 - 289, 290
3C 227 - 234
3C 268.4 - 288
3C 273 - 241
3C 275.1 - 288
3C 309.1 - 288
3C 345 - 288
3C 390.3 - 234
3C 455 - 288
4C 32.33 - 288
CTA 102 - 288
DA 240 - 234
DDO 72 - 299
DDO 125 - 299
FORNAX A - see NGC 1316
GALAXY - 12, 14, 31, 33, <u>37</u>, 61, 65, 70, 74, 99, <u>139</u>, 179, 227, 232, 236, 240, <u>247</u>, 263, 273, 293
HOLMBERG I - 200, 264, 299
HOLMBERG II - 264
HYDRA I CLUSTER - see ABELL 1060
IC 10 - 200, 264
IC 342 - 36, 47, <u>191</u>
IC 1613 - <u>251</u>, 264, 280
IC 1727 - <u>24</u>
IC 2233 - 90
IC 2951 - 84
IC 4296 - 49

LARGE MAGELLANIC CLOUD - 17, <u>131</u>, 142, 143, 247, <u>251</u>
LOCAL GROUP - 191, <u>227</u>, <u>251</u>, 263, 280, 294, 299
M31 - 2, 12, 14, <u>23</u>, 33, 37, 38, 42, 43, 47, 63, <u>73</u>, 99, 128, 132, 135, 137, 142, <u>149</u>, <u>159</u>, <u>163</u>, <u>169</u>, <u>175</u>, <u>183</u>, <u>191</u>, <u>195</u>, <u>198</u>, <u>227</u>, <u>247</u>, <u>251</u>, 263, 268, 297
M32 - 164, 247, 253
M33 - 6, 8, 37, 73, 122, 127, <u>132</u>, <u>149</u>, <u>191</u>, <u>197</u>, 227, 247, <u>251</u>, 264, 297
M51 - 1, <u>34</u>, <u>45</u>, 54, 63, 73, 79, 89, 93, <u>94</u>, 98, 100, 102, 132, 135, 153, 276
M63 - 132
M77 - see NGC 1068
M81 - <u>23</u>, 29, <u>37</u>, 63, 89, 99, 101, <u>105</u>, 122, 123, 126, 127, 132, 135, <u>207</u>, 224, 240, 247, 256, 257, <u>264</u>, 269, 289
M82 - 17, 65, 112, <u>131</u>, <u>207</u>, 224, 227, <u>230</u>, 240, 247, 257, 264, 271, <u>289</u>, 301
M83 - 79, 132, 135, 200, 231, 264
M87 - 7, 227, 234, 241, 244, 247, 258, 299
M101 - 1, 29, 30, 36, 73, 79, 89, 99, 143, 264, 265, 281, 290
MAGELLANIC CLOUDS/STREAM - 273, 289, <u>293</u>
MARK 1 - 288, 289
MARK 3 - 288
MARK 6 - <u>287</u>
MARK 231 - <u>288</u>
MARK 296 - <u>300</u>

NGC 55 - 289
NGC 128 - 6
NGC 147 - 251
NGC 157 - 73
NGC 185 - 251
NGC 205 - 164, 251
NGC 210 - 6
NGC 224 - see M31
NGC 253 - 131, 231, 232, 289
NGC 300 - 6, 8, 289
NGC 598 - see M33
NGC 672 - 24
NGC 891 - 28, 40, 63, 64, 180, 195, 216
NGC 925 - 24, 73
NGC 972 - 73
NGC 1023 - 274, 290
NGC 1042 - 54
NGC 1052 - 54, 221, 224
NGC 1068 - 65, 132, 206, 231, 232
NGC 1079 - 94, 95
NGC 1084 - 73
NGC 1268 - 84
NGC 1275 - 217, 227, 233, 234, 289
NGC 1291 - 14, 94, 95, 116
NGC 1300 - 99
NGC 1316 - 49, 50, 259, 260
NGC 1326 - 14, 94
NGC 1505 - 218
NGC 1510 - 274, 275
NGC 1512 - 274, 275
NGC 1533 - 53
NGC 1553 - 5, 6
NGC 2146 - 213
NGC 2366 - 264
NGC 2403 - 23, 73, 234
NGC 2535 - 282
NGC 2536 - 282
NGC 2623 - 17, 281
NGC 2655 - 207
NGC 2685 - 65, 268
NGC 2768 - 90
NGC 2841 - 37
NGC 2903 - 73
NGC 2911 - 50
NGC 3031 - see M81
NGC 3034 - see M82
NGC 3067 - 287
NGC 3077 - 224, 269, 289, 290

NGC 3079 - 207
NGC 3109 - 25, 264, 299
NGC 3115 - 6, 90
NGC 3190 - 65
NGC 3227 - 233
NGC 3256 - 281
NGC 3312 - 94
NGC 3379 - 244
NGC 3504 - 206
NGC 3556 - 42, 195
NGC 3557 - 22
NGC 3593 - 24, 73
NGC 3623 - 24, 65, 224, 289
NGC 3626 - 24
NGC 3627 - 224, 289
NGC 3628 - 224, 267, 289
NGC 3718 - 65
NGC 3860 - 84, 85
NGC 3921 - 282
NGC 4038 - 269, 279
NGC 4039 - 269, 279
NGC 4111 - 6
NGC 4138 - 288
NGC 4145 - 274
NGC 4151 - 124, 231, 274, 290
NGC 4214 - 300
NGC 4216 - 65
NGC 4236 - 99
NGC 4244 - 28
NGC 4258 - 37, 211
NGC 4274 - 54
NGC 4278 - 20, 53, 54, 221, 224
NGC 4324 - 24
NGC 4374 - 49 50
NGC 4378 - 91
NGC 4388 - 49
NGC 4402 - 49
NGC 4406 - 49, 50
NGC 4911 - 88
NGC 4425 - 49
NGC 4435 - 49
NGC 4438 - 49
NGC 4472 - 50
NGC 4517 - 24
NGC 4527 - 24
NGC 4559 - 24, 25
NGC 4565 - 24, 65, 195
NGC 4579 - 207
NGC 4594 - 6, 7, 17, 65, 207
NGC 4622 - 85, 86

INDEX OF GALAXIES AND CLUSTERS OF GALAXIES

NGC 4627 - 224, 275, 276
NGC 4631 - 24, 27, 28, 42, 214, 218, 224, 273, 275, 276, 295
NGC 4636 - 54
NGC 4651 - 288
NGC 4656 - 24, 27, 224, <u>273</u>, 295
NGC 4676 - 281
NGC 4698 - 24
NGC 4725 - 274
NGC 4727 - 274
NGC 4736 - 37, 214, 218
NGC 4762 - 6, 7
NGC 4826 - 65
NGC 4945 - <u>131</u>, 289
NGC 5005 - 65, 214
NGC 5033 - 214, 218
NGC 5055 - see M63
NGC 5101 - 6
NGC 5102 - 53, 54, 267
NGC 5128 - 17, 20, 21, <u>49</u>, 132, 134, 227, 230, 234, <u>241</u>, 259, 267, 268, 289
NGC 5194 - see M51
NGC 5195 - 276
NGC 5236 - see M83
NGC 5253 - 268
NGC 5364 - 77, 78
NGC 5383 - 65, 100, 276, 277
NGC 5474 - 264
NGC 5477 - 264
NGC 5548 - 233

NGC 5832 - 288
NGC 5846 - 53
NGC 5907 - 28, 42, 195
NGC 6212 - 288
NGC 6503 - 73, 288
NGC 6621 - 281
NGC 6622 - 281
NGC 6814 - 233
NGC 6822 - 247, <u>251</u>, 263, 264, 299
NGC 6902 - 13, 53, 275
NGC 6946 - <u>34</u>, 47, 48, 63
NGC 7252 - 17, <u>282</u>
NGC 7305 - 288
NGC 7331 - 24, 37, 65
NGC 7413 - 288
NGC 7640 - 195
PERSEUS CLUSTER - 84, 258
SCULPTOR GROUP - 293, 294
SEXTANS A - 264
SMALL MAGELLANIC CLOUD - 142, <u>251</u>, 299
U 8877 - 277
VIRGO CLUSTER - 49, 227, 258, 299
WLM - 264
ZW 0642.0+7334 - 287
VII ZW 303 - 12
1506+37 - 289
1749+701 - 288